COMPREHENSIVE ANALYTICAL CHEMISTRY

ELSEVIER SCIENCE PUBLISHERS B.V.
Sara Burgerhartstraat 25
P.O. Box 211, 1000 AE Amsterdam, The Netherlands

Distributors for the United States and Canada:

ELSEVIER SCIENCE PUBLISHING COMPANY INC.
655 Avenue of the Americas
New York, NY 10010, U.S.A.

ISBN 0-444-87376-7

Printed in The Netherlands

COMPREHENSIVE ANALYTICAL CHEMISTRY

Contributors to Volume XXV

Professor Yu.A. Zolotov and Professor N.M. Kuz'min
Vernadsky Institute of Geochemistry and Analytical Chemistry,
U.S.S.R. Academy of Sciences, Kosygin Str., 19, 117975 GSP-I, Moscow
V-334, U.S.S.R.

Translated from the Russian by
Prem Kumar Dang

Wilson and Wilson's

COMPREHENSIVE ANALYTICAL CHEMISTRY

Edited by

G. SVEHLA, Ph.D., D.Sc., F.R.S.C., F.I.C.I.

Professor of Analytical Chemistry
University College, Cork, Ireland

VOLUME XXV

PRECONCENTRATION OF TRACE ELEMENTS

by

Yu.A. ZOLOTOV and N.M. KUZ'MIN

ELSEVIER
AMSTERDAM–OXFORD–NEW YORK–TOKYO
1990

WILSON AND WILSON'S

COMPREHENSIVE ANALYTICAL CHEMISTRY

VOLUMES IN THE SERIES

Contents

Editor's preface

In *Comprehensive Analytical Chemistry* the aim is to provide a work which, in many instances, should be a self-sufficient reference work; but where this is not possible, it should at least be a starting point for any analytical investigation.

It is hoped to include the widest selection of analytical topics that is possible within the compass of the work, and to give material in sufficient detail to allow it to be used directly, not only by professional analytical chemists, but also by those workers whose use of analytial methods is incidental to their work rather than continual. Where it is not possible to give details of methods, full reference to the pertinent original literature is given.

The present volume describes a large number of methods for the preconcentration and determination of trace elements. Depending on the matrix, different procedures are presented and the reader's attention is drawn to the index which is structured according to the analyte. Most of the methods presented here have been tested by the staff of the *Vernadsky Institute of Geochemistry and Analytical Chemistry, Moscow, U.S.S.R.,* where the authors work.

G. Svehla

Authors' preface

Without the development of the theory and practical methods of trace elements preconcentration, it would have been difficult for analytical chemistry to solve many important problems presented to it with the advances made in nuclear technology, electronics, metallurgy, medicine, agriculture and environment studies. Large-scale use of direct instrumental methods of analysis does not lower, as it is sometimes thought, the significance of preconcentration methods. On the contrary, their new merits are disclosed. The combination of preconcentration and subsequent determination techniques has increasingly given rise to a new group of promising methods (known as hybrid methods). Physical and chemical methods of preconcentration are being developed. Reasonable competition between these methods and mutual penetration of different fields have resulted in rapid development of preconcentration methods which occupy an important place among the methods of analytical chemistry. Interesting concepts about this are found in the article by G. Tölg [1]: "We are becoming unaccustomed to one-way methodical thinking. At least in the coming years it will not be obviously uncommon to raise such questions as: Will preference be given to a physico-instrumental method as opposed to a chemical method, or should one use mass spectrometry of solid bodies rather than activation analysis. It is quite possible that success will come to those who will develop a large number of independent chemical or physical methods free from systematic errors and aimed at solving a problem, or to those who, knowing the 'weak points' of their methods, would try to compete with other specialists in finding an optimal alternative. Even the most perfect instrumental analytical methods will not lead to any success if the knowledge of the substance is lacking from the view point of classical analytical chemistry".

Special monographs [2–4] are devoted to analytical preconcentra-

XVII

tion. General aspects of preconcentration of trace elements have been considered in many other books as well as reviews. Important aspects of different methods of preconcentration and their combination with other methods of determination, mainly specific procedures demonstrating the role of preconcentration in the analysis of environmental samples, mineral raw materials, pure substances and other inorganic materials, metals and alloys, organic substances, biological samples, and soils, are considered in hundreds of original publications.

The authors of the present volume, using their experience and published material [2], have tried to describe the concentration methodology, fundamentals of most commonly used methods of preconcentration, problems of their combination with subsequent methods of determination, and, finally, have tried to throw light on the use of preconcentration in analysis of various substances. The book deals mainly with preconcentration of trace elements in inorganic analysis; preconcentration of organic trace components and gas impurities has not been included.

We had to face certain difficulties in selecting the literature. Nevertheless, we tried to cover all of the known preconcentration techniques, and their application in various fields, and to determine when and where a preconcentration method can be better applied.

We tried to illustrate the material with those procedures to which, in our opinion, preference should be given in performing analysis in laboratories. In total we had about 15000 works at our disposal. About 8% of the examined publications are cited.

Suggestions and comments made by specialists at different stages of preparation of the manuscript enabled us to comprehend numerous materials and to write this volume on general problems of trace elements preconcentration.

The authors wish to express their sincere thanks to Dr. L.A. Okhanova for her assistance in collecting material for this book. Thanks are also due to all colleagues who critically read the original manuscript and offered many suggestions.

The authors will be grateful to readers for their comments irrespective of their nature.

Yu. A. Zolotov
N.M. Kuz'min

References

1   G. Tölg, *Naturwissenschaften*, 63 (1976) 99.
2   Yu.A. Zolotov and N.M. Kuz'min, *Preconcentration of Trace Elements*, Khimiya, Moscow, 1982 (in Russian).
3   J. Minczewski, J. Chwastowska and R. Dybczinski, *Separation and Preconcentration Methods in Inorganic Trace Analysis*, Ellis Horwood, Chichester, 1982.
4   A. Mizuike, *Enrichment Techniques for Inorganic Trace Analysis*, Springer-Verlag, Berlin, Heidelberg, New York, 1983.

Chapter 1

General characteristics of preconcentration

1.1. Concentration and separation

The methods of analytical chemistry can be divided into two large groups: (1) methods of separation and concentration of components (isotopes, elements, molecules, phases), (2) methods of determination (detection) of components of the material to be analysed. Some new methods – hybrid methods – have also appeared at the boundary of these groups. All these methods have their roots in various fields of chemistry and physics (inorganic, organic and physical chemistry, electrochemistry, spectroscopy, nuclear physics, etc.), but are also based to a significant extent on the theoretical foundations of analytical chemistry. In this book the methods of determination will not be considered. We shall describe only the possibilities and special features of separation and preconcentration methods. Hybrid methods will also be considered to some extent.

Separation and preconcentration are described by at least three terms: "separation", "preconcentration" and "isolation". Separation is a process in which the components constituting the starting mixture are separated from each other. Preconcentration is a technique by which the ratio of concentration (or the amount) of trace components to the concentration (or the amount) of macrocomponent is increased. In separation, the components constituting the mixture may or may not differ in concentration from each other. In preconcentration, the components that have significantly different concentrations are treated. It is difficult to suggest the best meaning of the term "isolation".

Separation and preconcentration are usually effected by the same methods: extraction, precipitation, co-precipitation, sorption, crystallization, electrochemical and volatilization methods, sublimation, flotation and others. Some methods, *e.g.* paper and thin-layer chromatography, are more suitable for separating components than for preconcentration. On the contrary, fire melting (fire assay) is used only for preconcentration. The successful application of each method depends on how correctly the conditions ensuring quantitative transfer of the desired (or the disturbing) component into one of the phases have been chosen, or, in a homogeneous system, on accumulation of the component in a definite part of the system.

Preconcentration can be absolute or relative (Fig. 1.1). In absolute concentration the trace components are transferred from the large mass of sample into a small mass, and the concentration of the trace components increases. An example of this is preconcentration by evaporation of matrix in the analysis of waters, mineral acids and organic solvents. Relative concentration increases the ratio of trace to the main interfering macrocomponents; the solvents, in this case, are not grouped with the latter.

Relative concentration can be regarded as a particular case of separation with the difference that in the considered case the concentrations of components differ significantly. Here the ratio of initial and final masses of the sample is not of great importance. The main aim of relative concentration is to exchange the matrix for a suitable collector (generally of smaller mass) to prevent its interference in the analysis. For example, in the analysis of gallium arsenide for copper and zinc, the matrix elements (after dissolving the sample in hydrochloric acid containing bromine) can be extracted with active oxygen-containing solvent; thereafter, the trace elements can be determined by any suitable method.

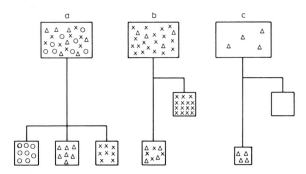

Fig. 1.1. Separation (a), relative (b) and absolute (c) concentration.

In practice, relative and absolute preconcentrations are often combined; the initial matrix is replaced by another organic or inorganic one and the concentrate of trace components is "squeezed" by giving additional treatment, for example by evaporation, until a required mass is obtained.

1.2. Significance of preconcentration and fields of application

Preconcentration of trace elements holds an important place among the techniques used in modern analytical chemistry. An analytical cycle usually includes sampling, treatment of the sample prior to determination, determination proper, and processing of results. Preconcentration is a constituent part of the sample treatment (preparation) stage. Other stages of pretreatment can be the decomposition of the sample, dissolution, masking and simple separation of its components. The choice of operations at the sample pretreatment stage depends mainly on the problem being solved, the nature of the sample and the methods of subsequent determination. There are no universal procedures for this.

The development of instrumental methods for direct analysis – spark-source mass spectrometry, atomic absorption, neutron activation analysis and other methods – has not reduced the significance of preconcentration techniques. On the contrary, the requirements for concentration have increased, particularly because of new combinations with methods of determination. The introduction of new instrumental methods has not restricted the sphere of application of preconcentration, but has revealed its new possibilities.

Earlier, the main, if not the only, advantage of preconcentration was considered to be that it enables the relative detection limits of trace components to be lowered. Of course, such a decrease is in fact achieved in many cases, for example using stripping voltammetry, atomic emission spectrometry with preliminary concentration and extraction photometric analysis. Successful use of preconcentration has placed these methods on a par with the most sensitive instrumental methods. Later on it was found that preconcentration also lowered the absolute limit of detection of trace components in a number of cases, particularly in extraction atomic absorption analysis. The introduction of a combustible extract of trace elements into a flame often produces a good effect on the pulverisation conditions and the atomization process.

Preconcentration has increased the possibilities of many analytical determinations by eliminating the matrix effect which often significantly worsens the detection limits and other metrological parameters of the procedure and which can sometimes prevent the determination of one or other trace elements.

Preconcentration may significantly ease calibration. Removal of matrix by preconcentration makes it possible to use one series of standards with the same matrix for different analytes (aqueous solution in spectrophotometry and luminescence, graphite powder in atomic emission spectrometry, etc.).

Another advantage of preconcentration is that it enables large representative samples to be treated and thereby reduces the sampling error. In fact, most modern physical methods of determination, having quite low limits of detection, give information about analysed samples of very small mass, sometimes containing a few milligrams of the material. This, on the one hand, is their advantage. On the other hand, it is a significant drawback of these methods because a small sample may not be very representative. Preconcentration ensures isolation of trace elements from a large weighed sample. To take a classic example, the fire assay technique is used to concentrate noble metals, which are non-homogeneously distributed in ores, rocks, etc., from very large samples.

Besides, it may be mentioned that at the preconcentration stage it is convenient to introduce internal standards, if necessary, or measured amounts of trace elements if quantitative analysis is carried out by the method of additions. Preconcentration with complete separation of the matrix proves to be useful in the analysis of toxic, radioactive and expensive materials.

Often preconcentration is an integral part of the preliminary preparation of the sample for analysis [1]. For example, on decomposing samples of soils or minerals with mixtures containing hydrofluoric acid, silicon is distilled off as volatile tetrafluoride. Water is distilled off in preparing samples of natural and waste waters; vegetable materials, animal tissues and soils are almost always mineralized by dry or wet methods.

There are some drawbacks which have to be mentioned. Preconcentration complicates the determination and increases the determination time. It may also lead to losses of trace elements to be determined or to contamination. The number of trace elements determinable is sometimes lower than in direct analysis. Special working procedures and reagents of high purity are necessary.

The availability of very sensitive physical methods of analysis, which enable very small amounts (up to 10^{-15} g) of trace elements to be determined, encourages chemists to adapt preconcentration to them. It often widely increases their possibilities. This can be illustrated by the flameless atomic absorption method which was earlier developed as a technique not requiring the removal of matrix. Soon it became clear that non-selective disturbances appearing in the presence of matrix increase the absolute detection limit of trace elements. Preconcentration removes this restriction. Another example of this is spark source mass spectrometry – a method ensuring, in principle, low absolute limits of detection for at least 70 elements, but, due to the small amount of the sample to be analysed, it gives relative detection limits no better than 10^{-5}–10^{-7}%. In this case also, preconcentration proves useful as it widens the possibilities of the method.

1.3. Quantitative characteristics of preconcentration

Every preconcentration technique has its own quantitative characteristics, but at least three of them are used to describe a preconcentration method: recovery (R), concentration coefficient (K) and separation coefficient (S).

The recovery (R) is a dimensionless value showing what fraction of the absolute amount of trace element is present in the concentrate, and is expressed as

$$R = \frac{q_c}{q_s} \tag{1.1}$$

where q_c and q_s are respectively the quantities of trace element in the concentrate and in the sample. It is usually expressed as a percentage

$$R(\%) = \frac{q_c}{q_s} \cdot 100 \tag{1.2}$$

The knowledge of this characteristic enables one to correct the determination results for a systematic error associated with the loss of trace element as a result of incomplete separation. For example, if R is equal to 0.8, the determination result should be corrected by dividing the obtained value by 0.8.

R is usually determined by analysis of standard reference materials or specially prepared synthetic samples containing a known amount of a trace element. The possibility of loss of trace elements to be determined and of contamination, particularly when working with small concentrations of trace elements [2], should be taken into consideration as it can significantly distort the value being determined. This is the reason why such experiments are conducted under strictly controlled conditions accurately imitating the actual process of preconcentration. Other conditions being equal, R can depend on the concentration of trace element and its state. The nature of this dependence is usually ascertained at the development stages.

The concentration coefficient (K) is the ratio of absolute amounts of trace elements and the matrix in the concentrate, divided by a similar ratio in the starting sample:

$$K = \frac{q_c}{Q_c} \bigg/ \frac{q_s}{Q_s} = R\frac{Q_s}{Q_c} \qquad (1.3)$$

where Q_c and Q_s are the amounts of the matrix in the concentrate and the sample, respectively. As $q_s \ll Q_s$ and $q_c \ll Q_c$, the values of Q_s and Q_c are usually taken as being equal to the total mass of the sample and the concentrate, respectively. If $R = 100\%$, then

$$K = \frac{Q_s}{Q_c} \qquad (1.4)$$

The concentration coefficient is taken into consideration for constructing calibration curves showing the dependence of analytical signal on the mass or concentration of trace elements. A correction for K is also applied in quantitative analysis by other methods, *e.g.* by the method of additions.

For large degrees of absolute preconcentrations it is difficult to measure the mass of the concentrate with the required accuracy. In this case the concentration of a trace element (X) in the sample is calculated from the equation

$$X(\%) = \frac{q_c}{RQ_s} \cdot 100 \qquad (1.5)$$

The separation coefficient (S) is expressed by

$$S = \frac{Q_c}{q_c} \bigg/ \frac{Q_s}{q_s} \qquad (1.6)$$

From this it is evident that S is the reciprocal of K.

1.4. Selective and group preconcentration

In most cases an analyst has to carry our multi-element determination. For example, impurities of many heavy metals are determined in environmental samples; several trace elements are also determined in soils, plants and animal tissues; different detrimental impurities and alloying trace elements are found in the materials used in the electronic industry. The study of an unknown, recently synthesized substance is always associated with the task of obtaining review information about its element composition. The methods of multielement analysis are therefore finding wide application. Prior to such a determination it is necessary to concentrate all the elements to be determined.

However, it is not always possible to determine the composition of a substance in a single stage of analysis. Besides, it is often necessary to determine single elements (for example, gold in ores and concentrates; mercury in the air of production areas; daughter element in parent element, etc.). Unfortunately, most of the monoelemental methods, say, spectrophotometric and luminescence methods, give distorted results if the sample contains several components or they do not determine the concentration of the desired trace component at all. Therefore, it is necessary to eliminate the difficulties associated with the multicomponent nature of the sample, to increase the determination selectivity, to mask the matrix and other interfering components, to vary the oxidation state of the matrix element or the trace element, *i.e.* to create conditions where the matrix or other trace components will not affect the results. This is achieved by preconcentration. The practical chemical analysis thus demands that the preconcentration techniques must ensure group as well as individual (selective) separation of trace elements.

Selective preconcentration is a procedure by which one trace element or several trace elements are separated in succession. It is suitable for single-element determination methods like photometry, fluorimetry and atomic absorption spectrophotometry.

Group preconcentration ensures separation of several trace elements in one stage. It is convenient for all multielement determination methods, *e.g.* atomic emission, X-ray fluorescence, spark-source mass spectrometry, etc.

Preconcentration is based on the difference in chemical and physical properties of macro- and trace elements: solubility, sorption, electrochemical characteristics, boiling and sublimation temperatures, the

TABLE 1.1

Relationship between the material analysed and the preconcentration method used

Preconcentration-method	Water	Mineral raw material	Pure substances, inorganic materials	Metals and alloys	Organic substances	Plants, soils, animal tissues
Extraction methods	+	+	+	+	+	+
Sorption methods	+	+	+	+	+	+
Precipitation and co-precipitation methods	+	+	+	+	+	+
Electrochemical methods	+	+	+	+	+	
Volatilization methods including those involving chemical transformation	+	+	+	+	+	+
Crystallization methods			+	+		+
Fire assay		+		+		
Flotation methods	+	+	+		+	
Partial dissolution of the matrix			+	+		+

aggregate state and the dimensions of ions or compounds, the difference in their charge or mass, etc. Compared to group concentration, selective concentration is a refined and complicated process. Use is made not only of different properties of trace elements and the matrix, but also of the difference in the properties of various trace elements; if necessary, the difference in the properties is created artificially. The selectivity of preconcentration increases if it is effected in several stages. In this respect various chromatographic methods and zone melting have proved to be successful.

1.5. Removal of matrix and separation of trace elements

Preconcentration can be effected by two methods: by removing the matrix and by separating the trace elements. Both methods are successfully used in practice, and preference cannot be given, in general, to either of them without reference to the substance to be analysed and the subsequent method of determination.

In fact, the selection of the method depends strongly on the nature of the material being analysed, of course. If the matrix is simple, *i.e.* contains one or two elements, it is easy to remove the matrix. If the matrix contains several elements (complex minerals and alloys, soils), it is better to separate the trace elements.

The selection depends on the preconcentration method used. For example, co-precipitation is suitable for the separation of trace elements but not for the removal of the matrix because the trace elements can partially co-precipitate with the matrix. Volatilization can be conveniently used for separating the matrix of comparatively simple composition, or homogeneous, highly volatile materials: natural waters, volatile halides, acids, organic solvents.

Removal of the matrix and separation of the trace elements are used for group preconcentration; for selective concentration preference is given to the separation of trace elements.

Compared to the separation of trace elements, removal of the matrix demands large amounts of reagents and time and incurs losses of the trace elements being concentrated.

In Table 1.1 an attempt has been made to show the relationship between the substance analysed and the preconcentration method used. Here, " + " stands for the decision most often taken by an analyst; the left-hand section of vertical columns refers to a decision to remove the matrix, and the right-hand section to the decision to separate the trace elements.

References

1 R. Bock, *A Handbook of Decomposition Methods in Analytical Chemistry*, International Textbook Co., Glasgow, 1979.
2 Yu.A. Zolotov and N.M. Kuz'min, *Ekstraktsionnoe Kontsentrirovanie (Extraction Preconcentration)*, Khimia, Moscow, 1971.

Chapter 2

Methods of preconcentration

2.1. Classification and general characteristics of methods

Numerous methods are used for analytical preconcentration of trace elements. A large majority of them had previously been employed only for separation. Probably evaporation was first utilized for preconcentration; precipitation, extraction, electrochemical and other methods gained recognition at a later stage. Fire assay is one of the oldest methods of preconcentration. The availability of a large number of methods has a drawback also: sometimes it is difficult to decide which method should be utilized. These methods should therefore be classified and compared.

By the nature of the separation methods used, they can be classified into (1) chemical and physico-chemical and (2) physical methods.

With the first group may be classed extraction, sorption, precipitation and co-precipitation, partial dissolution of matrix, flotation, volatilization after chemical transformations, chemical transport reactions, fire assay, electrochemical methods and dialysis. The second group may include volatilization, crystallization, freezing out, filtration and gel filtration, and ultracentrifugation.

Even in this classification, combining chemical and physico-chemical methods into one group, certain assumptions have been made. Thus, evaporation, which is grouped with physical methods, can be attended by a change in the chemical state of trace elements and the matrix, and the extraction of simple compounds, for example I_2 and $AsBr_3$, can be tentatively deescribed as "physical distribution" of species between two immiscible phases. On the whole, however, this classification is considered to be satisfactory.

The methods of preconcentration can also be classified according to the number and nature of phases involved in the separation process. Methods based on the distribution of species between two immiscible phases are the most important; in these one of the phases can concentrate the trace elements. The classification by the principle that two phases exist in a system was proposed in particular by Sandell and Onishi [1]; its variants are also known [2, 3]. A classification based on the same principle was proposed by Semov [4]. Taking into consideration the earlier schemes, Zolotov [5] proposed a classification which, after additions and improvements, is given in Table 2.1 and encompasses most of the preconcentration techniques.

However, there are methods where components are separated in one phase, e.g. electrodialysis, electrophoresis, diffusion and thermo-diffusion methods. These are not completely covered by the proposed classification. We can, however, resort to the use of an analogy: under the influence of external energy the components of a system get divided into two parts which can be separated from each other, e.g. by a semipermeable membrane during dialysis. Conventionally the components can be assumed to be distributed between "two phases".

Sometimes more complex systems are also encountered. Thus, in extraction preconcentration of trace elements the extracts may separate into two phases with the formation of a three-phase system. For example, when extraction is carried out with diantipyrylmethane in a mixture of benzene and chloroform [6, 7], the third phase represents a saturated solution of an organic solvent in diantipyrylmethane dithiocyanate, the intermediate layer is a saturated solution of organic salt in the solvent, and the top layer is an aqueous solution of inorganic salts and acids.

These methods of concentration are used to different degrees, and this indicates the importance of individual methods [8]. Each field of application of analytical chemistry has its own distinctive set of methods: in the petrochemical industry preference is given to chromatographic methods; in toxicological chemistry, to extraction and chromatography; in electronics, to volatilization and extraction, etc.

The selection of a preconcentration method is dictated by (i) the practical problem being solved, the nature of the material to be analysed, trace elements to be determined, the specified metrological parameters of the technique; (ii) the origin and previous history of the material to be analysed (this relates first of all to the substances and industrial materials obtained and purified through the use of a simple

12

TABLE 2.1

Classification of methods by the phase state of systems during preconcentration and by the final state of concentrate

System's phase state during preconcentration	Final phase state of concentrate	Method of preconcentration
Liquid–liquid	Solid	Fire assay Extraction in molten phase
	Liquid	Liquid–liquid extraction
Liquid–solid	Solid	Precipitation and co-precipitation Sorption Freezing out Partial dissolution of matrix Electrodeposition, electrodissolution Cementation Wet mineralization Volatilization with preliminary chemical transformation Filtration Flotation Oriented crystallization Zone melting
Liquid–gas	Liquid or gas	Evaporation
	Liquid	Wet mineralization
	Liquid or solid or gas	Sorption
		Volatilization with preliminary chemical transformation
Solid–gas	Solid	Dry mineralization Filtration of gases Chemical transport reactions
	Solid or gas	Volatilization with preliminary chemical transformation Sublimation
	Solid or liquid	Sorption
Liquid	Liquid	Dialysis Gel filtration Electrodialysis Electrodiffusion Electroosmosis Electrophoresis
Gas	Gas	Thermodiffusion
Liquid–liquid–liquid	Liquid	Three-phase extraction

method of separation; therefore, the preconcentration method must be different from the method used for purification of the substance); (iii) the combination of the selected method of preconcentration and subsequent method of determination of trace elements in a concentrate; (iv) the simplicity, the availability and the duration of the method; (v) the equipment available in the laboratory of the scientist and in those laboratories which will use the method; (vi) the specialization and qualification of the researcher developing the technique and of the analysts of the laboratories where this technique is to be employed; (vii) the need to ensure safe working conditions.

When selecting a preconcentration method, consideration is given to the question whether the method ensures distinct separation of macro- and trace components. A priori information, or intelligent suggestions concerning the species in which trace elements are present in the substance to be analysed, also aid correct selection of the method. There is one more factor which affects the selection, *i.e.* mutual effects of the matrix and trace elements in the process of sample treatment. For example, extraction is often accompanied by co-extraction, precipitation is followed by co-precipitation, and distillation is complicated by the formation of azeotropic mixtures.

Below we shall give the general characteristics of some main preconcentration techniques.

Solvent extraction is an effective and widely used method of preconcentration. It can be applied both for the removal of matrix and for the selective, group, or subsequent separation of trace elements. One of its variants – extraction chromatography – is an effective way of separating substances with almost similar properties and ensures high efficiency of preconcentration.

Sorption techniques, *e.g.* ion exchange, prove to be most effective in those cases when the subsequent method of determination may suffer from mutual interfering effects of trace components. It is used in combination with spectrophotometry, voltammetry and neutron activation methods. The assortment of sorbents used at present has greatly increased, particularly due to the synthesis of chelate-type and liquid sorbents which are deposited on inert solid carriers. Co-precipitation on inorganic and organic collectors ensures high efficiency of absolute concentration. It yields good results when used in combination with the determination methods designed for obtaining analytical signals from solid samples, particularly with X-ray fluorescence or atomic emission spectrometry. Electrochemical techniques ensure high efficiency of preconcentration. One practical method involving electrolytic deposition of trace elements on solid graphite electrodes proved good for atomic emission or atomic absorption determination.

Practically no substance of vegetable or animal origin can be analysed without ashing. Use is made of simple, practicable, and often "reagentless", distillation methods, in particular fractional distillation. The latter is extensively used in atomic emission analysis. It is expected that in the near future analysts will more often use chemical transport reactions which are highly effective and represent practically reagentless purification of various compounds (an instance of the use of a technological method is already known; this is zone melting).

There are some remarks which may be made on technical implementation of preconcentration. Batch and continuous processes, single- and multi-stage preconcentration are finding application. Which is selected depends on the task facing the analyst and how different are the properties of the macro- and trace elements involved. As mentioned earlier, the multi-stage methods, *e.g.* chromatographic methods, are convenient for selective preconcentration. If the concentration coefficient is 30–50 or more, then use can be made of a single-stage method.

2.2. Solvent extraction

2.2.1 CHARACTERISTICS OF SOLVENT EXTRACTION AS A METHOD OF PRECONCENTRATION

Solvent extraction is probably the most extensively used method of preconcentration. The theory of this method is being constantly developed. The extraction systems have been classified. Much is being done to ascertain the composition, the structure and properties of the compounds to be separated, the dependence of extractability of metals on the composition of the aqueous phase, the type of extractant, the concentration of the element to be separated, and temperature. The processes of mutual effect of elements in different systems have also been studied. More and more emphasis is being given to rational combination of extraction and subsequent methods of determination.

Extraction methods are suitable for absolute and relative preconcentration. They may be used for (selective or group) separation of trace elements in the extract or for matrix separation in analysis of diverse industrial and natural materials. The important advantage of extraction methods is that they are universal with respect to the type of elements to be isolated and to their concentration. Conditions have been found for extraction of practically all elements from different systems. Simplicity and rapidity are other advantages of these methods. Usually, solvent extraction ensures high efficiency of preconcentration and can be combined with different methods of subsequent determination. There is hardly any laboratory specializing in the determination of trace elements that cannot utilize solvent extraction methods and does not use these methods in its day-to-day work.

Extraction as a method of preconcentration has been reviewed in ref. 9. The knowledge gained in this direction in subsequent years is summarized in several books, reviews and general papers [10–25].

Solvent extraction is a method of isolation, separation and concentration of substances; it is based on the distribution of dissolved substance between two immiscible liquid phases. Most commonly water forms one phase and the organic solvent the other. Extraction, being a heterogeneous process, is governed by the phase rule

$$N + F = Z + 2 \tag{2.1}$$

where N is the number of phases, F denotes the degrees of freedom and Z stands for the number of components. If the number of phases equals two and one substance is distributed, then at constant temperature and pressure the system is univariant. In equilibrium conditions the ratio of concentrations of the substance distributed in both phases is a constant value independent of total concentration of the substance. This value, designated as distribution constant, is described by

$$\frac{C_{I}}{C_{II}} = K_{D} \tag{2.2}$$

Here, K_{D} is the partition constant; C_{I} and C_{II} are the equilibrium concentrations of the substance distributed in both phases in one and the same form, if this form does not vary with concentration. This is an ideal and rarely occurring case in inorganic analysis. Numerous processes, *e.g.* dissociation and association, solvation and hydrolysis, formation of polynuclear complexes, oxidation and reduction, etc., occurring in both phases complicate the extraction equilibrium. However, the effect of such processes is not always significant, and certain extraction systems can be described by the partition law.

Various systems are used in extraction of trace elements. Correct selection of a system largely determines the success of extraction preconcentration. Therefore, classification of extraction processes and of the compounds to be extracted has importance for an analyst. Having taken the compound that goes into the organic phase as the basis of the classification, the following extraction systems can be singled out.

(i) Coordinatively unsolvated neutral compounds with covalent bonds, *e.g.* $GeCl_4$, AsI_3 and OsO_4. The distribution of such substances is well described by the partition law, they being extracted by different types of solvents; their extraction with inert solvents is very selective. The halides of mercury(II), antimony(III), arsenic(III) and germanium-(IV) are easily extracted.

(ii) Chelates, formed by metal ions with reagents containing at least two atoms which are capable of coordinating with the metal. Atoms of oxygen, nitrogen, sulphur and some other elements play the role of these atoms. One of the active groups of such a reagent usually contains a mobile hydrogen atom which, upon complex formation, is substituted by metal; the second (third and other groups) can be acidic, or, often, basic. Chelates are one of the widespread classes of compounds used for preconcentration of trace elements. Extensive use is made of such reagents as dithizone, dithiocarbamate, 8-hydroxyquinoline, oximes, β-diketones, N-benzoyl-N-phenylhydroxylamine, 8-mercaptoquinoline, etc. Compounds formed upon extraction of metals by organophosphorus acids or carboxylic acids are also classed with chelates.

(iii) Coordinatively solvated neutral (mixed) complexes of the type $ScCl_3(TBP)_3$ or $UO_2Br_2(TBPO)_2$ where TBP stands for tri-*n*-butylphosphate and TBPO denotes tri-*n*-butylphosphine oxide. Here the inorganic anions present in the aqueous phase as well as the molecules of the extractant, appear in the internal coordination sphere of the metal atom. Compounds of this type are better extracted by highly donor-active solvents which have the ability to coordinate with the metal.

(iv) Coordinatively unsolvated ion associates. These include tetraphenylarsonium, tetraphenylphosphonium, tetraalkylammonium and similar large hydrophobic cations with perrhenates, perchlorates and other large anions which do not have coordination bonds with water molecules. The salts of large complex cations like FeL_3^{2+} (where L stands for 1,10-phenanthroline) extracted in the presence of suitable counterions, *e.g.* perchlorate, and also the compounds formed by complex anions of the type $SbCl_6^-$ with basic dyes and other bulky cations can also be placed in this group.

(v) Mineral acids (hydrohalic acids, nitric, sulphuric, perchloric acids and others) which are separated by polar solvents with quite high basicity, *e.g.* by ethers and esters, ketones, alcohols, amines and their oxides, salts of quaternary ammonium bases.

(vi) Complex metalloacids of general formula $H_m MX_{m+n}$, where m is usually equal to 1 or 2, *e.g.* $HFeCl_4$ or H_2CdI_4. These compounds are extracted only with active solvents capable of adding protons in acidic medium, and also by salts of quaternary ammonium bases. They are well extracted in both macro- and micro-amounts; this enables the trace elements and the matrix to be transferred into the organic phase.

Heteropoly acids can also be placed into this group. They are extracted with oxygen-containing solvents; extraction is carried out for concentration of arsenic, molybdenum, germanium, tungsten, vanadium and some other elements.

(vii) Different compounds: external spherical halides and pseudo-halides of the type $[Ca(TBP)_m^{2+}](SCN)_2$, mixed complex compounds containing molecules of chelate-forming reagent, *e.g.* $Sn(Ox)_2Cl_2$, in which Ox is the anion of 8-hydroxyquinoline, coordinatively solvated complex acids of the type $H[UO_2Cl_3(TBP)]$, and others.

For a quantitative assessment of extraction preconcentration of trace elements, besides the characteristics common to all the methods of preconcentration, use is made of the distribution coefficient (D) equal to the ratio of total analytical concentration of the element in the organic phase (C_o) to its total analytical concentration in the aqueous phase (C_w), the form of existence of the element in these phases being not taken into consideration.

The separation factor for elements A and B is provided by the equation

$$S = D_A/D_B \tag{2.3}$$

D_A here being greater than D_B.

Recovery of a substance (in %) is the degree of extraction of the substance into an organic phase. It may be expressed by the distribution ratio of the element

$$R(\%) = \frac{100D}{D + v_w/v_o} \tag{2.4}$$

Here, v_w and v_o are the equilibrium volumes of the aqueous and organic

18

phases. If $v_w = v_o$, then

$$R(\%) = \frac{100D}{D + 1} \tag{2.5}$$

The extraction constant, K_{ex}, which is the equilibrium constant of a heterogeneous reaction of extraction, is of great importance for selecting concentration conditions. Thus, for a chelate whose extraction proceeds according to the equation

$$M^{n+} + nHA_{(o)} \rightleftharpoons MA_{n(o)} + nH^+ \tag{2.6}$$

the extraction constant is equal to

$$K_{ex} = \frac{[MA_N]_o\,[H^+]^n}{[M^{n+}]\,[HA]_o^n} \tag{2.7}$$

The development and utilization of highly sensitive instrumental methods in modern analytical chemistry placed stringent requirements upon concentration, in general, and on extraction concentration, in particular. This initiated the search for, and synthesis of, new effective extractants, the development and introduction of progressive methods of extraction preconcentration, and simplification of the analytical cycle by eliminating additional treatment of the concentrate. Significant successes have been made in developing methods for separating noble, rare and scattered elements in rocks and ores, and toxic elements in environmental materials.

2.2.2 EXTRACTION OF TRACE ELEMENTS

Extraction separation of trace elements is possible if the matrix elements do not react with the reagent used. Examples of such samples are natural and waste waters and alkali and, often, alkaline-earth metals salts, compounds of metals that exist in the anionic form during extraction. Other cases are also possible when the matrix elements are "removed" by masking or by other methods, say preliminary volatilization.

2.2.2.1 Chelates

For extracting trace elements chelates are often used. These are compounds of metals with organic polydentate reagents. The general formula of most common chelates is MA_n, where A stands for the anion of the reagent which is a weak acid ($pK_a = 3$–13) and n represents the metal ion charge. Examples of such chelates are copper(II) diethyldithiocarbamate, or aluminium acetylacetonate.

Chelates are characterized by high distribution coefficients in extraction despite their low solubility in organic solvents. In water, chelates dissolve to an insignificant extent. Owing to poor solubility in organic phases, the capacity of the extracts is very low but sufficient enough for the extraction of trace elements. Chelates are often coloured compounds and this permits their use in photometric determination of elements. Certain chelates are thermally stable and volatile. This enables their extraction to be accomplished jointly with gas chromatographic separation and determination of metals.

The extraction theory of chelates is quite well developed. The extraction equation for a typical chelate, MA_n, is given earlier (eqn. 2.6). From the extraction constant equation (eqn. 2.7) the following conclusion can be drawn. Under conditions when all forms of a metal in the aqueous phase can be disregarded, excepting the M^{n+} ion, the ratio $MA_{n(o)}/M^{n+}$ represents the distribution coefficient. We have

$$D = K_{ex} \frac{[HA]_o^n}{[H^+]_w^n} \tag{2.8}$$

$$\log D = \log K_{ex} + n\log [HA]_o + n pH \tag{2.9}$$

This is the main equation that describes extraction of chelates.

The extraction constant depends on the following: stability constant, β_n, of the complex being extracted, reagent dissociation constant, K_{HA}, distribution constants of complexes, $K_{D,MA}$, and the reagent distribution constant, $K_{D,HA}$. This dependence can be expressed by

$$K_{ex} = \frac{\beta_n K_{D,MA} K_{HA}^n}{K_{D,HA}^n} \tag{2.10}$$

Thus, the better the extraction, the higher the stability of the complex and the greater its distribution constant. The greater the acidity of the reagent and the less it goes into the organic phase, the greater will be the extraction.

20

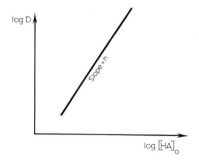

Fig. 2.1. Distribution coefficient of metal in the form of a chelate *vs.* equilibrium concentration of reagent in organic phase at constant pH of aqueous phase.

The relationship between extraction and reagent concentration is important for choosing preconcentration conditions. At constant pH of the aqueous phase the dependence of the logarithm of the metal distribution coefficient on the reagent's equilibrium concentration is linear (see eqn. 2.9). The equilibrium concentration of a chelate-forming reagent can be taken equal to initial concentration, provided very small amounts of metal are extracted with a reagent having a high distribution constant. Such a dependence is shown in Fig. 2.1. In other words, the degree of extraction increases with reagent concentration, other conditions being equal.

However, the control of hydrogen ion concentration is still more important. It is seen from eqn. 2.9 that at constant equilibrium concentration of reagent in an organic phase log D depends linearly on pH, the slope being equal to n (Fig. 2.2).

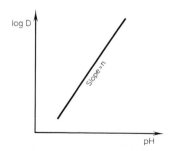

Fig. 2.2. Distribution coefficient of metal in the form of chelate *vs.* equilibrium value of pH at constant equilibrium concentration of reagent in organic phase.

References pp. 179–203

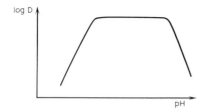

Fig. 2.3. Distribution coefficient of metal *vs.* aqueous phase equilibrium pH at constant equilibrium concentration of reagent in the organic phase.

The extraction–pH dependence is only partially described by eqn. 2.9 and Fig. 2.2. As the pH is increased, new metal forms, besides M^{n+} ion, begin to appear in the aqueous phase, for example MA^{+}_{n-1} and finally MA_n. That is the reason why the dependence starts deviating from the straight line and appears on the plateau in the region where MA_n exists in both phases. At still higher pH values, metal hydrolysis manifests itself; sometimes the formation of MA^{-}_{n+1} metal ion complexes takes place and extraction decreases. Fig. 2.3 shows a typical curve for the dependence on pH.

In analytical practice the values of distribution coefficients are not very often used. Of greater significance are the degrees of extraction expressed in percent. Their dependences on pH are S-shaped curves with different slopes depending on the metal ion charge and, conse-quently, on the composition of chelates (Fig. 2.4). Such a dependence enables pH control to be applied in separating different metals with one and the same reagent. Similar dependences are also employed for selecting conditions for group extraction of elements. A few of them are shown in Figs. 2.5–2.7.

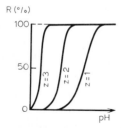

Fig. 2.4. Recovery of metal chelates *vs.* aqueous phase pH for metal ions with different charge (z).

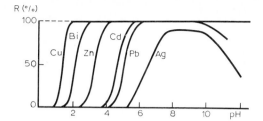

Fig. 2.5. Recovery of metal chelates of 8-hydroxyquinoline *vs.* aqueous phase pH. A 0.1 *M* solution of oxine in chloroform was used.

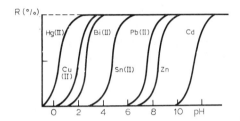

Fig. 2.6. Recovery of metal chelates of dithizone *vs.* aqueous phase pH.

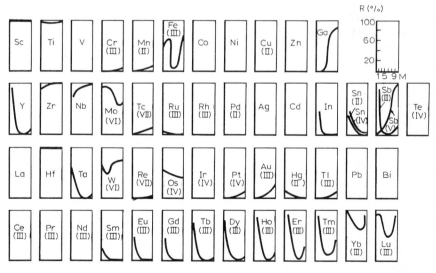

Fig. 2.7. Recovery of metal chelates of di(2-ethylhexyl)phosphoric (D2EHP) acid from HCl solutions. 0.75 *M* solution of D2EHP was used [from I.H. Qureshi, L.T. McClendon and P.D. LaFleur, *Modern Trends in Activation Analysis*, NBS Special Publication 312, vol. 1, 1969; *Radiochimica Acta*, 12 (1969) 107].

Chelates can be coordinatively saturated and unsaturated. In the latter, the central atom of metal is capable of adding neutral ligands, such as water, into the inner coordination sphere. This is why the nature of the solvent has a marked effect on the extraction of chelates. Coordinatively saturated complexes are extracted with different solvents. Aluminium acetylacetonate is an example of such a complex. Every molecule of acetylacetonate occupies two coordination positions in the inner coordination sphere. The coordination number of Al equals 6; that is the reason why in AlA_3, where A stands for acetylacetonate anion, there are no free coordination positions. Such complexes are known as coordinatively saturated complexes.

Dithiocarbamates, particularly sodium diethyldithiocarbamate (I) and ammonium pyrrolidinedithiocarbamate (II), dithizone (III), 8-hydroxyquinoline (IV) and 8-mercaptoquinoline (V) are chelate-forming reagents which are extensively used for group concentration.

The ammonium pyrrolidinedithiocarbamate–methylisobutyl ketone system is of great importance in the extraction of trace elements, particularly in atomic absorption determination of small amounts of elements in biological and environmental materials. Unlike other dithiocarbamates including diethyldithiocarbamate, ammonium pyrrolidinedithiocarbamate is more stable in acid medium. Hexamethyleneammonium hexamethylenedithiocarbamate possesses a similar property [26–28].

Many techniques are based on the use of dithizone; among them is the technique for determining cadmium in zinc by employing extraction of cadmium dithizonate with carbon tetrachloride [29]. Cadmium is determined in the extract by the flame atomic absorption method; 10^5-fold amounts of zinc are no hindrance to determination.

24

Acylpyrazolones, in particular 1-phenyl-3-methyl-4-benzoyl-pyrazolone-5 (VI), are other known reagents.

$$
\begin{array}{c}
H_3C-C=N \\
| \quad\quad\quad N-C_6H_5 \\
H_5C_6-C-CH-C \\
\;\;\|\quad\quad\;\| \\
\;\;O\quad\quad O
\end{array}
$$

VI

Reagents of this class extract at least 50 elements including alkaline earth metals. They are readily available and can be comparatively easily synthesized. They can be used for extracting trace elements with high distribution coefficients, equilibrium in the system being attained rapidly. Using reagent VI, trace elements in alkali metal halides [30], phosphorus [31], arsenic [32] and sodium compounds [33, 34], highly pure salts and waste waters [35] are determined by atomic emission methods. It can also be used for selective extraction of some elements prior to photometric [36–38], luminescence [39], polarographic [40] and atomic absorption [41] determination. More than 150 papers have been written on the application of acylpyrazolones. Most of them have been summarized in ref. 19.

Group extraction preconcentration of trace elements pursues the objective of isolating a maximum number of elements in a single step using a minimum number and amount of chelate-forming reagents which can be easily purified and do not go into the concentrate. Therefore, it is necessary to collect the maximum analytical information with the minimum of disturbance in the course of concentration. This is achieved by different methods and techniques. For instance, a procedure has been developed for preconcentration of trace elements by using a mixture of cupferron and sodium diethyldithiocarbamate [42]. At pH values more than 7 these reagents exist as anions and therefore transfer into an extract only as chelates of extractable elements. Besides, small values of distribution coefficients of reagents in the form of anions permit their purification by the same extraction method. The addition of trioctylphosphine (TOPO) into the organic solvent (chloroform) as a donor active additive encourages the extraction of alkaline earth metals which form coordinatively unsaturated chelates as also do heavy metals; TOPO remains in the extract but does not interfere with subsequent spectrochemical determination of trace elements. Extraction of Ag, Al, As, Bi, Ca, Cd, Co, Cr, Cu, Fe, Hg, Mg,

Mn, Mo, Ni, Pb, Sb, Sn, Ti and Zn is carried out successively at several pH values. Trioctylphosphine oxide is not used if the matrix contains an element forming coordinatively unsaturated chelates.

The composition of the aqueous phase significantly affects the extraction efficiency of chelates. Large amounts of salts (electrolytes) produce changes in the dielectric permeability of the system. Such changes affect the constants which characterize extraction equilibria. A particular case of such an effect is the salting out of the extracting compounds and the extraction of charged forms of chelate-forming reagents in the presence of suitable counterions in the aqueous phase. The anion of the salt can combine with trace elements into non-extracting complexes (masking). Large hydrophobic anions can promote extraction of cationic chelates (along with neutral compounds) which are not extracted in the presence of the salt. On the contrary, the cations of the matrix element favour in some cases the extraction of anion compounds. At small concentrations of the chelate-forming reagent, the extraction of elements in the form of simple or complex salts can become significant with the anions present in the aqueous phase, and at comparatively high concentrations of the acid the extraction of complex metalloacids ($HFeCl_4$, $HInBr_4$, etc.) is appreciable. The macro-component present in the aqueous phase can encourage coprecipitation of certain trace elements. Some salts catalytically speed up the process until equilibrium is attained (for example, alkali metal halides in the extraction of chromium(III) chelates [43]. This must be taken into consideration in developing definite analytical techniques. Thus, the ion partners of the aqueous phase can be used for increasing the number of extractable elements or for improving the selectivity of preconcentration. For instance, it has been shown [44] that cationic chelate of gold(III) with 8-mercaptoquinolinate is extracted well in the presence of trichloroacetate ions. In this case, Ag, Al, Co(II), Fe(II), In, Ni, Sb(III), Tl(III) and Zn are not extracted. Use of this method has been made for selective extraction of gold in analysis of indium, nickel chloride and aluminium nitrate.

2.2.2.2 Neutral coordinatively solvated compounds

For extraction concentration of trace elements use is also made of neutral coordinatively solvated compounds. Here we talk about mixed complexes containing inorganic ligands and neutral extraction reagents, which are of the type $MX_n L_m$, where M stands for metal, L is an extraction reagent, and X represents an inorganic anion, say chloride. Such compounds are extracted only with reagents capable of entering into the inner coordination sphere of the complex. Depending on the type of metal, reagents with different active atoms are used. For alkaline earth and rare earth metals, zirconium, uranium and other hard metals, use is made of oxygen-containing extraction reagents; for soft metals – platinum metals, mercury or bismuth – use is made of sulphur-containing reagents. Both these types of reagents can be used for transition d-elements.

The extraction equation can be described by

$$M^{n+}_{(w)} + nX^-_{(w)} + mL_{(o)} = MX_n L_{m(o)} \tag{2.11}$$

If the activity coefficients are ignored, then

$$K_{ex} = \frac{[MX_n L_m]_o}{[M^{n+}]_{(w)}[X^-]^n_{(w)}[L]^m_{(o)}} = \frac{D}{[X^-]^n_{(w)}[L]^m_{(o)}} \tag{2.12}$$

$$D = K_{ex}[X^-]^n_{(w)}[L]^m_{(o)}; \quad \log D = \log K_{ex} + n\log[X^-]_{(w)} + m\log[L]_{(o)} \tag{2.13}$$

The last equation, as in the case of chelates, can be used for determining the composition of the compounds to be extracted. It also characterizes the effect of basic parameters of the system, for example concentration of the inorganic anion, X. For extracting trace elements by this mechanism use is made of several neutral extractants, for example sulphur-containing compounds, triphenylphosphine, aromatic amines and tributylphosphate.

Among sulphur-containing extractants, derivatives of thiourea and organic sulphides are of significance [25, 45]. Use has been made of these reagents for concentrating trace amounts of noble metals. Platinum-group elements are not very suitable for extraction, particularly due to their existence in the form of kinetically inert complexes. The following are examples of using derivatives of thiourea for extraction of noble metals.

A method involving the use of diphenylthiourea (DPTU, VII) has been suggested for extraction concentration of silver and platinum-group metals after they have been labelled with tin(II) chloride [46]. DPTU makes it possible to concentrate noble metals and to separate them from non-ferrous metals. Addition of DPTU to acetone, heating and boiling ensure quantitative separation of Pd, Pt, Rh, Ru, Ir, Ag and Au with chloroform. The separation of noble metals is selective. The point is that neutral sulphur-containing compounds are selective extractants owing to their ability to enter into the inner coordination sphere of only a few metals. Besides, these reagents are weakly protonized and do not extract anion complexes as a rule. The concentrate of trace elements can be analysed by atomic emission, atomic absorption, and other suitable methods. Hexamethylenephenylthiourea (VIII) can also be effectively used for solvent extraction preconcentration. Petrukhin *et al.* [47] have studied the mechanism of extracting Os, Pd, Pt, Rh and Ru from sulphate solutions using DPTU and chloroform; the procedure involved prelabelling of metal complexes with tin(II) chloride. Under these conditions the metals are extracted as coordinatively solvated complexes of composition $MCl_n(DPTU)$.

$$H_5C_6-NH$$
$$\diagdown$$
$$C=S$$
$$\diagup$$
$$H_5C_6-NH$$

VII

$$H_2C-CH_2-CH_2 \quad HN-C_6H_5$$
$$| \qquad\qquad\qquad | $$
$$\qquad\qquad\qquad N-C=S$$
$$| \qquad\qquad\qquad \diagup$$
$$H_2C-CH_2-CH_2$$

VIII

Solution of diphenylthiourea in chloroform has been suggested for extraction of silver, mercury, gold and, partially, of copper [48, 49]. This procedure can be applied to atomic absorption determination of silver in rocks [50]. The limits of detection for 1% absorption are not less than 0.1 μg/ml Ag in the nebulization method, and no less than 0.0005 μg in the furnace-flame method. Silver down to 0.01 g can be determined in 1 tonne of rock. Chloroform solution of diphenylthiourea has also been used by Shaburova *et al.* [51] for extraction separation of copper, silver and thallium in halides of some alkali and alkaline earth elements, and also in halides of Cd, Co and Ni. For atomic absorption determination the extract was nebulized into an air–acetylene flame. The detection limits were 10^{-5}–10^{-7}% depending on the substance of analysis and the weighed amount.

Rakovsky *et al.* [52] used di-o-tolylthiourea for solvent extraction of Pd, Pt, Rh and Ir from chloride solutions containing tin(II) chloride. Dichloroethane was utilized as organic solvent. Conditions have been determined for solvent extraction preconcentration of the enumerated elements [53]. Based on the difference in the reactivity of initial chloride complexes of platinum metals, a procedure has been worked out for their extraction separation: palladium is extracted in the absence of $SnCl_2$ and without heating; platinum and rhodium are isolated only in the presence of $SnCl_2$ or on heating; iridium is extracted only after heating in a single-phase system containing di-o-tolylthiourea and SnCl$_2$.

Organic sulphides (individual sulphides and a mixture of petroleum sulphides) are extensively applied for selective extraction of gold, silver, palladium and mercury [54–58]. They are readily available, stable, practically insoluble in aqueous solutions, combine well with different organic solvents, and are characterized by high metal uptake. Sulphides are quite convenient for selective concentration of trace amounts of Au, Ag and Pd in the analysis of low-grade ores and also for extraction of matrix in the analysis of these metals. Gilbert *et al.* [59] employed extraction of gold and palladium with sulphides from chloride solutions for their simultaneous neutron-activation determination in natural and industrial substances. At a neutron flux density of $2 \cdot 10^{13}$ neutrons cm^{-2}s^{-1}, the detection limit of cadmium was $2 \cdot 10^{-10}$ and of gold $5 \cdot 10^{-12}$ g. In analysis of high-purity palladium the matrix was extracted with di-n-amylsulphide [60], atomic emission analysis being the final procedure; 26 elements were determined, the detection limits were between $5 \cdot 10^{-6}$ and $2 \cdot 10^{-8}$%. This same reagent was used for analysis of high-purity gold [61] and silver [62]. Extraction with petroleum sulphides which are mainly mixtures of mono-, bi- and tricyclic sulphides was employed for assaying a mixture of Au, Ir, Pd and Pt [63]. Making use of different reactivities of chloride complexes of these elements, the authors extracted gold with dilute solution of sulphides in dichloroethane by bringing the phases into contact for a few seconds. Palladium was extracted with $0.15 M$ solution of the reagent in dichloroethane in 5 min; platinum was extracted in the presence of SnCl$_2$; iridium was extracted by heating in the presence of SnCl$_2$.

Triphenylphosphine (IX) containing phosphorus(III) reacts mainly

H$_5$C$_6$ \
H$_5$C$_6$ —P \
H$_5$C$_6$

IX

with the ions of metals which are soft acids. It is especially suitable for extracting silver(I) from halide solutions with benzene; silver is extracted in the form of the compound, AgX(Ph$_3$P) [64]. The authors [64] believed that effective extractants for silver and mercury would be the reagents with vacant orbitals which play the role of both donors and acceptors. Using these data as the basis, an atomic absorption technique has been developed for determining silver in rocks, ores and minerals, the detection limit being 0.01 g per tonne [65]. Gold in the form of coordinatively solvated compound, $Au(Ph_3P)Cl$, is extracted well with benzene solution of triphenylphosphine from cyanide solutions in the presence of hydrochloric acid [66]. Here, triphenylphosphine also acts as reducing agent. A method has been developed for extraction atomic absorption determination of gold in industrial solutions obtained after treatment of ores and minerals.

Senise and Levi [67] have shown that palladium can be extracted with triphenylphosphine. Mojski [68] has developed a method for extracting platinum metals from HCl solutions using TFF solution in 1,2-dichloroethane. Palladium, platinum and osmium separate out at an acid concentration of over 6 M; at a much lower acidity only palladium separates out. TFF, as an extractant for concentration of noble metals, has been compared with DFTM [69] with the object of effective separation of noble metals for their subsequent determination by atomic emission spectrometry and inductively coupled plasma (AES–ICP) method. The conditions at which concentration was effected were not optimal. Besides, these authors did not succeed in determining the extractable noble metals (Ag, Au, Pd and Pt) directly in the extract.

Aromatic amines are capable of forming coordinatively solvated compounds. True, in solutions with high concentration of hydrogen ions they are in a position to protonize and often form ionic associates. p-Octylaniline (OA, X) has proved to be a very effective extractant for separation of platinum-group elements from chloride solutions [70] even in the absence of labilizing additions. A special property of this reagent is its high extractivity in reference to Ir, Rh and Ru in any oxidation state. Non-noble metals, $e.g.$ Bi, Ca, Cu, Fe, Mn, Ni, Sn and others remain almost completely in the aqueous phase; small amounts of these metals, which go into the extract, are removed from it on washing. The separation factor for platinum and other metals attains values from 10^3 to 10^4. A method has been developed for concentration of platinum-group metals in natural materials and industrial products. A cheaper extractant of this type – technical p-alkylaniline – is also available [71]. Ir, Pd, Pt, Rh, Ru and also Au were extracted from 3 M HCl solutions using toluene solution of alkylaniline which was first transformed into hydrochloride; the concentrate after a suitable treatment was analysed by the atomic emission method.

The authors of ref. 72 compared extraction of platinum-group metals using toluene solutions of hydrochlorides of n-octylaniline and tri-n-

octylamine. Octylaniline extracts Ir, Rh and Ru best of all, and Pd and Pt poorly. With TOA, floating precipitates may appear at the interface. Knowing this, preference was given to octyl- or alkylaniline for analysis of industrial materials. The determination of trace elements was completed by atomic emission or atomic absorption methods.

$$\text{X} \qquad\qquad \text{XI} \qquad\qquad \text{XII}$$

2-Octylaminopyridine (XI) is an interesting extractant used for concentration of small amounts of platinum-group metals [73]. It makes it possible to extract easily iridium and rhodium, the elements which are usually extracted with great difficulty, iridium being extracted best of all other platinum-group metals [74, 75]. 4-Octylaminopyridine (XII) has proved still more effective for concentration of iridium [76]. Generally speaking, separation of platinum-group metals, gold, and silver is a very difficult task. It is successfully solved by employing extraction.

2.2.2.3 Ionic associates

This is a very important group of extracted compounds which are applied for concentration of trace elements. Of these, many are extensively used for separation of complex metallo-acids, for instance metal halides. The condition for their extraction is the ability of the extractant to protonize in an acidic medium or to exist in the cationic form.

First of all we will discuss extraction of compounds of the type $H_m MX_{n+m}$, where M stands for metal, X represents a single-charged electronegative ligand (for example, fluoride, chloride, cyanide, nitrate), and n signifies the metal ion charge. Examples of these compounds are: $HFeCl_4$, $HNbF_6$ and $HInI_4$. Complex acids are extracted only with highly basic oxygen-containing extractants, ketones, ethers, esters, etc., or with amines. Inert organic solvents like benzene or chloroform do not extract them.

The extraction equation is described by

$$m H^+_{(w)} + M^{n+}_{(w)} + (m + n) X^-_{(w)} \rightleftharpoons H_m MX_{m+n\,(o)} \qquad (2.14)$$

Then

$$K_{ex} = \frac{[H_m MX_{m+n}]_{(o)}}{[H^+]^m_{(w)} [M^{n+}]_{(w)} [X^-]^{m+n}_{(w)}} \qquad (2.15)$$

$$\log D = \log K_{ex} + m\log [H^+] + (m + n) \log [X^-] \qquad (2.16)$$

In practice, extraction of metal-containing acids is employed for separation of Nb and Ta from fluoride solutions; Fe, Ga, Au and Sb from chloride solutions; Au and In from bromide solutions; Bi and Cd from iodide solutions; and Fe and Co from thiocyanate solutions.

In addition, acids of the type $HReO_4$ or H_2MoO_4 are extracted by a similar mechanism.

We shall restrict ourselves to some examples. Extensive use is made of high-molecular-weight amines (including aromatic amines), salts of quaternary ammonium bases and organic N-oxides for preconcentration of trace elements. Spivakov et al. [77] have extracted 16 elements from iodine–sulphate medium (1 M in KI or NH_4I and 2.3 M in H_2SO_4) with trioctylamine (TOA) sulphate solution in o-xylene and methylisobutylketone. Under these conditions, Ag, As, Au, Bi, Cd, Cu, Hg, In, Pb, Sb, Sn, Te, Tl and Zn are extracted quantitatively. Iron practically remains in the aqueous phase. On the strength of the results obtained, an atomic absorption technique has been worked out for determining Bi, Cd, Cu, Pb, Sb and Zn in steels, the detection limits being 10^{-4}–$10^{-5}\%$.

Of the oxygen-containing extractants, diantipyrylmethane (XIII) and its analogs are of interest.

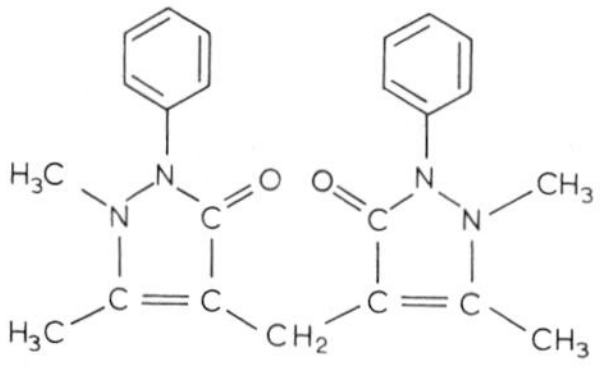

XIII

Petrov [78] has reviewed these very useful and available extractants as to their analytical application.

Yudelevich et al. [79] have made extensive use of 2,2′-dichlorodiethyl-ether (chlorex), a readily available compound; it is the by-product of certain processes and is mainly used to remove the matrix element (see elsewhere). Examples of using this compound in extraction of trace components are also known. In the analysis of cadmium and its high-purity salts, Au, Fe, Ga, In, Sb and Tl were first extracted from halide solutions with chlorex and then determined by atomic emission spectroscopy [80].

Oxides of amines are interesting. Torgov *et al.* [81] compared trioctylamine, N-oxide of TOA, and α-n-nonylpyridine oxide (α-NPO) as extractants for W, Mo and Re (0.1–0.2 M solutions of the extractants in benzene and toluene). The most significant advantage of amine oxides over amines, as described in ref. 81, is the ability of trioctylamine oxide to recover W, Mo and Re from solutions with much higher pH values, and also to extract selectively rhenium from solutions with pH > 7. Group extraction of W, Mo and Re with trioctylamine oxide and extraction of molybdenum with α-NPO was employed for atomic absorption determination of these metals in complex technological solutions [82], and also to remove the matrix in the atomic emission analysis of high-purity rhenium [83, 84].

Certain anions of acids are highly charged and therefore they are poorly extracted. Among them are not only metal-containing acids, but also more common acids, for example, phosphoric or arsenic acid. Organotin compounds, for instance dinonyl- and di-octyltin dinitrates (XIV) of tin have a unique ability to extract such (oxygen-containing) multicharge anions.

$$\begin{matrix} R \\ \diagdown \\ \diagup \\ R \end{matrix} Sn(NO_3)_2 \qquad R = C_8H_{17},\ C_9H_{19}$$

XIV

Using chloroform, methyl isobutyl ketone, butyl acetate or mixtures of any organic solvents with tributyl phosphate, trioctylphosphine oxide or high-molecular-weight alcohols as diluents, gives the following unusual sequence of extractability of anions: AsO_4^{3-}, PO_4^{3-} > SeO_4^{2-}, SeO_3^{2-} > WO_4^{2-}, MoO_4^{2-} > SO_4^{2-} > ReO_4^{-} [85–89]. They extract phosphate and arsenate best of all (distribution coefficients equal 10^2–10^3), sulphate, selenite and selenate not quite so well although satisfactorily, and extract single-charged anions relatively poorly. This sequence is, of course, unusual for anion-exchange extraction which in the given case increases with the charge of the anion. The composition and structure of the compounds to be extracted were studied by an extraction method as well as by infrared and Mossbauer spectroscopy. The composition of the compounds being extracted depends on the nature of oxo-anion, the type of donor-active additive, the concentration of acids, and the salt composition of the aqueous phase. Extraction formally proceeds as per the anion-exchange mechanism, but compounds are formed in which tin is bound with the oxygen of the anion to be extracted by a coordination bond, the molecules of the donor-active additives being a part of the compounds being extracted. The obtained data was used for working out extraction photometric techniques for determining phosphorus and arsenic in vanadium, and for neutron-activation determination of phosphorus in aluminium [80] and for developing several other techniques. Recently, a résumé of works on the use of organotin extractants has appeared [91].

Of course, several other extractants and extraction systems are used for preconcentration of trace elements. A wide range of extractants enables an analyst to choose the best reagent, on the one hand, and poses the very difficult task of ascertaining which of the available reagents is most suitable, on the other hand. Unfortunately, there are almost no publications giving systematic comparison of the properties of the new reagents and those already in use by discussing examples of typical substances of analysis and typical determinable trace elements.

In the review [92] dedicated to the determination of trace elements in geological samples, the following conclusion was drawn: "Natural selection of reagents has lead to the allotment of a relatively small group of reagents for the determination of trace elements in geological samples." Thus, wide use is made of basic dyes, dithizone and sodium diethyldithiocarbamate in the laboratories of the USSR Geological Service. It can be assumed that a similar situation prevails in other fields: non-ferrous and ferrous metallurgy, chemical, electronic, nuclear and other industries (possibly with a slightly different assortment of reagents). It seems that much attention should be paid to "artificial selection" of reagents.

2.2.3 EXTRACTION OF MATRIX

We have already discussed to some extent the extraction of matrix for obtaining a concentration of trace elements in the aqueous phase. This is a widely used technique and is of great importance in analytical practice. The capacity of the organic phase has to be high and the extraction has to be sufficiently selective for the trace elements to remain completely in the aqueous phase. Metal chelates are less convenient for the purpose because of their moderate solubility in the organic phase, but various ion-associates, coordinatively solvated and coordinatively unsolvated non-ionized compounds are quite suitable.

Extraction of matrix proves useful only when the analysed sample has a comparatively simple macroelement composition. This technique is therefore extensively employed in analysis of metals and alloys, simple salts and oxides. Matrix extraction is rarely used in analysis of natural materials, e.g. rocks or biological samples.

In analysis of samples of relatively simple composition the matrix is extracted from halide, thiocyanate, and nitrate media using tri-n-butylphosphate, di-2-ethylhexylphosphoric acid, 2,2'-dichlorodiethyl ether, or methyl isobutyl ketone. Many methods of extracting matrix for the purpose of concentrating impurities are given in the book by Yudelevich et al. [79]. For extraction of matrix elements use is often made of the same extractants as for the extraction of trace elements. Recently, some chelate-forming reagents have also been used for extracting large amounts of metals.

Of the examples of using 2,2'-dichlorodiethyl ether for the removal of macroelement, we shall cite two: concentration of 17 impurities in high-purity cadmium [93] and concentration of 21 non-extracting trace elements in analysis of indium and gallium [94]. In the former case, impurities were determined by the atomic emission method; in the latter case, by the atomic absorption method. Earlier, the same extractant was used to remove indium [95]. Dioctylsulphoxide in xylene was employed to remove uranium in the atomic absorption determination of Cd, Co, Cu and Ni [96].

Among the reagents recommended by the authors of this monograph, mention may be made of O-isopropyl-N-ethylthiocarbamate (XV). The reagent is liquid, miscible with organic solvents and has high efficiency with respect to macro- and microamounts of silver [97]. Only mercury, palladium and, in part, gold are extracted together with silver from $1\,M$ nitric acid; other elements, e.g. Al, As, Ba, Be, Bi, Ca, Cd, Cr, Co, Cu, Fe, Ga, In, Mg, Mn, Pb, Sn, Te, Tl, V, Zn and Zr remain in the aqueous phase. A single extraction of silver with $2\,M$ reagent in chloroform is practically 100% complete. A similar extraction has been used in determining trace impurities in high-purity silver. The detection limits were $5 \cdot 10^{-6}$ to $2 \cdot 10^{-8}$%. O-Isopropyl-N-methylthiocarbamate (XVI) is a more convenient reagent for extraction and is readily available. Methods involving the use of this reagent have been worked out for extraction of silver [98, 99] and also of gold in determination of impurities in native gold [100].

(i) C_3H_7—O
$\quad$ C=S
C_2H_5—NH

XV

(i) C_3H_7—O
$\quad$ C=S
CH_3—NH

XVI

Of great importance for preconcentration and separation of elements is the mutual effect of elements in their extraction, for example, in the form of complex metalloacids from hydrohalic and pseudo-hydrohalic acids (see for example refs 101–107). The variation in the distribution coefficients of trace elements, depending on the concentration of the extracting macro-element (co-extraction and suppression of extraction), is determined by polymerization (association) of the extracting compound in the organic phase, dissociation of the metalloacids in the organic phase, polymerization in the aqueous phase, saturation of the organic phase due to limited solubility of the extractable compound, and by the decrease in equilibrium concentration of free extractant due to its bonding in the extractable compound. The mutual effect of elements is observed also in extraction from halide solutions with amines and quaternary ammonium bases [108–112]; in most cases, the distribution coefficients of trace elements in the presence of extracting macrocomponent are decreased, *i.e.* suppression of extraction takes place.

The mutual effects of elements are observed in other extraction systems also. In ref. 113 a study was made of the effect of neutral trihalide on the extraction of trace elements forming complex metalloacids, namely In, Ga and Sb(V). Possible causes of variation in the distribution coefficients of trace elements owing to the decrease in the activity of free extractants, to hydrolysis of macrocomponent, etc. have been suggested. A study has been made [114] of extraction of trace amounts of indium in the form of bromindate of brilliant green $(R^+ InBr_4^+$, where R^+ stands for the cation of the dye) in the presence of macroelements Cd, Fe(III), Ga, Hg, Tl(III) and Zn having different extractabilities under conditions of optimum extraction of indium. Benzene and its mixtures with 1,2-dichloroethane, chloroform, 2,2′-dichloroethyl ether were used as organic solvents. It has been shown that the better the extraction of a macrocomponent in the form of an ion associate the more it suppresses the extraction of indium due to dissociation of the extracting ion associates in the organic phase (common ion effect).

Here mention must be made of a review concerning data on suppression of extraction of one element by another [115], and of a work [112] that gives a mathematical description of the mutual effect of elements.

2.2.4 EXTRACTION CONCENTRATION TECHNIQUES

The following solvent extraction methods have gained wide recognition for preconcentration of elements: batch extraction in which the substance to be extracted goes from one phase into another on merely mixing the phases; continuous extraction, in which the extractant continuously flows through the solution to be analysed or *vice versa*; counterflow extraction and extraction chromatography. These methods are reviewed in our monograph entitled "Extraction Preconcentration" [9].

Extraction is most often carried out in a water–organic solvent system using a separating funnel (batch method). There are, however, other specific procedures also.

Extraction with melts attracts widespread attention [116–118]. Two groups of organic compounds are used in such processes: substances which continuously act as reagent and (if melted) solvent, for example 8-hydroxyquinoline, and substances playing only the role of easily fusible solvents, *e.g.* naphthalene, biphenyl, benzophenol. In some cases, extraction carried out with easily fusible extractants has an advantage over usual extraction because it is easy to create very high concentration of the reagent. Therefore, the degree of formation of less stable complexes increases. For this reason and due to the application of elevated temperature the rate of extraction increases. It is of significance in the extraction of compounds which are formed slowly, *e.g.* from kinetically inert complexes. Besides, extraction with melts is more convenient than the usual solvent extraction in combination with X-ray fluorescence spectrometry and other methods of determination demanding the use of solid samples. The use of this extraction can be illustrated with several examples.

Japanese researchers have worked out procedures for extraction spectrophotometric determination of Bi, Cd, Co, Cu, Fe, In, Mg, Mn, Ni, Pb, U and Zn [119–123] and for polarographic determination of In, Ni and Pb [124, 125]. Using molten 8-hydroxyquinoline, the authors of ref. 126 extracted Ag, Al, Au, Co, Ga, Mn, Ni, Pb and Sn from aqueous solutions of monosodium phosphate at pH 5–8; extraction was carried out at 80–85°C. The process was completed with atomic emission analysis; the detection limits were 10^{-3}–10^{-5}%. In the analysis of natural waters and organic materials, Lobanov *et al.* [127] extracted Co, Cr, Cu, Fe, Ni and Zn also with molten 8-hydroxyquinoline and determined the impurities by X-ray fluorescence spectrometry. The advantages of extraction with melts show up in quantitative extraction of metals which readily hydrolyse in aqueous media, *e.g.* zirconium; at pH 5–7 the hydroxyquinolinate of this element completely transforms into a melt. Chromium(III) was extracted by using 8-hydroxyquinoline with the aim of spectrophotometric determination [128]. *n*-Dichlorobenzene was used by this author [129] as fusible solvent; it is superior to naphthalene and biphenyl by virtue of its large precipitate volume.

Mercury, after the decomposition of small amounts of organomercury compounds, was separated from sulphuric acid solution at 60°C using a fatty acid melt; it was then determined by the X-ray fluorescence spectroscopy method [130]. From the solutions of salts, chromium was isolated with molten diethylammonium diethyldithiocarbamate [131]. A complex of copper with 3-(4-phenyl-2-pyridyl)-5,6-diphenyl-1,2,4-triazine and tetraphenylborate was extracted with molten naphthalene. Solidified concentrate was removed, dissolved in dimethylformamide and copper was determined by the AAS method. Using stearic alcohol melt, heavy metals (Bi, Cd, Co, Cu, Fe, Ni, Pb and Zn) were extracted as complexes with ammonium pyrrolidinedithiocarbamate, and were determined by the X-ray fluorescence spectrometry method [132].

High absolute concentrations are achieved in three-phase extraction, *e.g.* in systems containing diantipyrylmethane and its homologues (see, for example, [78, 133–138]). The authors of these works – Zhivopistsev and co-workers – have observed the formation of three phases (one aqueous and two organic phases) in the extraction of melts from halide and thiocyanate solutions using chloroform or other organic solvents in the presence of reagents of the type diantipyrylmethane (DAPM). In particular, by extracting elements from thiocyanate media with DAPM solution in benzene and chloroform, one can see the formation of two phases; one of them is diluent containing a small amount of DAPM thiocyanate while the other is DAPM thiocyanate with the diluent dissolved in it. High concentration of thiocyanate ion and DAPM in the second organic phase of small volume, which is known as the third phase, creates convenient conditions for the extraction of elements that form thiocyanate complexes. Up to 95% of extractable elements concentrate in the third phase, *e.g.* Ag, Au, Co, Fe, Ge, Mo, Sb, Sn, Ti, Th, W, Zn, Zr and others. In this case the concentration coefficient of trace elements can be increased by one to two orders due to effective segregation of the extract into two phases [137]. The obtained concentrate can be conveniently analysed by atomic emission and atomic absorption methods.

There are papers dedicated to the so-called "homogeneous" ("single-phase") extraction. This is a convenient method for increasing the rate of attaining equilibrium and/or for raising the concentration coefficients of impurities. As extractant, Japanese researchers [139] used propylene carboxylic acid which forms a homogeneous system with the solution of extractable elements (in the given case, with a solution of molybdenum(VI)) at a temperature exceeding 70°C. A two-phase system is formed on cooling to room temperature. A single extraction of metal is 97% complete. The second phase can be formed by decreasing the mutual solubility in the water–organic solvent system due to salting out. Thus, isopropyl alcohol mixes with water in all proportions under ordinary conditions; if the aqueous phase containing this solvent is saturated with sodium chloride, two phases are formed [140]. This fact has been used in isolating Au(III), Ti(IV), V(IV) and other elements. The addition of surfactants increases extraction. For decreasing the equilibrium attainment time in the extraction of iron(III) thenoyltrifluoroacetone, HTTA was introduced into the aqueous phase in the form of isopropyl alcohol solution [141]. The mixture was divided into layers by adding sodium nitrate. Acetone is a quite convenient solvent for atomic absorption analysis, but it mixes with water in all proportions; Matkovich and Christian [142] extracted with acetone the dithizonates, 8-hydroxyquinolinates and pyrrolidinedithiocarbamates of Co, Cu, Fe, Mn, Ni, Ti, V and Zn from the aqueous phase containing calcium chloride.

Similar to these systems are the systems with polyethylene glycol (PEG) which is a solid and dissolves well in water. Upon adding large amounts of a number of salts, for instance, ammonium sulphate, the water–PEG system separates into layers. The solution of PEG in water forms one phase and that of water in PEG the other, the water content being significant in both phases. Such a system is suitable for extraction of elements and has essential merits. It enables extraction of strongly hydrated compounds including multicharge complexes and complexes with several SO_3H groups, which under ordinary conditions are difficult to extract. Besides, the system does not contain any combustible, strong smelling, or toxic organic solvents. PEG is completely harmless and is readily available. Other water-soluble polymers may also be used. Refs. 143–145 discuss extraction of metals in the PEG–water system.

In recent years there has been intensive development of different methods of membrane extraction, as applied to separation and isolation of inorganic compounds [146]. In membrane extraction the substance is redistributed in a three-phase system in such a way that the layer of one of the liquids – membrane – separates two solutions which are immiscible with it and have different compositions. Its disadvantage is that it is time consuming. Several ways have been suggested to get over this disadvantage: attempts are being made to intensify extraction of ionic compounds by applying an electric field (by the so-called electroextraction). Also of interest is membrane extraction in multiple emulsion. This is probably the most important method for technological application of extraction.

In between extraction and sorption lies the concentration method involving the use of extractants coated on polyurethane foams [147–151].

2.2.5 EXTRACTION CHROMATOGRAPHY

Extraction chromatography is an effective method of concentration and separation [16, 152, 153]. It can be regarded as a peculiar variant of continuous extraction – the compound to be extracted is distributed between two liquid phases, one of which is fixed on a solid inert carrier placed in a column, while the other travels along the column. The chemical nature of the process remains extraction, but the technique of carrying out the process becomes chromatographic. In a number of cases, column extraction chromatography has advantages over the usual solvent extraction [154, 155]. Elements with almost similar properties can be separated by repeating many times the elementary events of extraction. The implementation of the process is no more complicated than in extraction. Use can be made of organic solvents which form stable emulsions in normal extraction. The distribution coefficients of components can be determined from elution curves; this is important in those cases when it is difficult to determine these values by normal methods. Other advantages of extraction chromatography include the high degree of absolute concentration and the possibility of carrying out experiments in "sterile" and isolated conditions and, hence, of lowering the value and the fluctuations of the blank experiment correction compared to conventional extraction. Extraction chromatography can be easily controlled to achieve group or selective concentration. Of no less significance is the ease with which this process can be automated.

One of the founders of extraction chromatography, Siekersky [16], believed that this method could compete with ion-exchange chromatography in both selective separation and experimental techniques. The possibilities of extraction chromatography show up most vividly when elements with almost similar properties are separated, *e.g.* lanthanides, actinides, zirconium–hafnium, niobium–tantalum, etc. Good results can be obtained by correct selection of the extraction system and the composition of the aqueous phase, particularly by adding masking substances, oxidisers, reducers, and also by controlling the separation conditions. The latter depend on the parameters of the chromatographic column, *e.g.* height of the theoretical plate (HTP) and the number of such plates. HTP depends on the column contents, the nature and granulation of the carrier, the elution rate, temperature, the type of extractant and its amount, its degree of dilution with neutral solvents and the composition of the aqueous phase.

The theoretical principles of extraction chromatography, which are common to all types of chromatography, are described in some fundamental works [156–158]. In these and also in refs. 16, 159 and 160 are given the methods of calculating parameters of chromatographic columns.

The distribution coefficients determined through the use of conventional extraction and extraction chromatography are usually in good accord with each other. But this fact, which has been confirmed experimentally, cannot be extended to all possible systems and cases. One of the reasons is that in the chromatographic column the activities of the extractant and the extractable complex can vary due to the reaction of immobile phases with the carrier, and extraction can take place under non-equilibrium conditions (the time is not sufficient for establishing extractional equilibrium).

The total retained volume of any component can be calculated provided the distribution coefficient, D, and the volumes of aqueous (mobile), V_w, and organic (immobile), V_o, phases are known.

$$V_{\max} \;=\; V_w + D V_o \qquad\qquad (2.17)$$

This is an important equation because it combines conventional (static) and dynamic extractions.

To separate three or more components it is sometimes advisable to use eluents of varying composition with a view to minimizing the separation time and to reducing the volumes of analysed solutions, *i.e.* performing gradient elution. If elution is carried out successively with several solutions (1, 2, . . . , n) for which the distribution coefficients D_1, D_2, ..., D_n can be taken from the data on conventional extraction, and the first eluent (volume V_1) washes off only the trace elements having minimum distribution coefficient, D_1, the second (volume V_2) washes off another trace element, and so on, then eqn. 2.17 characterizing the retained volume of the nth component takes the form [161]

$$V_{\text{max},n} = (V_{\text{w}} + DV_{\text{o}}) \prod_{1}^{n-1} \left(\frac{V_{\text{max},i} - V_i}{V_{\text{max},i}} \right)$$
(2.18)

Different extractants are used as immobile phase, *e.g.* neutral organophosphorus compounds, amines, their oxides and salts of quaternary ammonium bases, chelate-forming reagents, and others.

Some special requirements are placed on the carriers of immobile phases [16]. They should be well wetted by the immobile phase and must retain it in sufficient amounts. Besides, the immobile phase must not be detached from the carrier by the mobile phase. The carrier should be chemically inert and should not swell, it should not dissolve in the extractant nor sorb the mixture components to be separated. It must be uniform in particle size, should have a sufficiently developed surface and should ensure the flow of mobile phase through the column at the desired rate. The dimensions of particles, the density, mechanical strength and porosity are of no less importance. Finally, the carrier should be readily available. Hydrophobic silica gel, polyfluoro-chloroethylene, polytetrafluoroethylene (PTEE, Teflon), polyethylene and others have found extensive use as carriers. Block-type porous Teflon is an interesting carrier which gives columns with stable parameters [162–164].

The distinctive characteristics of extraction chromatography as a method of concentrating trace elements are considered in the review by Korobeinikova *et al.* [152]. Bolshova and Shapovalova [165] have gathered data that show how the extracting and non-extracting macroelements affect the behaviour of trace elements under the conditions of extraction chromatography.

Here we shall consider some examples of using extraction chromatography for concentration of trace elements. Often, trace elements are first concentrated in columns and then separated under such con-

42

ditions as to maximise the difference in the distribution coefficients (at least more than ten times as high). Then the matrix elements pass through the column, practically without being absorbed, while the trace elements are retained by the phase.

Glukhov *et al.* [166] isolated gold from complex composition solutions (particularly from natural waters) containing 21 cations with 0.5 M benzene solution of petroleum sulphides, on the pellets of porous Teflon. Only gold and palladium were absorbed from hydrochloric acid solutions. Determination was carried out by the neutron activation method. Alimarin *et al.* [167] used a system containing toluene solution of tridecylamine and hydrochloric acid solutions for concentration (and subsequent separation) of indium and cadmium in the analysis of pyrites. As a carrier of the mobile phase, they used Teflon. Indium and gallium present in different fractions of the eluent were determined by the polarographic method. Using systems with di-(2-ethylhexyl)-phosphoric acid and tributylphosphate, rare-earth elements were isolated for their determination in mineral raw materials [168].

Extraction chromatography is employed when the matrix has to be extracted. The volume of the immobile (organic) phase is very small, of course, and this imposes restrictions on obtaining high values of concentration coefficients and retaining the swollen organic phase on the carrier. But sometimes it is more important to achieve high selectivity of separation. Thus, analyzing high-purity tellurium by the neutron activation method, Grazhulene *et al.* [169] separated the matrix element in a column packed with tributylphosphate (on polytetrafluoroethylene) and HCl; they succeeded in determining eleven trace elements.

In neutron activation analysis of antimony, gallium and gallium antimonide – strongly activating matrices – for 35 elements-impurities, Gilbert *et al.* [170] passed the samples of mass 0.3–0.6 g (after dissolving them in 3–5 ml 9 M HCl solution) through chromatographic columns containing graphite powder, silica gel or Teflon and 2,2′-dichlorodiethyl ether as immobile phase. The matrix in this case was separated from the trace elements with a separation factor of 10^2; the radioactivity of the matrix elements did not exceed the background level. The solution after the elution of trace elements with 120–150 ml 9 M HCl was subjected to gamma-spectrometric analysis.

In another work [171] gold and silver were isolated for determining impurities in them by the NAA method (after irradiation). Dioctylsulphide on Teflon powder was used as immobile phase; HNO_3 and HCl solutions served as mobile phase for silver and gold, respectively.

Earlier, these authors concentrated impurities by extracting gold and silver after irradiating the samples in a reactor [172, 173]. Three to four extractions were carried out. However, extraction separation of a highly active matrix was not found to be an easy process; use had to be made of heavy face shields. Extraction chromatography is more convenient in this respect.

2.2.6 AUTOMATION

Even in the 1960's, Butlet *et al.* [174] had automated an atomic absorption procedure for determination of gold in cyanide solutions after its preliminary extraction with methyl isobutyl ketone. At the same time the AKUFVE device was developed [175] to automate the process of studying extraction. The device has a mixer, a centrifuge for separating phases, different detectors that enable radioactivity, optical density, refraction, pH and other properties of the phases to be determined immediately after the separation of phases. The productivity of the analyst increases in this case by 10–100 times.

A two-stage extraction system was employed for automatic determination of trace amounts of metals by the atomic absorption spectrometry method. Using a Technicon autoanalyzer [176], Cd, Co, Fe and Pb were determined after separating them as pyrrolidinedithiocarbamates with methyl isobutyl ketone. The output of the method is about 60 samples per hour. An automated extraction device is described in ref. 177, which, in combination with a 6-channel atomic fluorescent spectrometer, was used for determining traces of Co, Cr, Cu, Fe, Mn and Zn in sea water. The trace elements were extracted as diethyldithiocarbamates with a mixture of butyl alcohol and methyl isobutyl ketone. The output of this device was 25 samples per hour.

Flow injection analysis and similar systems of analysis offer significant possibilities for automating extraction concentration.

Kraak [178] in his review article dedicated to automatic preparation of samples with the use of extraction and sorption methods, has mentioned that the concentration-and-separation stage is the most labour intensive. Therefore, it is this stage that limits the creation of automatic systems for flow analysis. Also, he is of the opinion that only half of the automated systems described in the literature are suitable for continuous extraction by the flow injection analysis method. He has given a flow diagram of the device for single-stage automatic extraction; this device is intended for use in combination with high-perfor-

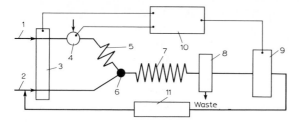

Fig. 2.8. System for flow injection analysis (solvent extraction version). 1 and 2 = Flows of aqueous and organic phases; 3 = pump; 4 = sample injection; 5 = mixing coil; 6 = segmentor; 7 = extraction coil; 8 = separation of phases; 9 = detector; 10 = microprocessor; 11 = regeneration of an organic phase.

mance liquid chromatography (HPLC), gas chromatography (GC) and AES–ICP. The concentration coefficient attained in this system does not exceed 4.

Fig. 2.8 shows a diagram of the flow injection analysis system for extraction. Here, various selective analyzers, including atomic absorption and atomic emission spectrometers with inductively coupled plasma, are used as detectors [179]. Nord and Karlberg [180] have described the flame atomic absorption determination of copper in solutions after its extraction as pyrrolidinedithiocarbamate with methyl isobutyl ketone. Copper concentrates by 5.5 times. An interesting method of concentrating Cd, Co, Cu, Ni and Pb as pyrrolidinedithiocarbamates in trichlorotrifluoroethane (Freon-113) in a flow system is described in ref. 181. In the proposed device, extraction is carried out at pH 6, and back-extraction is accomplished with Hg(II) solution in nitric acid. The back-extract goes to the graphite atomizer of the atomic absorption spectrometer. For highly accurate automated extraction–atomic absorption determination of heavy metals in sewage waters, Vatanabe et al. [182] have suggested a procedure for flow injection analysis. The procedure involves mixing of sample with reagents and an organic solvent, introduction of reference samples of determinable elements, mixing and separation of phases and transfer of extract into the spectrometer crucible.

2.3. Sorption methods

2.3.1 GENERAL CHARACTERISTICS OF METHODS

In these methods of preconcentration, based on sorption of gases and of dissolved substances with solid or liquid sorbents, use can be made of diverse sorption mechanisms: adsorption (sorption of substances on the surface of a solid or liquid body), absorption (sorption of gases, vapours or substances dissolved in the volume of a solid or liquid phase), chemical sorption (sorption of substances by solid or liquid sorbents with the formation of chemical compounds) and capillary condensation (formation of a liquid phase in the pores and capillaries of solid sorbent during sorption of vapours of the substance). In practice however, it is difficult to encounter separately any of the above mentioned types of sorption: they are generally used in combination with each other. Thus, adsorption usually precedes chemisorption. Adsorption and chemisorption methods are widely used for concentration of trace elements. Among the latter extensive use is made of ion exchange and sorption which is accompanied by complex-formation, say, on chelate-forming sorbents.

A classification based on the technique employed can also be considered. Here, we shall single out static and dynamic (particularly chromatographic) methods. Using the classification by geometric parameters of immobile phase, chromatographic methods can be divided into column and plate methods [paper and thin-layer chromatography (TLC)].

Sorption methods of concentration, in particular ion-exchange methods, are often employed when the mutual interfering effect of trace elements present in the sample can show up in subsequent determination of the trace elements. Sorption concentration generally ensures good selectivity of separation and high values of concentration coefficients. Concentration can be relatively easily controlled; it does not call for high temperatures and sophisticated instruments. Therefore, sorption methods can be conveniently used in field conditions; they are quite convenient for group concentration also.

For concentration of trace elements use is made of different sorbents which, besides having good sorption power and selectivity, should have the ability to be easily regenerated, and be chemically and mechanically stable. The following sorbents are in general use: activated charcoals, normal and modified cellulose, ion-exchange and chelate synthetic sorbents and different inorganic sorbents.

Many papers are devoted to sorption methods, the majority being experimental rather than theoretical [20, 183–192].

For quantitative estimation of sorption, use is made of the degree of separation and the distribution coefficient. Here the distribution coefficient, D, is generally the ratio of concentration of the sorbed substance in a solid sorbent (c_{solid}) to its concentration in solution (c_{sol}):

$$D = \frac{c_{solid}}{c_{sol}} \tag{2.19}$$

With consideration of the specificity of phases, D can be more accurately expressed as

$$D = \frac{\text{amount of the sorbed substance (in g of sorbent)}}{\text{amount of the sorbed substance (in ml of solution)}} \tag{2.20}$$

Sometimes, c_{solid} is expressed in volume of the sorbent, for instance, in millilitres.

Under static conditions the distribution coefficients of the component between the liquid phase and the sorbent can be determined as follows. A weighed amount of the sorbent is taken in a flask, a definite volume of solution containing a known amount of the element to be sorbed is added, and the contents are shaken until equilibrium is attained. After the separation of liquid and solid phases, an aliquot of the solution is taken and analysed. Concentration of the element in the sorbent is found from the difference. In chromatography, the distribution coefficient is the ratio of the total amount of the substance absorbed on every theoretical plate (stage) (see elsewhere) to its total amount in that portion of the solution which corresponds to the free volume of the theoretical plate after equilibrium has been established. The distribution coefficients are computed from C or F diagrams.

We shall consider in detail the sorption concentration technique. The static method is employed when the distribution coefficients of the sorbed substance are large enough. A weighed amount of the pulverized sorbent is brought into contact with a known volume of the analyzed solution; usually this is done in flasks by mixing the contents for a definite period. The concentrate is then filtered or decanted, washed (by different methods), and usually dried. Drying is not necessary when the sorbent decomposes or desorption is employed. Desorption with suitable solutions is carried out in the same manner as sorption. The sorbent is necessarily decomposed in irreversible sorption, as is done, say, in the case of platinum metals or, more often, when using determination methods that involve working with solutions. A typical case is flame atomic absorption spectrometry. The sorbent can be decomposed by treating it with acids or oxidizers. Decomposition of the concentrate is not obligatory and may even be unnecessary when X-ray fluorescence, neutron activation, and electrothermal atomic absorption methods of determination are used. Thus, for XRFS analysis it is sufficient to press the concentrate into a tablet.

The sorption filter method consists of filtering the analysed solution through a thin layer of the sorbent coated on a substrate, through a layer of paper having sorption properties, or through a specially made membrane. This is a very simple technique of concentration, but it can be employed only when the distribution coefficients are large and, mainly, when sorption is carried out quickly. In any event, the solution must not be passed through the filter quickly. After concentration, the absorbed substance may be desorbed or the filter is decomposed or directly used for analysis. As with the static variant, the latter method is preferred because it is simple and does not bring in additional impurities when the concentrate is treated chemically.

Sometimes the filtering layer is made thick; in this case, the separation method approximates a chromatographic method.

Column chromatography is widely used for sorption concentration. It is very convenient for group concentration of elements, including micro- and macroelements, with not very different properties.

There are three types of chromatography: elution, frontal and displacement. These differ in the experimental procedure and also in utility. Quantitative separation of a mixture of components is accomplished by elution chromatography as a rule. Elution chromatography is convenient because for series analysis of samples it is not necessary to regenerate the column.

The experimental procedure may be slightly changed depending on what elements – macro or micro – are better absorbed on the column. For example, when the matrix elements are weakly sorbed, the column need not be large; also, its volume is unimportant. When the matrix element is absorbed and the traces are relatively readily eluted, the volume of the sorbent contained in the column should be sufficient. The task is more complicated in those cases when sorption of macro- and different trace components is significant and elution has to be performed quite skilfully. For instance, it becomes necessary to use large columns, to choose a narrow range of particle size, and so on.

The theory of chromatography is well developed and has been dealt with in special literature, for instance in refs. 183, 186 and 192. As in other cases, efficiency, selectivity, symmetry of peaks and separation rate must be ensured in concentrating elements by the chromatography

48

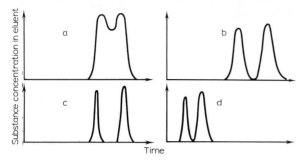

Fig. 2.9. Chromatograms characterizing efficiency and selectivity of separation. a = Insufficient separation; b = improved selectivity; c = improved efficiency; d = more rapid separation.

method. In chromatography, efficiency is measured by the narrowness of peaks on a chromatogram; selectivity means non-overlapping of peaks (Fig. 2.9). The peaks are symmetrical when sorption isotherms are linear and the separation rate, among other things, is characterized by the distance between (the well-resolved) peaks. A sorption isotherm is the dependence of the substance equilibrium concentration in one phase on that in the other phase (Fig. 2.10). Its shape, concave (a), linear (b), or convex (c), determines the form of a chromatographic peak.

There are two theories of chromatography which to a different

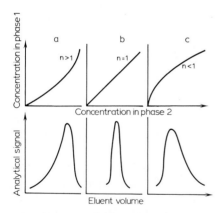

Fig. 2.10. Chromatographic peak form vs. sorption isotherm character. a = Concave isotherm, peak with a stretched front; b = linear isotherm, symmetric peak; c = convex isotherm, peak with a tail.

extent help in selecting conditions for sorption concentration in columns: the theory of theoretical plates or stages, and the kinetic theory. They operate on many common concepts and there is no distinct boundary between them.

In the theory of theoretical plates a column is visualized as a totality of discrete touching horizontal layers. Equilibrium is assumed to be established in every layer between the mobile and immobile phases. Such layers are identified as theoretical plates. The movement of substance and solvent is considered as displacement from one stage to another. The separation efficiency increases with the number of equilibria, that is, with the increase in the number of theoretical plates. In other words, the number of theoretical plates, N, is the measure of column efficiency. The other quantity is the height, H, equivalent to a theoretical plate (HETP). If L denotes the column length, then

$$N = L/H \tag{2.21}$$

The plate height, H, decreases with increasing column efficiency.

The theory of theoretical plates describes the migration rate of the dissolved substance. It contains equations for determining the peak shape. Nevertheless, it does not explain the effect of the elution rate or of the properties of the column packing on the peak width and the values of H and L. Therefore, this theory was rejected in favour of the kinetic theory. However, H and N appear also in the kinetic theory as efficiency parameters.

The second, kinetic, theory explains the effect of different process conditions on the peak width and the time of its appearance.

Here, the retention time, t_R, i.e. time needed for a peak to reach the column end, is introduced. The substance migration rate is expressed by the relationship

$$v = L/t_R \tag{2.22}$$

For a solvent, we have

$$v_n = L/t_n \tag{2.23}$$

The retention index is an important parameter in chromatography. It is the ratio of the flow-rate of the dissolved substance to that of the solvent:

$$R = \frac{L/t_R}{L/t_n} = t_n/t_R \tag{2.24}$$

50

In planar chromatography, R_F is an important characteristic.

Use is also made of the term, retained volume, V:

$$V = t_R F \tag{2.25}$$

Here, F stands for the flow-rate of the mobile phase.

For a solvent

$$V_n = t_n F \tag{2.26}$$

It can be shown that

$$R = V_n / V \tag{2.27}$$

The physical meaning of R is that it characterizes the time a molecule of the dissolved substance remains in the immobile phase.

Usually, after column preconcentration the concentrate is in the form of solution (eluate). It is then either directly analysed by some method or subjected to further treatment, say evaporated. In the last few years, a different approach typical of modern chromatographic analytical methods has been noted. It involves continuous detection, *i.e.* determination of components present in the eluent in the course of the process. Of such techniques, as applied to inorganic analysis, the most interesting is ion chromatography where detection is effected by a conductometric, or other, detector.

Next we shall consider how traces of elements are concentrated with the use of some important sorbents.

2.3.2 ACTIVATED CHARCOALS

Activated charcoals have found extensive use in concentration of trace elements, concentration being effected principally by molecular adsorption processes (though sorption also plays a part in other mechanisms, say ion exchange). Therefore, we shall not discuss in detail adsorption as a process.

In adsorption, substance concentrates on the surface of the phases to be separated due to molecular forces acting on the surface of the sorbent. An adsorption system consists of an adsorbent (a substance with a large specific surface) and an adsorbate (a substance whose molecules are adsorbed). Several physical (forces of attraction) and chemical (formation of a compound) factors are responsible for the interaction between the molecules of the adsorbate and the surface of the solid body. Physical adsorption can be easily reversed.

For describing different sorption processes, use can be made of the Langmuir, Polyany, Freundlich, and other empirical equations which connect the amount of sorbed substance (a) and the equilibrium concentration of the substance in the solution or the gaseous phase (C). All these give rise to a relationship for the sorption isotherm, *i.e.* $a = f(C)$. Thus, the Freundlich equation is written as

$$a = bC^n \tag{2.28}$$

where b and n are constants.

Depending on the values of n, the curves corresponding to this equation – sorption isotherms (*cf.* Fig. 2.10) – can be convex ($n < 1$), linear ($n = 1$) and concave ($n > 1$), the system with convex isotherms being the most favourable [193]. The values of b and n increase with the boiling point of the adsorbate and decrease with the increase in the adsorption temperature. Each adsorbent has its own isotherm which is the main characteristic of its adsorption power.

When adsorption is carried out in a flow, the preconcentration efficiency depends not only on the type of adsorption isotherm, but also on the mass exchange rate. Sorption takes place in three stages: the substance to be sorbed is transferred to the outer surface of particles (outward diffusion); the molecules penetrate into the particles and the pores of the sorbent (inward diffusion); attachment of molecules to the inner surface of the pores of the sorbent (adsorption itself), the latter being the fastest stage of adsorption. The effect of outward and inward diffusion depends on the conditions under which the process is carried out and on the system. In most cases, the total adsorption rate is determined by the process of transferring the substance to the inner surface of the pores. Therefore, in adsorption procedures special emphasis is placed on the study of the dynamics of the process.

Adsorption on activated charcoals is effected first of all by the dispersion forces acting over very small distances, unlike heteropolar sorbents such as silica gel, alumogel and zeolites, with electrostatic forces being the reason for sorption on them. Therefore, the efficiency of activated charcoals depends first of all on the chemical structure of micropores with radius less than 1 nm. Three types of oxides exist on the surface of activated charcoals [194]:

The presence of acidic and basic oxides is responsible for the amphoteric nature of activated charcoals when acids and bases are adsorbed. Physical adsorption may be accompanied by cation- or anion-exchange processes.

Several papers on the use of activated charcoals as effective adsorbents of trace elements have been published [195]. For determination of gold and silver in rocks and ores, Shugurova [196] concentrated these elements on activated charcoals or sulphonated coal in columns made of carbon electrodes which are used in atomic emission analysis. This enabled the sample preparation to be simplified to the maximum extent for subsequent analysis because, after drying at 105°C, the column was used as the lower electrode. A procedure has been worked out for absorbing elemental lead from atmospheric air [197], which involves filtration of sample through a graphite disc and then through a layer of activated charcoal where lead is sorbed. The sorbent was burnt at 1500°C and lead was determined by the atomic absorption method. The procedure for determining gaseous bromine and iodine in atmospheric air [198] involves their concentration on activated charcoal, besides electrostatic precipitation. Thereafter, the sorbent is irradiated with a neutron flux and treated with sodium hydroxide solution containing a silver salt as carrier and a small amount of sodium hypochlorite. The precipitates of silver bromide and iodide are removed and their β-radioactivity is measured.

The impurities (Ag, Bi, Cd, Co, Cu, Fe, Hg, In, Mn, Ni and Pb) present in sodium perchlorate were concentrated, after dissolving the salt in water, by filtering the solution through a filter paper coated with 50 mg of activated charcoal [199]. The metals were then desorbed with concentrated nitric acid and determined by flame AAS; mercury was not desorbed and was determined in a different way.

Oxidized activated charcoals occupy a special place. They are selective multi-functional cation exchangers. Unlike synthetic cation exchangers, they can be easily prepared, readily regenerated and are quite stable to chemical, thermal and radiation effects. The monograph by Tarkovskaya [200], dedicated to activated charcoals, contains structures of surfaces of such charcoals (Fig. 2.11). An oxidized activated charcoal like a sorbent exhibits predominantly ion-exchange properties, namely, it first of all acts as a cation exchanger. Therein [200] are given examples of using such charcoals for concentration of, say, alkaline-earth metals in the presence of alkalis.

Tarkovskaya et al. [201] concentrated trace elements selectively with oxidized charcoals to determine Cr, Mo and V in water, brines of

TABLE 2.2

Examples of concentrating trace elements as complexes on activated charcoals

Trace elements concentrated	Substance analysed	Reagent used	Method of determination	Remarks	Ref.
Ag, Cd, Co, Cu, In, Ni, Pb, Tl, Zn	Tungsten	Sodium diethyldithiocarbamate	AAS, XRFS,	Desorption HNO_3	202
Cd, Co, Cu, Fe, Ni, Pb, Zn	High-purity selenium	Sodium diethyldithiocarbamate	AAS	Desorption HNO_3	203
Ag, Bi, Cd, Co, Cu, In, Ni, Pb, Tl, Zn	Chromium(III) salts	Hexamethylene-dithiocarbamate Hexamethylene-ammonia	AAS	Sorption filter, desorption HNO_3	204
Bi, Cd, Cu, In, Pb, Tl	Minerals, cement, metallic iron	Ammonium diethyldithiophosphate	AAS, AES-ICP	Sorption filter, desorption HNO_3	205
Bi, Cd, Cu, In, Pb, Tl	Gallium and high-purity aluminium	Ammonium diethyldithiophosphate	AAS	Sorption filter, desorption HNO_3	206
Cd, Co, Cu, Pb	Aqueous solutions of salts (NaCl, $MgCl_2$, $CaCl_2$)	Dithizone, diphenylcarbazide, 8-hydroxyquinoline, anthranilic acid	AAS, AES	Sorption filter, desorption HNO_3	207
Ag, Al, Bi, Cd, Co, Cr, Cu, Fe, Hg, Hf, In, Mn, Mo, Nb, Ni, Pb, Pd, Re, Se,Sn, Ta, Ti, Tl, W, Zn, Zr, REE*	—	8-Hydroxyquinoline	XRFS	Concentration coefficient $1 \cdot 10^4$	208

Elements	Material	Reagent	Method	Procedure	Ref.
Cd, Co, Cr, Cu, Fe, Hg, Hf, Mn, Ni, Pb, Re, Zn, REE	Natural waters	8-Hydroxyquinoline	Photometry, AAS, XRFS, NAA, γ-activation analysis	—	209
Ag, Bi, Cd, Co, Cu, Fe, Hg, In, Ni, Pb, Tl	Zinc and zinc nitrate of high purity	Potassium xanthate	AAS, AES	—	210
Bi, Cd, Co, Cu, Fe, In, Ni, Pb, Tl, Zn	Manganese and its compounds	Potassium xanthate	Flame AAS	Sorption filter, desorption with acid	211
Bi, Co, Cu, Fe, In, Pb	Metallic silver and thallium nitrate	Xylenol orange	AAS	Sorption filter, desorption with HNO_3	212
F	Natural waters	Alizarin S	Photometry	Activated charcoal impregnated with zirconium nitrate, sorption filter, desorption with NaOH	213

* Rare earth elements

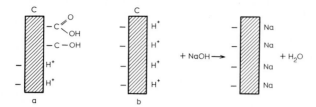

Fig. 2.11. Surface of oxidized charcoal (a) and the ion exchange mechanism (b).

mercury electrolysis, zinc and nickel salts. The concentrate was then directly burnt in the channel of the graphite electrode for AES.

Most often, on activated charcoals are adsorbed the complexes of metals with specially introduced reagents; the latter are selected in such a manner that adsorption of a macrocomponent is much less than the adsorption of complexes of trace components. The efficiency of concentrating complex compounds is determined by their stability constants, nature and structure of ligands, and the charges of complexes. It is possible to introduce complex-forming agents into the analysed solution, or the process can be carried out directly on activated charcoals. Examples of this concentration are given in Table 2.2. Figure 2.12 shows the filtration device (activated charcoal on a paper disc) used in ref. 203.

Fig. 2.12. Apparatus for the filtration of a sample through activated charcoal.

56

On synthetic ion exchangers separation takes place mainly by the ion-exchange mechanism. Ion exchangers form complex compounds with adsorbable ions. Equipment and procedures for working with ion exchangers are described in the books by Marhol [190], Senyavin [193], Rieman and Walton [183] and others.

The application of ordinary high-molecular-weight synthetic ion exchangers is hampered by the need to handle large volumes of solutions. However, these sorbents are employed in many laboratories, particularly when use is made of microcolumns. Trace elements, as well as the matrix, can be separated. If the trace elements are sorbed, then for their determination they are desorbed or the sorbent is ashed. If very pure sorbents are available, direct atomic emission (or any other) analysis of the sorbent concentrate is possible.

Synthetic ion exchangers are usually made of copolymers of styrol and divinylbenzene having divinylbenzene cross-links. The degree of cross-links determines many important properties of the sorbent, in particular, swelling ability and sorption rate. Use is mostly made of strongly cross-linked sorbents (up to 10 or more percent of divinylbenzene) which comprise fine grains (100–400 mesh). Care is taken that the grains have a spherical shape and the fraction used contains grains of very similar diameter.

The main thing is, however, that acidic $(SO_3^- H^+)$ or basic $(N(CH_3)_3^+ OH^-)$ groups, *i.e.* ionogenic groups, attach to this polymeric matrix. Ion exchangers with acidic groups can exchange cations, and are called cation exchangers; sorbents with basic groups can exchange anions, and are known as anion exchangers. Examples of ion exchangers are given in Table 2.3. In common use are the sorbents with highly acidic or highly basic groups, about which we have mentioned earlier. Elsewhere are given the structures of the cation-exchange resin KU-2 (XVII) and of the anion-exchange resin AV-17 (XVIII).

XVII   XVIII

TABLE 2.3

Some synthetic ion exchangers

Type	Ionogenic group	Trade name	Exchange capacity (m equiv./g)
Cation exchanger	$-SO_3H$	KU-2	4.9–5.1
		Dowex 50	5
		Amberlite IR-120	5
	$-SO_3H, -OH$	KU-1	4.5–5.1
	$-COOH$	KB-4	10
Anion exchanger	$-N(CH_3)_3Cl$	AV-17	4.3
		Dowex 1	3
		Amberlite IRA-400	3
	$=NH, \equiv N$	AN-2F	10.6
	$-NH_2, =NH$	AN-1	4

Ion exchangers of the above forms exist not only as fine-grain powder, but also as membranes and paper; macroporous granules are also known. In ion chromatography use is made of spherical ion exchangers, say, in the form of a thin layer coated on the surface of a ball of other material.

Commercial ion exchangers are, as a rule, contaminated with trace elements and, therefore, call for purification. For instance, cation-exchange resins are purified by washing with acid solutions.

The essence of an ion-exchange process is to replace the sorbent counterions (for instance, H^+ in the case of $SO_3^- H^+$) with the ions of the elements to be concentrated, which are present in the analysed solution.

Ion-exchange reactions can be treated as normal chemical reactions; in particular, use can be made of constants of chemical reactions known as exchange constants. The point is that ion-exchange processes are usually reversible. In the general form, a cation-exchange reaction is expressed as

$$X^- A^+ + B^+ \rightleftharpoons X^- B^+ + A^+$$

Here, X^- stands for polymeric matrix with a dissociated functional group. For a concentration exchange constant (which does not account for the activity coefficient), we shall have

$$K_{ex} = \frac{[X^- B^+][A^+]}{[X^- A^+][B^+]} = \frac{[B^+]_{solid}[A^+]}{[A^+]_{solid}[B^+]} \tag{2.29}$$

In this expression, $[B^+]_{solid}$ and $[A^+]_{solid}$ are the concentrations of cations in the ion exchanger.

In an ion-exchange method, use is often made of complexes of elements present in the analysed solution or especially obtained prior to preconcentration. Complex formation enables the difference in sorption behaviour of separated elements to be increased, sometimes very significantly; in particular, one of the elements to be separated is transferred into an anion complex and the other is kept as cation. An example of this is the separation of iron(III) and aluminium by adding ammonium thiocyanate. Only iron forms an anionic thiocyanate complex, $Fe(SCN)_4^-$. Fe(III) and aluminium are easily separated using an ion exchanger. Most often use is made of organic complex forming reagents. Water–organic solutions often aid in increasing the ion-exchange concentration efficiency.

Ion exchangers are employed to sorb trace elements and to absorb the matrix elements.

An example of solving the difficult task of concentrating rare earth elements present in high-purity yttrium oxide is found in the work of Cheng Jai-Kai *et al.* [214]. All the determinable rare earth elements and yttrium are sorbed in a long (50 cm) ion-exchange column using a cation exchanger. First, yttrium is eluted with $0.15\,M$ solution of α-isobutyric acid, and then La, Ce, Pr, Nd, Sm and Eu, with $0.5\,M$ solution of the same acid. The concentration coefficient equals $1 \cdot 10^4$. AES is employed to determine rare earth elements in the concentrate. For determining inorganic forms of mercury in natural waters [215], mercury was sorbed on 200 mg of an ion exchanger under static conditions. Then the sorbent was filtered out and added into the flask containing a solution of tin(II); here mercury was reduced to elementary mercury and the vapours of the latter were removed from the flask by a current of air into a closed system, and the mercury was determined by the AAS method.

Other examples are listed in Table 2.4.

Ion chromatography is another useful technique for preconcentration. Absolute concentration may be effected, in particular, in a concentration column. Preconcentration enables the detection limits of ions to be lowered (sometimes by two to three orders of magnitude). Thus, by using a concentration column, Zolotov *et al.* [226] succeeded in lowering by 10^3 times the detection limit of selenium present in natural waters in the form of SeO_4^{2-}. Without concentration the minimum determinable concentration of selenium was $10\,\mu g/l$. By using a

References pp. 179–203

TABLE 2.4

Examples of concentration of trace elements on synthetic ion exchangers

Determined elements	Substance analysed	Ion exchanger used	Sample preparation and concentration conditions	Method of determination	Ref.
Sorption of trace elements					
Ag, Au, Ba, Bi, Cd, Co, Cr, Cu, Fe, Mg, Mn, Ni, Pb, Zn	High-purity arsenic and its trichloride	Cation exchanger KU-2-8	Oxidizing with a mixture of HNO_3 and HCl, neutralization up to pH 0.9–1.0; column chromatography	AES	216
Ag, Ba, Ca, Co, Cr, Cu, Fe, Mg, Ni, Pb, Zr	High-purity germanium	Cation exchanger KU-2-8	Dissolving in a mixture of HF and HNO_3, evaporating to dryness, dissolving in HF, diluting to pH 0.9–1.0; column chromatography	AES	217
Al, Bi, Cd, Co, Cr, Mg, Mn, Sb	High-purity gallium and gallium arsenide	Cation exchanger KU-2-8 and Anion exchanger EDE-10P	Dissolving in a mixture of HCl and HNO_3, evaporating to dryness, dissolving in 0.75–1.0 M HCl; column chromatography	AES	218
Cd, Co, Cu, Ni, Pb, Zn	Natural waters	Cation exchanger Dowex A-1	pH 6.9–7.1; sorption under static conditions	AAS	219

Elements	Matrix	Ion exchanger	Procedure	Method	Ref.
Ce, La, Pr	Carbon steels	Anion exchanger Bio-Rad 1-X10	Dissolving in HNO_3; reducing Ce(IV) to Ce(III) with H_2O_2, mixing with methanol and ethanol; column chromatography	XRFS	220
Ca, Cd, Cu, Fe, Mg	–	Anion exchanger Bio-Rad Ag MP-1	Forming chelates with thyron; column chromatography	AES–ICP	221
Bi, Pb	Steels	Anion exchanger EDE-10M	Dissolving in HNO_3 and then in HCl; evaporation; dissolving in $2\,M$ HCl; column chromatography	AAS	222

Sorption of matrix

Elements	Matrix	Ion exchanger	Procedure	Method	Ref.
Cd, Cu, Pb	High-purity bismuth	Cation exchanger KU-2-8	Dissolving in HNO_3, evaporation, adding thyron and ethylenediamine; column chromatography	SV	223
As, Cd, Co, Cu, Fe, Ga, Hg, Sb	High-purity silver	Cation exchanger KU-2	Dissolving in $3\,M$ HNO_3, adding EDTA, pH 2, column chromatography	NAA	224
Co, Cr, Cs, Fe, Ga, K, La, Mn, Na, Rb, Zn		Anion exchanger AV-17	Dissolving in HNO_3, eluent $0.1\,M$ HNO_3		
Pr	Oxide and sulphide of lanthanum	Anion exchanger Dowex 1X8	Dissolving in $10\,M$ HCl, mixing with eluent ($0.5\,M$ $LiNO_3 + 0.1\,M$ HNO_3 in methanol) and $4\,M$ LiOH, column chromatography	NAA	225

concentration column this limit was lowered to 0.01 µg/l, and this enabled selenium to be determined in natural water. In a like manner, by effecting concentration in a column, small amounts of anions present in rain water were determined [227]. The chromatograph was equipped with ion-selective electrodes (for F⁻ and Br⁻), a UV detector for determining nitrate ions, a minicomputer for continuous (round-the-clock) analysis control, and a supplementary concentration column packed with the ion exchanger Zipax SAX, into which the sample (4 ml) was introduced. The device can analyse twelve samples per hour.

2.3.4 COMPLEX-FORMING SORBENTS

Upon using weakly acidic cation exchangers containing, for instance, a COOH group, the ion-exchange process is accompanied by the formation of a complex of metal with the functional group. The complex-formation process is further enhanced if use is made, say, of the $PO(OH)_2$ group or, better, of amino-diacetate. This leads us to the currently popular group of sorbents which form complex compounds, most often (but not always) chelate compounds. These sorbents can be divided into at least three groups: (1) sorbents with complex-forming groups inoculated to the polymeric or inorganic matrix; (2) the so-called polymeric heterochain sorbents; (3) sorbents modified or impregnated with complex-forming reagents. There are some very diverse compounds in this latter group.

2.3.4.1 Complex-forming sorbents with inoculated groups

(i) On a polymer base. The great number of analytical reagents available and the extensive information on their reactions with elements make it possible to prepare polymer-based sorbents with various selective and non-selective groups of such reagents. Such chelate-forming sorbents have proved to be very effective sorbents for concentration of trace elements in analysis of different natural and industrial samples. Their high selectivity is mainly due to the nature of "inoculated" function groups. Refs. 191, 228 and 229 contain detailed characteristics of these sorbents and consider the possibilities for their application.

The main merit of such sorbents is indeed their inexhaustable diversity. Nevertheless, it is not always easy to synthesize sorbents. Sorbents retain a number of properties of normal ion exchangers, say swelling ability; as in the case of ion exchangers, the swelling ability depends on the degree of stiffness of the polymer skeleton. Sorbents can be prepared as grains, fibres and even fabrics. Sometimes, fine grains of sorbents are added to an inert material which can be conveniently processed; such sorbents are known as filled sorbents.

Extensive use is made of the sorbent Chelex-100 (XIX) containing an iminodiacetic acid group.

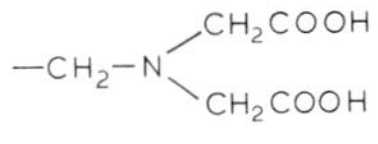

$$-CH_2-N\begin{array}{c} CH_2COOH \\ CH_2COOH \end{array}$$

XIX

Thus, in the sodium form (100–200 mesh) this sorbent has been used [230] for concentrating thirteen trace elements present in urea, the concentration coefficient being 50; the trace elements are determined by the X-ray spectrometry method with proton excitation (PIXE). Conditions have been found for simultaneous concentration of Cd, Co, Fe, Mo and Ni (pH 4.7–5.5) with this sorbent in analysis of biological samples [231].

In the analysis of natural waters for uranium content [232] the Chelex-100-based membranes were placed at the bottom of the vacuum filtration device, covered with a filter to protect the membranes from suspension particles, and 250 ml of the sample to be analysed was passed through them. The membrane was then dried and analysed by the X-ray fluorescence method. Analysis including preconcentration was completed in 15 min. The detection limits of uranium were $5 \cdot 10^{-8}\%$. Van Grieken et al. [233] have shown that heavy metal ions can be concentrated from water using Chelex-100 filters. The volume of the sample was 200 ml, pH 7–8, and pressure 200–300 kPa. Concentration time was 20 min. The filters after concentration were analysed by the X-ray fluorescence method.

Sorbents with several other complex-forming groups have been synthesized and used. Grote and Kettrup [234] have recently reported the use of P–D sorbent containing inoculated dithizone (XX) for isolation of noble metals:

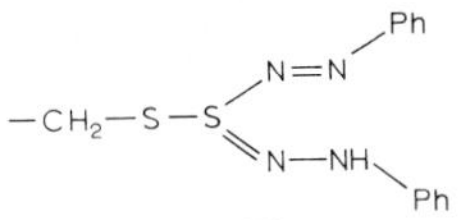

$$-CH_2-S-S\begin{array}{c} N=N-Ph \\ N-NH-Ph \end{array}$$

XX

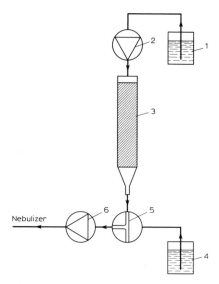

Fig. 2.13. Schematic diagram of a dynamic system combining liquid chromatography and atomic absorption spectrometry. 1 = Solution to be analyzed; 2 and 6 = pumps; 3 = column with a sorbent; 4 = solution of a multi-elemental reference sample; 5 = three-way stopcock.

Polystyrol serves as matrix. The sorbent was used for concentration of noble metals. It is interesting to note that the authors studied the sorption properties using a dynamic system that incorporates liquid chromatography and AAS (Fig. 2.13).

Myasoedova and Savvin [191] have synthesized a number of important sorbents, particularly for concentration of noble metals. Among them is the sorbent PVB–MP containing group XXI.

$$-N\begin{array}{c} \diagup CH{=}CH \\ \diagdown \\ N{=}\underset{|}{C}{-}CH_3 \end{array}$$

XXI

The polymer matrix is made of copolymers of styrol with (8–20%) divinyldibenzol and has a microporous structure. The sorbent exhibits selectivity towards noble metals and has a fairly large capacity (600 mg/g for gold). By IR spectroscopy it has been shown that a nitrogen atom of the pyrazole group participates in the complex-formation process. The sorbent has been recommended for concentration of Ag, Au and Pt metals from solutions containing Al, Co, Cu, Fe, Ni and other elements [235, 236]. Also, it is used to determine noble metals in copper–nickel ores and other substances [237]. Some sorbents of the series Polyorgs®, obtained by the same authors, are also used for the purpose. These include, for instance, Polyorgs® VI with group XXII.

XXII

The matrix is of polyvinyl fibre. The sorbent finds a number of specific applications (see, for example ref. 238). The chelate-forming sorbent suggested in ref. 235 was used for selective concentration of noble metals by thin-layer chromatography. Finely ground sorbent was mixed with an equal amount of cellulose powder in the presence of chloroform, the mixture was placed on a support and dried. Up to 0.01 ml of the analysed solution was placed on the start line and was left overnight; the plate was immersed in 1 M HCl; after the mobile phase had risen up to about 5 cm the plate was dried. The sorbent zone along the start line was removed, decomposed by an acid, and noble metals were determined in the residue.

Generally the plane (two dimensional) variants of chromatography are not very convenient for concentration, though they are simpler and more economical than column chromatography. These concentration techniques combine poorly with the instrumental determination methods. Besides, they do not always enable the components appreciably differing in concentration to be separated. Determination has to be made in a small volume of the sample, and this adversely affects the values of concentration coefficients.

As matrix, use is made not only of synthetic polymers but also of natural ones, particularly of cellulose. Thus, fibrous sorbent Milton-T is important for concentration of platinum metals. It contains the $-CSNH_2$ group; a copolymer of cellulose with acrylonitrile having a fibrous structure serves as matrix. Orobinskaya et al. [239] have studied the ion-exchange sorption of platinum group metals under static conditions at 100°C. Pd, Pt and Ru were sorbed from 1.5–3 M hydrochloric acid, and Ir and Rh, from 1 M HCl. A procedure has been developed for concentration of traces of platinum group metals from dilute solutions. In another study [240], platinum group metals were also concentrated from waste solutions on a Milton-T fibre, and then determined by the AES method. Other works, for instance [241], are also dedicated to Milton-T. For determining Cu, Fe, Mn and Zn in soil extracts the trace elements were concentrated by passing the analysed solution through cotton precoated with iminodiacetyl cellulose [242]. The same sorbent has been used for concentrating traces of Cd, Cu and Pb in atomic absorption analysis of drinking waters, methanol, *n*-propyl alcohol and acetone [243], for separating out uranium from sea waters [243], for concentrating traces of different metals from citrate and acetate solutions [244], and also for concentrating traces of Bi, Cd, Co, Cu, Fe(III), Mn (II), Pb, Ti, V(V) and Zn in analyzing magnesium by the AES–ICP method [244]. Traces of Cu, Fe, Hg, Sr and Zn were concentrated from water using cellulose filters [245]. Preliminarily, a functional group of chromotropic acid was introduced into cellulose, and compound XXIII was formed:

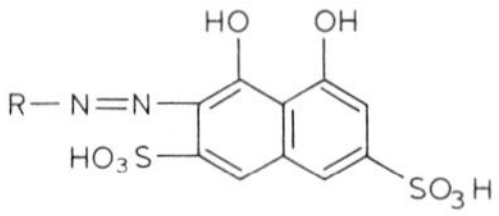

XXIII

The developed method excels in simplicity because cellulose filters are analysed by the X-ray fluorescence method without any additional treatment.

Lieser *et al.* [246] concentrated trace amounts of Cd, Co, Cr, Cu, Fe, Mg, Mn, Mo, Ni, Pb, U, V and Zn in analysis of water, concentration being effected with modified cellulose which contained 1-(2-hydroxy-phenylazo)-2-naphthol (XXIV) or 4-(2-pyridylazo)resorcin (XXV) [246].

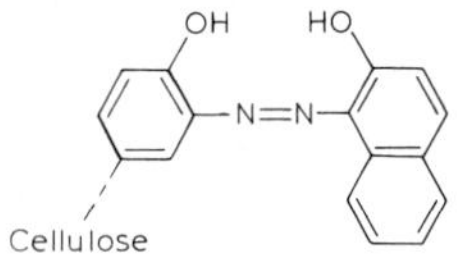

XXIV 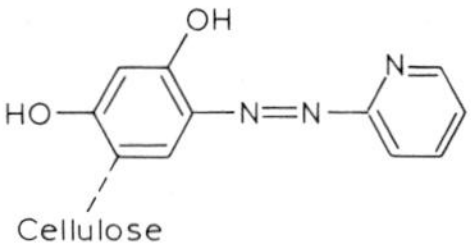   XXV

The exchange capacity of such resins equals 0.5 mmol/g. Two methods involving the use of modified cellulose were developed for analysing natural waters (column and filtration methods). In the column method, 1–5 l water was passed through a layer of sorbent and the trace elements were eluted successively with 50 ml of 1 *M* HCl and 30 ml of purified water. Thereafter, the eluate was neutralized with sodium hydroxide solution (pH up to 6–7) and mixed with a small amount of modified cellulose. The sorbent after drying was analysed by the X-ray fluorescence method. The filtration method, despite worse metrological

66

parameters, is attractive for its simplicity. In the case of the filtration method, 1–5 l of natural waters were passed through a filter of modified cellulose. Depending on the volume of water to be analysed, the concentration coefficients varied between 20 and 100.

Imai et al. [247] synthesized dithiocarbamate derivatives of cellulose by treating it with n-toluenesulphonyl chloride in pyridine, with subsequent reaction of tolyl cellulose and amines in dimethyl formamide, and by treating the obtained amino celluloses with carbon disulphide in the NH_4OH–CH_3OH medium. Such sorbents, depending on the type of the substituent, isolate Ag, As(V), Cd, Co, Cr(III, VI), Cu, Fe, Hg(II), Mn(II), Pb, Sb(V), Se(IV), Te(IV) and Zn from aqueous solutions. The sorbents were used to concentrate trace elements in the analysis of natural waters. Recently a method has been developed for concentrating uranium by the static or column method using cellulose containing two amino groups or one amino group and a dithiocarbon-containing group [248]; uranium can be extracted from acidified sea water.

Concentration of AsO_3^-, AsO_4^{2-}, CrO_4^{2-}, MoO_4^{2-}, SeO_3^{2-}, SeO_4^{2-}, VO_3^- and WO_4^{2-} from waters with the use of cellulose filters containing functional groups of 2,2-diaminediethylamine has been studied with a view to determine these anions by the XRFS method [249]. The extraction degree of anions decreases strongly in the presence of $\geqslant 0.01\,M$ NaCl.

Myasoedova et al. [250] used modified cellulose for concentration and determination of microgram amounts of silver by thin-layer chromatography. Cellulose containing the azorhodamine group was synthesized. Microgram amounts of silver give bright red colour to the sorbent layer. This method was utilized in the analysis of sulphide ores rich in Co, Cu, Fe and Ni; the ore was decomposed with nitric acid in the presence of perchloric acid. In chromatographic separation the non-noble elements predominating in the sample were shifted close to the front of the mobile phase with $0.01\,M$ ethylenediaminetetraacetic acid, while silver remained at the start. Silver was determined by the atomic absorption method with flameless atomization. The detection limits were $2.5 \cdot 10^{-11}$ g.

Anion complexes of Cd, Cu and Pb with humic and fulvic acids present in natural surface waters were concentrated [251] in a weakly acidic anion exchanger of Sephadex with diethylaminoethyl groups (Sephadexes are dextrin gels having cross-links). The column used for concentration is shown in Fig. 2.14.

An X-ray fluorescence method has been developed [252] for determin-

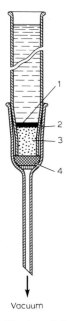

Vacuum

Fig. 2.14. Column for sorption preconcentration of cadmium, copper and lead from water.
1 = sorbent; 2 = ground joint; 3 = vitreous silica; 4 = vitreous silica wool.

ing uranium in natural waters after its preliminary separation on discs
of cellulose-phosphate paper, Whatman P-81. After concentration, the
discs are washed, dried and analysed. Toussaint *et al.* [253] separated
the suspended oxides of metals on Millipore filters and concentrated
the dissolved metals on the cation-exchange paper Reev Angel SA-2.
The separated trace elements were determined by the X-ray fluorescence method.

(ii) On an inorganic matrix. Complex-forming groups can be inoculated not only to organic but also to mineral matrices, such as silica
gel and glass beads. The obtained sorbents are unique hybrids of synthetic organic and inorganic sorbents (see elsewhere). These sorbents
do not swell; the technique of obtaining fine and even fractions of silica
gel is well developed.

Leyden *et al.* [254, 255] have developed procedures for preconcentration of ions on silica gel or small glass balls with complex-forming
groups immobilized on their surface. Such groups are shown below.

$$\begin{array}{l} \text{—O} \\ \text{—O—Si—}(CH_2)_2\text{—}N(CH_3)\text{—}C\underset{S^-}{\overset{S}{\diagup}} \qquad\qquad \text{XXVI} \\ \text{—O} \end{array}$$

$$\begin{array}{l} \text{—O} \\ \text{—O—Si—}(CH_2)_3\text{—}NH\text{—}CH_2\text{—}NH\text{—}C\underset{S^-}{\overset{S}{\diagup}} \qquad \text{XXVII} \\ \text{—O} \end{array}$$

$$\begin{array}{l} \text{—O} \\ \text{—O—Si—}(CH_2)_2\text{—}N\begin{array}{l} C\underset{S^-}{\overset{S}{\diagup}} \\ (CH_2)_2\text{—}NH\text{—}C\underset{S^-}{\overset{S}{\diagup}} \end{array} \qquad \text{XXVIII} \\ \text{—O} \end{array}$$

$$\begin{array}{l} \text{—O} \\ \text{—O—Si—}(CH_2)_2\text{—}NH\text{—}(CH_2)_2\text{—}NH_2 \qquad \text{XXIX} \\ \text{—O} \end{array}$$

The dithiocarbamate group ensures sorption of cations, and the diamine group that of anions. The sorption of trace elements is sufficiently complete under static conditions, but for isolating traces from a large volume the column method is preferred; in this case it is better to use glass balls rather than silica gel. After sorption, the glass balls or silica gel particles are mixed with an equal amount of powdered cellulose and pressed, and the tablet is analysed by the X-ray fluorescence method. Elements can be separated from very dilute solutions, *e.g.* selenium can be separated from 2 l of water at a concentration of 10 ng/ml. For determining trace amounts of oxyanions AsO_4^{3-}, $Cr_2O_7^{2-}$, SeO_4^{2-}, MnO_4^{2-}, WO_4^{2-} and WO_3^{-}, Leyden *et al.* used concentration on porous glass balls whose surface was treated with N-β-aminoethylaminopropyltrimethoxysilane. In so doing, 90% SeO_4^{2-} and MnO_4^{2-} were adsorbed in 2 min, $Cr_2O_7^{2-}$ in 4 min and AsO_4^{3-} in 8 min.

Active work is being carried out at the Moscow State University [256] and Paderborn University [257–260]. Table 2.5 lists some works of the latter.

2.3.4.2 Polymeric heterochain sorbents

Complex-forming sorbents can be obtained not only by inoculating complex-forming groups to any inert matrix. Active groups may constitute a component of the matrix proper; often the complex-forming centres enter into the composition of polymer chains as heteroatoms. An example of this is polymer thioester $(-CH_2\text{-S-})_x$. Here, the sulphur

TABLE 2.5

Some examples of concentration of trace elements with silica gel or glass-based complex-forming sorbents

Complex-forming reagent	Concentrated elements	Analytical application	Ref.
5-Methylene-2-(2′-thiazoleazo)-anizol	Pd	For separation from large amounts of copper	257
2-Amino-1-cyclopentane-1-dithiocarboxylic acid	Ag, Hg, Pd	Separation is possible from non-ferrous metals and Pt(IV)	258
β-Diketones	Cu, Fe(III), U(VI)	Concentration of U(VI) from sea water is possible	259
Formazines	Ag	Concentration is possible in the flow method	260
8-Hydroxyquinoline	Cd, Co, Cu, Ni, Pb, Zn	Determination by the AAS method in flow injection analysis	261

atom is involved in complex formation. In this case, the number of active atoms is determined by stoichiometry and turns out to be relatively large. These sorbents should have large capacity.

A number of techniques involving the use of polymer thioester have been developed for concentrating heavy and, particularly, noble metals. The sorbent selectively extracts noble metals from solutions with a complex composition. After sorption, the concentrate may be dissolved in nitric acid or may be directly analysed by the XRFS method; after the concentrate is dissolved, determination may be carried out either by flame AAS or by the AES–ICP method [262].

Polymer thioester has been used for concentration of heavy metals from natural waters and other substances. The metals in the concentrate are determined by the XRFS [263] or AAS–ICP method [264].

A polymer tertiary amine containing the above fragment proved to be an effective sorbent for noble metals. With this, platinum metals are extracted from chloride solutions without heating them and without introducing labelling additives of the type $SnCl_2$. A sorption-X-ray fluorescence method of determining platinum metals has been developed on this basis.

Other examples of using the two mentioned heterochain sorbents are listed in Table 2.6. Some metrological characteristics of the methods of determining noble metals, after concentrating them with polymer thioester, are given in Table 2.7.

With heterochain sorbents may also be grouped the sorbents that have macrocyclic compounds in the chain. An example of this is the sorbent containing 18-Crown-6 (XXX):

XXX

Synthesis and application of such sorbents are most completely presented in the works of Blasius and co-workers [271–273].

TABLE 2.6

Concentrating trace elements with polymer hetero-chain sorbents

Sorbent	Elements to be concentrated	Concentration conditions	Substance analysed	Method of determination	Ref.
Polymer thioester	Platinum metals (Ir, Os, Pd, Pt, Rh, Ru)	3 M HCl, static sorption in boiling solution	Industrial solutions	XRFS	265
	Platinum metals (Ir, Os, Pd, Pt, Rh, Ru)	1–3 M HCl, static sorption in boiling solution	Technological solutions	XRFS	266
	Platinum metals (Ir, Pd, Pt, Rh, Ru)	3 M HCl, static sorption in boiling solution containing $SnCl_2$	Copper–nickel slurry processing products	ETAAS	267
	Platinum metals (Ir, Pd, Pt, Rh, Ru)	3 M HCl, static sorption in boiling solution containing $SnCl_2$	Copper–nickel slurry processing products	AES–ICP	268
	Platinum metals (Ir, Pd, Pt, Rh, Ru)	3 M HCl, static sorption in boiling solution containing $SnCl_2$	Technological solutions and solid products	ETAAS	269
Polymeric tertiary amine	Platinum metals (Ir, Os, Pd, Pt, Rh, Ru)	1–3 M HCl, static sorption	Industrial solutions, concentrates and alloys	XRFS	270

TABLE 2.7

Metrological characteristics of the methods of determining platinum metals in polymer thioester-based solid concentrates

Element to be determined	Range of determined contents (μg)			Relative standard deviation referred to the mean of the range of determined contents		
	ETAAS	AES	XRFS	ETAAS	AES	XRFS
Pd	$4 \cdot 10^{-3}$–100	0.05–5	80–5000	0.10	0.11	0.030
Pt	$4 \cdot 10^{-3}$–100	0.05–5	100–5000	0.13	0.07	0.026
Rh	$4 \cdot 10^{-3}$–100	0.25–50	80–5000	0.08	0.05	0.032
Ru	$4 \cdot 10^{-3}$–100	0.10–10	80–5000	0.13	0.07	0.019
Ir	$4 \cdot 10^{-3}$–100	0.50–50	100–5000	0.18	0.09	0.032
Os	$1 \cdot 10^{-2}$–100		100–5000	0.20	–	0.036

2.3.4.3 Sorbents modified with complex-forming reagents

In general it is not necessary to chemically bind the complex-forming groups, or the reagents containing them, with the matrix by stable covalent bonds. Such reagents can be firmly fixed on the surface of an ion exchanger as a second layer, i.e. as a layer of counterions. Alternatively, when dissolved in an organic solvent or water, they fix as a thin layer on the surface of the porous carrier. Another variation involves mechanical "pressing-in" of complex-forming reagents in an inert matrix. This can be done in numerous ways, many of which have been realised.

Knapp et al. [274] have worked out a procedure for concentrating traces of metals, in particular of cadmium and mercury, with sodium diethyldithiocarbamate coated on Chromosorb W-DMCS. Japanese researchers [275] used dithizone and thenoyltrifluoroacetone gels coated on a polystyrene (2% divinylbenzene) bead for selective sorption of mercury from sea water. Ol'shanova et al. [276] used, as precipitants, disodium phosphate, dimethyl glyoxime, 8-hydroxyquinoline and alkalis for concentrating Cu, Cr, Fe, Pb, Ni and Zn from dilute solutions on the anion exchangers AV-17 and AV-18 in OH-form under static and dynamic conditions. Depending on the nature of ion and the type of precipitant, the concentration coefficient was 50–300.

Tanaka et al. [277] have worked out a method for spectrophotometric determination of Cd, Cu, Hg, Mn and Zn. The method involves the use of transparent membranes made up of 1-(2-pyridylazo)-2-naphthol, tricresyl phosphate and polyvinyl chloride. The optical density of a membrane immersed in the analysed solution is proportional to the concentration of determinable ion. To prepare a membrane, 20 mg of 1-(2-pyridylazo)-2-naphthol and 1 g polyvinyl chloride are dissolved in 1 ml tricresyl phosphate, the mixture is poured on a glass plate, the solvent is evaporated, and the film formed is pressed at 90°C under vacuum. In analysis of wastes of the platinum metal industry the noble metals were concentrated by ascending chromatography on a paper impregnated with thionalide solution in ethyl alcohol and containing acetic acid; the paper was then dried in air [278]. Mixtures of butyl and ethyl alcohols and acetone served as mobile phase. During chromatographic separation the platinum metals remained on the start line, while non-noble metals shifted upwards with the rise of the solvent front. Then the paper strips were dried in air; the yellow spot containing platinum group metals was cut out, mineralized, and analysed by the atomic emission method.

Different types of polyurethane-based sorbents have been obtained [147–150, 279–282]. These are highly porous materials with open pores. Therefore, polyurethanes have a very large specific surface; they can be easily saturated with complex-forming reagents or corresponding groups can be inoculated to the matrix. They have been used in a number of analyses. In a method for concentrating mercury on polyurethane foams treated with dithizone, acetone served as eluent [283]. For concentrating gold from 0.1 M HCl containing 3% thiourea and 1% sodium perchlorate, the solution was passed through a column packed with polyurethane foam, the foam being impregnated with tributylphosphate solution [284]. Gold was completely isolated from the column by dissolving the foam in hot nitric acid. Polyurethane membrane with open cells was used for selective separation of gallium [285] from chloride solutions in the form of $HGaCl_4$; the obtained concentrate was analysed by the atomic absorption method. Braun *et al.* [286] have studied the possibility of using polyester polyurethane foam containing 1-(2-pyridylazo)-2-naphthol as reagent for concentration and separation of Co, Fe and Mn under static and chromatographic conditions.

2.3.5 INORGANIC SORBENTS

With inorganic sorbents are grouped oxides and hydroxides of metals (silica gel, hydrated titanium oxide), several salts of metals (zirconium phosphate, sulphides), salts of heteropolyacids and others.

Their reaction mechanisms may differ, but ion-exchange and complex formation often take place. The advantages of such sorbents are: stability to heat, ionizing radiations and organic solvents, and at times high selectivity. Their disadvantages are that they do not always have a high capacity and their properties vary from lot to lot.

Small amounts of Ga, Ge, In, Mo, U, V and W were sorbed from sea waters with hydrated titanium oxide – an intermediate product obtained in the manufacture of titanium pigments from ilmenite [287]. Besides availability and low cost, this sorbent has exceptionally low solubility, heat-ageing resistance, and an acidic nature of the surface due to the presence of sulpho groups. Combining preconcentration with subsequent determination by atomic emission analysis ensures a detection limit of 10^{-7}–10^{-8}%. Chigetomi and Kojima [288] have studied adsorption of Fe(III), Ni and Cu(II) on a sorbent obtained from titanium hydroxide, acrylamide, and N,N'-methylenediacrylamide; maximum sorption was observed at pH 7–8.

As sorbents, Solovyeva et al. [289] used zirconium hydroxides and zirconium phosphates which have high selectivity, thermal and chemical stability, and high exchange capacity compared to ion exchangers and hydroxides of iron, titanium and aluminium. For neutron activation determination of trace elements in water and bottom sediments, Grancini et al. [290] sorbed As, P and W with aluminium oxide, and Ag, Ce, Co, Cr, Cs, Sb, Sc, Sr, Rb, Tb, U, Yb and Zn with hydrated antimony(V) oxide. This technique was used to study how trace elements are distributed in sea water at the surface and at different depths. In the analysis of sea water, hydrated oxides of titanium and zirconium were also used by Lieser et al. [291].

For extracting mercury vapours from air, Aleksandrov [292] used lead sulphide as sorbent; mercury was then determined by the AES method using a hollow iron cathode.

2.3.6 AUTOMATING THE SORPTION CONCENTRATION PROCESS

It is sometimes expedient to automate the sorption concentration process, particularly when samples of one type are to be prepared in large quantities, say, for analyzing natural or sewage waters or for clinical analysis. The sorption concentration technique allows automation. It is possible to automate not only the concentration operation, but also the transition to subsequent determination, and the determination proper. It has long been successfully tried in chromatographic methods. Under development is another approach involving the use of the flow injection analysis method.

Automatic atomic emission flame photometric determination of barium and strontium in water [293] was carried out after separating them in a chromatographic column packed with the cation exchanger Dowex 50 Wx8. Separation was effected in two Technicon devices equipped for carrying out two operations. The first operation involves transfer of sample into the column, addition of acetate buffer solution, elution of barium with a mixture of cyclohexane–1,2-dinitrilotetraacetic acid and ammonia, and washing of the column; strontium is eluted with the same reagent but at a different pH value. Thereafter, the column is washed. The second operation involves addition of sodium chloride solution to the eluate containing strontium and barium to improve ionization in the flame, and atomization of the solution into a flame. Reference solutions undergo the whole cycle of treatment. The output of the method is 40 samples in 110 min. A method has been suggested for automatic separation of Cu(II) from natural waters [294] using columns filled with an ion exchanger or glass packing immobilized with 8-hydroxyquinoline. It ends with atomic absorption analysis. The immobilized sorbent is convenient from the viewpoint of time spent on concentration.

A semi-automatic technique of determining phosphorus (as anions) in inorganic substances [295] from the emission spectrum of HPO molecule, which is excited in the cavity (MECA spectroscopy), involves preliminary separation of anions and subsequent introduction of solution into the graphite-filled cavity. After the solution has dried, the cavity is introduced into a hydrogen flame using a special device. An automatic device has been worked out for group chromatographic determination of Au, Ir, Pd and Pt from the irradiated sample solution in neutron activation analysis of geological and biological materials [296]. The irradiated sample, after oxidation fusion and dissolution in acids, was pumped with the aid of a peristaltic pump into the column filled with the ion exchanger Srafyon NMRR, and noble metals were sorbed selectively. After washing with hydrochloric acid the sorbent was taken out and subjected to gamma-spectrometric analysis. Five samples may be analyzed in 24 h; this does not include the irradiation time.

Ever increasing use is being made of sorption concentration as one of the stages of flow injection analysis. A flow injection analysis procedure has been worked out that involves preliminary concen-

76

tration (sorbent–phenolformaldehyde copolymer with inoculated functional groups of salicyclic acid) of Cd, Cu, Ni and Pb from tap, sea and sewage waters, and subsequent determination of trace elements by the flame AAS method [297, 298]. Here, use is made of an eight channel peristaltic pump, multiway stopcocks, and collectors which ensure introduction of the analysed solution, of eluent ($2\,M$ HNO$_3$) and buffer solution into two alternatively operating columns. Eluate is pumped into the spray unit of an AA spectrometer, and the column is regenerated by the buffer solution. This system ensures a 20–30-fold concentration of Cd, Cu, Ni and Pb; its output is 40 samples per hour.

For determining Cu, Mn and Pb in the aerosol particles of smoke that appear upon welding, the particles, after collecting them on a filter, were dissolved in a mixture of concentrated HNO$_3$ and HCl, and the solution was evaporated to dryness. After dissolving the residue in double distilled water the solution was injected into the flow injection analysis (FIA) system equipped with a microcolumn packed with chelate-forming sorbent Dowex A-1 [299]. Part of the eluate containing concentrated trace elements was introduced into an acetylene–air flame, and the elements were determined by the AAS method. The column with modified 8-hydroxyquinoline silica, built into the FIA–AAS system, was employed for concentration of trace elements, in particular of copper in the analysis of waters [300]. The flow-rate is limited by the consumption of solution in the AA spectrometer spray system. By preliminary sorption concentration, the detection limit is lowered by two orders. For concentrating Cd, Co, Cu, Ni, Pb and Zn, use was made also of the FIA–AAS system [261] (Fig. 2.15) using a microcolumn packed with 8-hydroxyquinoline immobilized on porous glass (XXXI):

XXXI

The system ensures a 500-fold concentration of trace elements in 25 min; it has been tested for determining copper in tap water.

A highly effective FIA–AAS system involving preliminary concentration on a chelate sorbent, Chelex-100, was utilized for determining trace elements in sea water [301]. In another work [302], such a system was used for determining Cd, Cu, Pb and Zn; the output of the

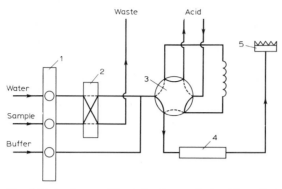

Fig. 2.15. A scheme of flow injection analysis with preliminary sorption concentration of trace elements. 1 = Peristaltic pump; 2 = pneumatic injection time-controlled valve; 3 = six-way valve; 4 = microcolumn packed with 8-hydroxyquinoline immobilized on porous glass; 5 = flame.

method is 30–60 samples per hour. Hartenstein *et al.* [303] have suggested a system for determining Al, Ba, Be, Cd, Co, Cu, Fe, Mn, Ni, Pb and Zn in waters; it enables the trace elements to be concentrated (at pH 9) in a microcolumn packed with Chelex-100 in the H^+ form. The trace elements are eluted with $2\,M$ HNO_3, the eluate is injected into a flow and analysed by the AES–ICP method. The system's output was raised by utilizing two columns, each being used in turn for sorption and desorption. Concentration lowered the detection limits of trace elements, excepting aluminium, by 10—80 times. The output of the system is 30 samples per hour.

A system combining FIA and AES–ICP was employed for determining phosphorus in steel [304]. With the aid of a measuring device 0.2–1.0 ml of the sample dissolved in 1% HNO_3 was taken; it was then injected into a flow of water (1.0 ml/min) and passed through a Teflon column packed with activated aluminium oxide. In so doing, the phosphate ions got sorbed in the column. Then 200 ml of $0.5\,M$ solution of potassium hydroxide was injected into a flow, and phosphorus was determined.

Ion chromatography is an automated method of concentration. A unique automatic analyzer, "Quik Chem" [305], determines the ionic composition of solutions containing anions and/or cations at a level lower and/or higher than the detection limit. It is intended for individual or successive simultaneous determination of ions in liquid or solid samples after dissolving them. The system combining FIA and ion chromatography constitutes a sampler, measuring device, two-four independent lines for FIA and one-two channels for ion chromatography, detectors (spectrophotometer, conductivity apparatus and atomic absorption spectrometer) which are connected to an individual or main-frame computer. Here, FIA and ion chromatography can be used simultaneously or in succession. The determinable components can be diluted or concentrated, separated selectively, and determined. The possibilities of the analyzer have been demonstrated in analyzing environmental samples and pharmaceuticals.

2.4. Precipitation and co-precipitation methods

Precipitation is one of the oldest methods of preconcentration and separation. Co-precipitation, which was earlier regarded as an unpleasant companion of precipitation, later on also became an effective method of preconcentration. These methods have many common points: the process of formation of precipitates in solution, factors determining the separation efficiency, experimental technique, etc. Therefore, it is expeditious to consider precipitation and co-precipitation in one section. Preconcentration takes place owing to different solubilities of the components in the aqueous, and less commonly, the aqueous–organic, solution. Depending on the purpose in hand, conditions have to be created such that the precipitate formed will carry with or, on the contrary, will not entrap the determinable trace components. The precipitate formation process is quite complicated and often it does not proceed instantaneously. Its progress depends on various factors, such as composition of the aqueous phase, pH, temperature, nature of the counter-ions forming the precipitate, sequence in which solutions are mixed, and the properties of the collector.

Precipitation, particularly co-precipitation with organic or inorganic collectors ensures a high degree of concentration. Co-precipitation is conveniently used in combination with subsequent determination methods demanding solid samples, *e.g.* with X-ray fluorescence or atomic emission spectroscopy.

Use is made of several groups of precipitates [306]. (1) Salts of strong acids. Partly soluble sulphates of Ba, Hg(I), Pb and Sr, chlorides of Ag and Hg(I), lead nitrate and others. (2) Salts of weak acids. Many of them are less soluble in water. The methods involving their application are not usually very selective. With them are classed sulphides, carbonates, phosphates and metal chelates. The selectivity of preconcentration can be increased by masking or using other methods. (3) Free acids (silicic, stannous, tungstic acids) and hydroxides of many metals. (4) Precipitates formed on reacting three components. For example, a metal can be transformed into a complex halide anion which can be precipitated after introducing an organic base. (5) Substances in the elemental state (sulphur, tellurium).

2.4.1 PRECIPITATION

2.4.1.1 Matrix precipitation

Concentration by precipitation can be carried out in two ways, *i.e.* by matrix precipitation and by precipitation of trace components. Selective concentration of the matrix by leaving behind in the solution the trace elements to be concentrated is a technique borrowed from gravimetric analysis [307]. Such a precipitation is associated with large consumption of reagents, prolonged duration of the process and co-precipitation: large mass of the precipitate may result in the loss of trace components. However, in some cases, most of these disadvantages can be avoided or reduced to a minimum. That is why matrix precipitation is still being used in analytical practice.

For determining Ag, Al, As, Bi, Ca, Cd, Co, Cu, Fe, Ga, In, Mg, Mn, Na, Ni, Sb and Tl in high-purity lead, Ustinov and Chalkov [308] precipitated the main mass of the macrocomponent in the form of nitrate. The concentration coefficient was 100–350. After separating lead nitrate, the solution was evaporated, the residue was ignited at 500–550°C, and analysed by the atomic emission method. The detection limits were 10^{-6}–10^{-8}%. The same technique was employed for determining 22 trace impurities in high-purity lead by the AES method [309].

Several methods involving precipitation of trace elements as nitrate, sulphate, or chloride have been developed by Jackwerth [310] for concentration of trace impurities present in lead.

Analysing high-purity tellurium for sodium content by the flame AES method, Szwabski and Dobrowolski [311] dissolved the sample in aqua regia. To remove the residues of nitric acid, the solution obtained was evaporated several times by adding concentrated HCl; the residue was diluted with water and the precipitate of tellurium dioxide containing sodium was separated by decantation. For concentrating Al, As, Bi, Cd and eighteen other trace elements, Tiptsova-Yakovleva *et al.* [312] separated the matrix in the elemental state from 2.0–3.0 M HCl with sulphur dioxide. Analysis was completed by the atomic emission method; the detection limits were 10^{-4}–10^{-6}%. Analysing mixed tellurides of cadmium and mercury, they separated the impurities from the solution containing cadmium by partial precipitation of tellurium in the form of hydroxide. This is how Al, As, Bi, Cr, Cu, Fe, Ga, In, Mo, Pb, Sb, Sn and Tl were determined with detection limits of 10^{-5}–10^{-7}%. Jackwerth et al. [313], in the analysis of high-purity tellurium by the AES method, separated the matrix from iodide solutions as TlI. Traces of Al, Bi, Ce, Co, Cu and twelve other elements remained in the solution containing EDTA.

2.4.1.2 Precipitation of trace elements

Organic precipitants are important for precipitating trace elements. Often, the organic reagents forming chelates which dissolve with difficulty in water have a role to play. Some examples of concentration of trace elements by precipitation are listed in Table 2.8.

2.4.2 CO-PRECIPITATION OF TRACE ELEMENTS

2.4.2.1 General

Co-precipitation of trace elements is much more extensively used than precipitation as a method of preconcentration. As is known, co-precipitation is the transfer of a substance into a precipitate of some compound if the substance does not form its own solid phase under the given conditions. Depending on the physical and chemical properties of the components and the experimental conditions, co-precipitation is owing to adsorption of the trace component on the surface of the collector, the formation of isomorphic mixed crystals, mixed chemical compounds, occlusion and mechanical inclusion of small amounts of other phases. Sometimes, all these factors act simultaneously to some extent.

Isomorphic co-precipitation and formation of mixed crystals are processes which have been studied relatively well; these are governed by the well known Bertelot–Nernst, Doerner–Hoskins and Khlopin relations. The theory explaining crystallization in such systems is discussed in a new light in the book by Melikhov and Merkulova [323]. The mechanism of adsorption that takes place on the surface of the forming and growing crystals has been fairly well studied; in par-

TABLE 2.8

Examples of concentration of trace elements by co-precipitation

Elements to be concentrated	Precipitant	Conditions	Method of determination	Detection limit (%)	Ref.
Ag, Bi, Co, Cu, Fe, In, Pd, Th, Ti, V, W	Alizarin blue	pH < 2	XRFS	–	314
Au, Pd	Dithizone	Membrane filter was used to separate out the precipitate	AAS	10^{-7}–10^{-8}	315
Cu, Fe, Ni, Zn	Pyrrolidine-dithiocarbamate	Membrane filter was used	XRFS	–	316
Co, Cr, Cu, Fe, Mn, Mo, Nb, Pb, Ta, Ti, V, W	Diethyldithio-carbamate and 8-mercaptoquinoline	Lanthanum and its preparations were analysed, pH 6, graphite powder was introduced	AES	ca. $1 \cdot 10^{-5}$	317
Co, Cu, Fe, Mo, Ni, Ti, V	Diethyldithio-carbamate, Cupferron, 8-hydroxyquinoline	Compounds of alkali metals were analysed, pH 4-5, graphite powder was introduced	AES	10^{-3}–10^{-6}	318
Ag, Au, Bi, Co, Cr, Cu, Fe, In, Mo, Ni, Pb, Sn, Ti, V	Diethyldithio-carbamate, dimethylglyoxime, 1-nitroso-2-naphthol	Compounds of zinc were analysed, pH 6, graphite powder was introduced	AES	$5 \cdot 10^{-5}$–$5 \cdot 10^{-7}$	319
As, Ga, Mn, Mo, Sn, and 19 other elements	Diethyldithio-carbamate	Precipitates are filtered through membrane filters	XRFS	–	320
Ag, As, Cr, Cu, Fe, Ga, Ni, Pb, V, Zn	Pyrrolidinedithio-carbamate	Precipitates are filtered through microporous filters	XRFS	–	321
Ag, Au, Ir, Pd, Pt, Rh, Ru	Diphenylthiourea	Industrial wastes are analysed	AES-ICP	*	322

ticular, the Fiens–Panet–Gan rule plays an important role. By this rule, adsorption increases with the growing surface area of the crystal and with the decrease in the solubility of the compounds of the trace element which it (the element) forms with oppositely charged ions of the crystal. Besides, according to this rule, the ions of the trace element are adsorbed on polar crystals if the crystal surface has a charge opposite to that of the ion of the trace component. Occlusion often consists of entrapping the solvent containing the trace element. This happens when precipitation takes place quickly and colloidial precipitates are formed.

On the strength of the theoretical basis of co-precipitation and the empirically accumulated factual material, it can be said that chemical and crystallochemical properties of compounds, the state of the co-precipitating element in solution, rate and sequence of adding reagents, precipitate ageing process, acidity of the solution, time, temperature, etc. can affect the co-precipitation of trace elements.

Thus, small crystals which are formed in the solution have a large number of active places on their surface, and they are more active than large crystals. The larger the surface of the precipitate and the more imperfect the form of crystals, the stronger is the adsorption of trace components from the solution. If precipitation takes place quickly and the obtained crystals are imperfect in their shape, then foreign ions get implanted in them; mechanical entrapment of trace components can also take place in this case.

The collectors (carriers) of trace elements must meet specific requirements. First of all, they must entrap the necessary element without picking up the matrix elements and, sometimes, certain interfering trace components. In general, preconcentration can be ensured by knowing the requirements that are placed for obtaining pure precipitates, for example in gravimetric analysis, and by doing just the opposite. Thus, large-surface amorphous flaky precipitates sometimes prove better than crystalline precipitates. Greater selectivity in co-precipitation can be attained by masking, changing the oxidation state of elements, and by applying other techniques.

The main requirement is that the collector should easily separate from the matrix solution; this is done by filtering and washing the precipitate. Other techniques used are centrifugal separation and flotation. It is desirable that the collector should be a pure and readily available substance. Requirements of the subsequent method of determination should also be taken into account. Thus, a particular requirement of atomic emission analysis is that use should be made of elements having simple spectra, say of hydroxides of aluminium and zirconium, but not of iron and manganese which have complex spectra. Sometimes, as a collector, it is convenient to utilize a compound of the element that may later be removed by evaporation or another simple method.

Co-precipitation may be effected by different techniques. The most common technique is that in which the element-carrier and a suitable precipitant (either inorganic or organic) are introduced into the analysed solution. For example, trace elements are co-precipitated on iron(III) hydroxide by adding a salt solution of this element and a solution of ammonia or sodium hydroxide.

Sometimes there is no need to specially introduce an element carrier, for it is already present in the solution to be analysed in micro- or macroamounts. In this case, the matrix is partially precipitated by regulating the amount of the added precipitant. This technique is, however, applicable only when the solubility product of the precipitated compound of the matrix element is more than the solubility product of the corresponding compounds of trace elements.

Concentration by partial precipitation of the matrix was employed by Chuiko and co-workers [324, 325], Jackwerth et al. [326], and other chemists. This is an unusually convenient technique when use is made of hydroxides. Chuiko et al. [327] have worked out a procedure for concentrating trace amounts of iron by partial precipitation of the matrix as phosphate in analysis of salts of Ni, Co, Cu and Zn. They were guided by the fact that, if the trace components formed much less soluble compounds with the reagent than the macrocomponent, they could be concentrated by partial precipitation of the matrix even in the absence of isomorphism. Iron is then separated from the collector by co-precipitation with aluminium hydroxide from ammonia solutions or with magnesium hydroxide from alkali solutions. The precipitate obtained is dissolved in an acid and iron is determined photometrically using 1,10-phenanthroline. In another work [324], traces of iron(III) were co-precipitated with Cd, Mn, Pb and Mg phosphates. In partial precipitation of these metals as phosphates, iron is quantitatively concentrated from cadmium and magnesium salt solutions.

Finally, there is one more co-precipitation technique applicable mainly to organic co-precipitants. Here, the precipitant is introduced as solution into an organic solvent which mixes with water, say acetone or ethanol. In the resulting water–organic solution where water

84

predominates, the introduced reagents go into the precipitate and carry with them the formed complexes of metals. This is how dithizone is introduced, for instance.

2.4.2.2 Inorganic co-precipitants

For co-precipitation of trace elements with inorganic collectors use is often made of amorphous precipitates with large active surfaces (hydroxides, sulphides, phosphates and others).

The mechanism of co-precipitation on such collectors has been thoroughly studied. The results of several investigations are given elsewhere in this volume.

Rudnev and Malofeeva [328] have considered co-precipitation on various inorganic precipitants from the viewpoint of transformations that take place in the solid phase, including formation of chemical compounds and solid solutions. The chemistry of co-precipitation depends on the location of elements in the Periodic Table. In particular, when the components have opposite donor–acceptor properties, a chemical compound is formed. This is one of the main causes of co-precipitation. Formation of solid solutions, which is typical of compounds having crystallochemical similarity, is of no less importance. It is difficult to draw a distinct boundary between the role of chemical compounds and solid solutions, because co-precipitation can be simultaneously caused by the action of both factors. What is more, solid solutions can be considered as continuous chemical compounds.

Sorption processes are of great importance. The formed precipitates have a large surface and porous structure, and therefore the whole volume of the finely dispersed precipitate is easily accessible to the ions present in the solution. These processes can be both physical and chemical, for example ion exchange observed in the co-precipitation of sulphides and hydroxides. Chuiko [329] has drawn a similar conclusion.

Considering co-precipitation of elements on hydroxides and sulphides, Rudnev et al. [330] inferred that co-precipitation on inorganic carriers was characterized by localization of the process on heterogeneous portions of the surface and its defects, and by an increase in the rate with time (autocatalytic reaction). Effective use of these features makes it possible, in particular, to concentrate 1 ng gallium on 1 mg silver sulphide, and to separate trace amounts of Co, In and Mn(II) from iron(III) hydroxide. The suggestions on the selection of collectors are interesting. The difference in acidic and basic properties of macro-

and trace components must be as much as possible. Chemical reactions that occur on the surface during co-precipitation are followed not only by the formation of stable, but also of intermediate, compounds. Proceeding from this, the authors recommend mixed collectors of increased sorptionability, capacity and selectivity, *e.g.* $Ag_2S-Ga_2S_3$, $Cu(OH)_2-Fe(OH)_3$ and $Cu(OH)_2-Zr(OH)_4$.

Several works on co-precipitation of elements with hydroxides have been done by Plotnikov and Novikov (see, for example refs. 331 and 332). They have studied co-precipitation of 83 elements with Fe(III), Zr, Mg, etc. hydroxides depending on the pH of the medium, composition of the aqueous phase, nature and concentration of oxidizers, reducers, complex-forming reagents, and other factors. Co-precipitation of cations starts at the trace component hydrolysis pH. The presence of complex-forming agents which do not form bridge bonds (NH_3, CO_3^{2-}) reduces co-precipitation if the complex ions coordinatively saturated with respect to the above ligands or water molecules are formed. In most cases, precipitation increases with the concentration of neutral salts. Data on co-precipitation of Be, In, Cd, Ce(III) and other elements with metal hydroxides have been systematized. According to these authors, neutral hydroxocomplexes of the type $Me(OH)_n$ are the main form of co-precipitating elements. In the pH range where a trace element is present in the charged form, co-precipitation is caused by the acid–base reaction of this element with the collector (provided there are no positively charged hydroxocomplexes of macrocomponents in the solution). Also, they believe [331, 332] that trace element hydrolyses in the surface layer of the collector and forms a neutral hydroxocomplex. The apparent loss of individual properties of trace components, which occurs in this case, shows up to a greater extent if partially hydrolysed ions of the element to be precipitated enter into the reaction. Maximum co-precipitation of the hydrolysing element is achieved with metal hydroxides at pH values corresponding to the formation of an independent solid phase by the trace element.

We shall give some examples of co-precipitation concentration on hydroxides. For determining trace amounts of Ag, Au, Cd, Co, Cr, Cu, Mn, Ni, Pb, Sn and Zn in gallium, arsenic and gallium arsenide with detection limits of $10^{-6}-10^{-7}\%$, Rudnev *et al.* [333] dissolved the sample in concentrated nitric and hydrochloric acids, evaporated to a small volume, and then antimony was removed as tribromide by adding hydrobromic acid. After neutralizing the solution with potassium hydroxide, bismuth nitrate was added, and the trace impurities were co-precipitated with the formed bismuth hydroxide which, having acidic and basic properties, is a convenient collector for precipitation of hydroxides of trace components with different properties. Here, formation of solid solutions or adsorption of impurities on bismuth hydroxide can take place along with the chemical reaction.

Lebedinskaya and Chuiko [334] concentrated trace amounts of Be, Bi, Cd, Co, Cu, Ni, Pb and Zn from natural waters using a mixed collector, *e.g.* magnesium and iron hydroxides. These elements are present in the analysed substance, therefore they need not be specially introduced. The trace elements forming less soluble hydroxides are quantitatively adsorbed by the colloidal precipitate of the collector. Trace elements are completely adsorbed if the macro- and trace components form compounds similar in chemical composition when there is excess alkali in the solution. The precipitate was centrifuged, dried, ground, mixed with graphite powder and analysed by the atomic emission method. The concentration coefficient was *ca.* 10^4. The detection limits were 10^{-7}–10^{-8}%. Other examples can be found in Table 2.9.

Wide use is made also of co-precipitation of trace components on metal sulphides. Formation of chemical compounds is one of the main causes of co-precipitation of sulphides. Trace element forms a solid solution. Preconcentration can be selective or group depending on the nature of the collector and the trace element. For examples, see Table 2.9.

Other inorganic collectors are also used. It has been shown, for instance, that palladium can be concentrated very selectively on silver cyanide [353]. Even 10^7–10^9-fold amounts of many elements, including Bi, Cd, Cu, Fe, Ni, Pb and Sn, do not affect the concentration result. Palladium is then determined by the photometric method. The technique has been used for determination of this element in nickel, copper and silver with a detection limit of $6 \cdot 10^{-7}$%. Other examples are listed in Table 2.9.

2.4.2.3 Organic co-precipitants

Organic co-precipitants were introduced into analytical practice in the 1950s mainly by Kuznetsov and co-workers [354–362]. Many examples of concentrating actinides with organic co-precipitants can be found in his monograph [360].

Organic collectors usually feature good selectivity. For example, from chloride solutions containing gallium and aluminium in the ratio

TABLE 2.9

Co-precipitation of trace elements with inorganic collectors

Collector	Precipitated elements	Substance analysed	Method of determination	Detection limit	Ref.
Hydroxides of Al, Fe(III) and others	Zr	Silicate rocks	AES–ICP	$0.32\ \mu g/g$ of rock	335
Fe(III) hydroxide	Bi, Sb, Se, Te, 14 REE*	High-purity copper Rocks, minerals, meteorites, lunar soil	AAS NAA	– 10^{-5}–$10^{-7}\%$	336 337
Aluminium hydroxide	As, Bi, Cr, Ge, In, Mn, Pb, Sb, Sn, Te, Ti, V	High-purity cadmium	AES	10^{-5}–$10^{-6}\%$	338
	Co	Solutions containing Cr, Cu, Mn, Ni, Pb	Photometry	–	339
Zirconium hydroxide	Sb	–	Square-wave voltammetry	$5\cdot10^{-7}\%$	340
Nickel hydroxide	Cr, Cu, Mg, Mn, Zn	High-purity aluminium	Flame AAS	$1\cdot10^{-4}\%$	341
Magnesium hydroxide	Be, Eu, Mn and 14 other elements	Brines	AES–ICP	0.02–$4\ \mu g/l$	342
Manganese(IV) oxide	As, Bi, Ga, In, Ni, Pb, Sb, Sn, Te, Ti	High-purity cadmium	AES	10^{-5}–$10^{-6}\%$	338

Substance	Elements	Matrix	Method	Detection limit	Ref.
Molybdenum sulphide	As	Soils, water	XRFS	$3 \cdot 10^{-5}\%$	343
Sulphides of a number of metals	Platinum metals		AES	$2 \cdot 10^{-4} - 1 \cdot 10^{-7}\%$	344
Sulphides in the presence of cellulose	As, Bi, Cd, Co, Cu, Hg, Se, Sn, Zn	Various substances	XRFS	$10^{-5} - 10^{-6}\ \mu g$	345
Lead chromate	Ba	Alkali-metal halides	AES	$6 \cdot 10^{-5} - 2 \cdot 10^{-6}\%$	346
Barium chromate and silver cyanide	SO_4^{2-}, Cl^-	Tap water	XRFS	$10^{-4} - 10^{-5}\ M$	347
Metal arsenates	Zr	Compounds of zinc and cadmium	Photometry	$1 \cdot 10^{-7}\%$	348
Elemental sulphur	Au, Hg, Pd, Pt Au	Silver, copper, lead Silver	– NAA	– –	349 350
Elemental arsenic	Se	Antimony and gallium	ac polarography	$10^{-5} - 10^{-6}\%$	351
Silver iodide in the presence of 1,10-phenanthroline	Bi, Cd, Co, Cu, Fe, Ni, Pb, Tl, Zn	Compounds of manganese	AAS	$1 \cdot 10^{-5} - 3 \cdot 10^{-7}\%$	352

* REE = Rare earth elements.

of $1:8 \cdot 10^9$, more than 90% gallium can be isolated by single co-precipitation with induline hydrochloride practically without adsorbing aluminium [361]. The effectiveness of organic co-precipitants is so great that a trace component can be isolated even when it is present in solutions in the ratio of $1:10^{15}$. Co-precipitants can be removed from the concentrate by simple combustion. The concentrate can be dissolved in an organic solvent. Organic co-precipitants often separate out in the precipitate on mixing the solution to be analysed with solutions of reagents. Concentration can be selective or group depending on the chosen conditions and the nature of the organic coprecipitant.

Elements can precipitate as different compounds. Thus, ion associates containing methyl violet cation and complex thiocyanate or iodide anion of the element co-precipitate, and barely soluble precipitates (methyl violet cation)SCN or (methyl violet cation)I serve as collectors. In this case, the element enters into the composition of the complex anion. Chelates form another group of compounds to be precipitated. If they are soluble in water (dithiocarbamates, dithizones, β-diketonates, etc.), they are co-precipitated by introducing an indifferent (chemically non-active) organic co-precipitant. Salts of heavy organic cations are added if the chelates dissolve in water. Thereafter, the chelates are co-precipitated together with the precipitates of the formed salts.

For separating elements from very dilute solutions, Kuznetsov [356] recommends the use of organic reagents with molecules in which the functional group reacting with the trace element is repeated, sometimes several times. Thus, arsenazo I is replaced by arsenazo II. Finally, the elements present in the solution as colloidal particles and the hydrolyzed polynuclear cations can be co-precipitated as per the colloidal-chemical mechanism. As reagents, Kuznetsov recommends tannin and basic dye, the tannate of which is barely soluble.

Indifferent co-precipitants are of interest [354, 358, 362, 363]. They do not contain reactive or complex-forming groups. Such co-precipitants are used in combination with any of the enumerated types of co-precipitants. Indifferent co-precipitants are introduced as ethanol or acetone solutions because they are insoluble in water. The precipitate formed absorbs suspensions of trace elements. Phenolphthalein, thymolphthalein, β-naphthol, 2,4-dinitroaniline and stilbene are used as indifferent co-precipitants.

Colourless co-precipitants hold promise for photometric analysis [364]. The reagent that reacts with the element to be concentrated by

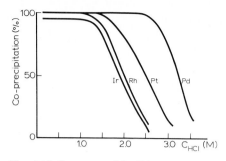

Fig. 2.16. Recovery of Ir, Pd, Pt and Rh in co-precipitation with thiobenzamide and diphenylamine precipitate *vs.* acidity of the medium. 10 µg each of Ir, Pd, Pt and Rh were dissolved in 100 ml.

forming coloured compounds is introduced into the solution, and then the required amounts of ingredients of an organic co-precipitant are added. The precipitates (usually readily forming) are filtered, dissolved in a suitable organic solvent, *e.g.* acetone or ethanol, and the solution obtained is measured with a photometer. Colourless organic co-precipitants in combination with subsequent spectrometric determination methods are convenient in the sense that the concentrate need not be ashed. For example, an express method has been developed [365] for concentrating uranium by co-precipitation of uranyl arsenazate in the presence of diphenylguanidinium chloride, a supplier of colourless organic cations. The barely soluble salt formed in this case was co-precipitated with naphthalene which was introduced as 10–20% ethanol or acetone solution. The method is suitable for photometric determination of uranium in natural waters with a uranium content of $\geqslant 5\,\mu g/l$.

Co-precipitation of small amounts of platinum group metals (Ir, Pd, Pt and Rh) with thiobenzamide and with diphenylamine precipitate in analysis of ores and rocks illustrates concentration of trace elements with indifferent co-precipitants [366]. The method was used to isolate elements from 0.1–2 M solutions of HCl, H_2SO_4 or $HClO_4$ obtained after the decomposition of the samples; it excels in high selectivity as Ag, Al, Cd, Co, Cr, Cu, Fe, Ni, Pd, Ti, Zn, alkali and alkaline earth elements do not form complexes with thiobenzamide and are not sorbed by diphenylamine. It is seen from Fig. 2.16 that platinum metals are quantitatively co-precipitated in the HCl concentration interval from 0.1 to 1 M. Fig. 2.17 shows the relationship between completeness of co-precipitation and the heating time of the solution after the addition of

TABLE 2.10

Co-precipitation of trace elements with organic co-precipitants

Elements to be concentrated	Complex-forming reagent	Indifferent co-precipitant or collector	Method of determination	Remarks	Ref.
Hf, Zr	8-Hydroxyquinoline	Phenolphthalein, β-naphthol, diphenylamine, and others	AES	Concentrate was mixed with graphite powder	367
Ag, Cd, Co, Cu, Hg, Mn, Zn	8-Mercaptoquinoline	8,8′-Diquinolyl disulphide	–	Natural waters and alkali-metal salts were analysed	368
Ag, Cd, Cr, Cu, Fe, Hg, Ni, Pb, Se, Zn	Diethylammonium diethyldithiocarbamate	Diethyldithiocarbamates of Co, Cu, Fe, Zn	AAS and AES-ICP	Aqueous solutions and organic substances were analysed	369
Cu, Mn, Pb, Zn, and others	Cobalt(III) pyrroli-dinedithiocarbamate	–	–	Difference waters were analysed	370
Fe, Ni, Sn	Hexamethylenedithio-carbamate	Hexamethyleneammonium, hexamethylenedithio-carbamate	XRFS	Zircalloy was analysed	371
U	Dibenzyldithiocarbamate	Iron dibenzyldithiocarbamate	XRFS, ETAAS	Natural waters were analysed	372
Ag, Bi, Cd, Co, Cu, Fe, Mo, Ni, Pb, Sn, V	Dibenzyldithiocarbamate	Dibenzyldithiocarbamates of silver and nickel	AES-ICP	Natural waters were analysed	373
Cd, Pb	Butylxanthate	Naphthalene	AAS	Concentrate was dissolved in dimethylformamide	374

Elements	Reagent		Method	Remarks	Ref.
Co, Ir, Ni	Tetrahydrofurfuryl-xanthate	Naphthalene	Photometry	Concentrate was dissolved in chloroform; alloys were analysed	375
Pd	Acenaphthequinone-dioxime	Naphthalene	AAS	Concentrate was dissolved in mixture of butyl amine and dimethylformamide	376
Large number of elements	Trioxylfluorones, particularly salicylfluorone	Salicylfluoronates of germanium and aluminium	AES	pH was varied step-by-step; the method is applicable for sea waters	377
Cd, Co, Cr, Cu, Mn, U, Zn	1-Nitroso-2-naphtholate	1-Nitroso-2-naphthol	NAA	Waters including sea waters were analysed	378
Ag, Au, Bi, Cd, Co, Cr, Cu, Fe, Hg, In, Zn	Thiocyanate or iodide, crystalline violet	2,4-Nitroaniline	NAA	Red phosphorus was analysed; concentrate was mineralized	379
Zn	Thiocyanate	Thiocyanate of antipyrine dye	AES	—	380

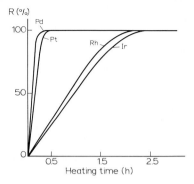

Fig. 2.17. Recovery of Ir, Pd, Pt and Rh in co-precipitation with thiobenzamide and diphenylamine precipitate *vs.* solution heating time. 5 μg each of Ir, Pd, Pt and Rh were dissolved in 100 ml (0.25 M in HCl).

thiobenzamide. Platinum and palladium are quantitatively co-precipitated after 10–30 min of heating, and rhodium in 2 h; iridium is completely isolated on heating the solutions for 2.5–3 h. Optimum conditions for group concentration of platinum group metals are: 0.25 M HCl, heating time 3 h, 30–50 mg thiobenzamide. Other examples are listed in Table 2.10.

2.5. Electrochemical methods

Electrochemical methods of preconcentration (electrodeposition, cementation, electrodialysis and others) are used for analysis of various natural and industrial materials. Techniques involving electrochemical concentration (ECC) make it possible to vary the elemental composition of the concentrate by changing the conditions under which electrochemical processes proceed. They do not require large amounts of chemical reagents (here electricity is the main "reagent") and are accessible practically for any laboratory.

2.5.1 ELECTRODEPOSITION

This is the most commonly used method of electrochemical concentration. The behaviour of an element during electrolysis is determined by its electrochemical potential which depends on the nature of the element, its chemical form, the concentration of the component to

94

be deposited, and the general composition of the electrolyte as well as on current density, material and design of electrode, and the constructional features of the electrochemical cell. In the electrolysis of multicomponent solutions, the voltage applied at the cell is more negative (for cathode processes) or more positive (for anode processes) than the values of the equilibrium potential of the corresponding oxidation–reduction systems calculated by the Nernst equation. As a result of the shift in electrode potentials from the equilibrium potential (polarization), some processes take place at the electrodes, which may be accompanied by the separation of one or several components on the surface or in the volume of the electrode. The cathode processes are of great significance for preconcentration, which can be used for both separation of macrocomponents and concentration of trace elements. In the latter case, electrolysis can pursue the objective of selective as well as of group concentration. Here, one must be guided first of all by the electrode potential of the elements to be concentrated.

Electrolysis at controlled potential (potentiostatic conditions of polarization) is a most commonly used variant of electrodeposition. It makes it possible to achieve practically complete deposition of the determinable component or components on the electrode without co-deposition of the accompanying elements even when the difference in electrodeposition potentials is insignificant (ca. 0.10–0.15 V). Electrolysis at a constant current (galvanostatic conditions) also often ensures quantitative separation and concentration of elements. As element is isolated from the solution, the current has to be gradually decreased according to the change in the cathode potential during electrolysis. Galvanostatic conditions are often justified in the separation of macrocomponents, particularly in those cases when the discharge potentials of the matrix elements are more positive than the charge potential of the trace impurities to be determined, e.g. in determination of trace elements in copper, bismuth or lead.

Most often the trace elements (platinum, graphite and others) are deposited on a solid cathode. Large values of the separation factor for micro- and macrocomponents is a necessary condition. This condition is fulfilled when the main constituents are electrochemically active (e.g. NaCl or $AlCl_3$) or when the analysed substance does not contain any macrocomponents that can separate out on the cathode (natural waters, organic acids). In a number of cases, deposition of matrix elements can be avoided by choosing a suitable medium and composition of the electrolyte. The effectiveness of electrodeposition of trace impurities depends on the potential, material of electrode and the area of its working surface, duration of electrolysis, properties of the electrolyte (composition, viscosity), temperature, and the mixing rate. Since trace elements are electrodeposited from very dilute solutions, the rate at which the components are transported to the surface of the electrode and the electrode potential become important factors.

In the simplest case, the amount of the ith component deposited on the electrode can be expressed by the relation [381]

$$Q_i = \frac{Z_i FSD_i C_i t_e}{\delta_i} \tag{2.30}$$

where Q_i is the amount of the ith component on the electrode, and is expressed in terms of the electrochemical equivalent; F is the Faraday constant; Z_i stands for the number of electrons participating in the electrode process; S denotes the electrode surface; D_i is the coefficient of diffusion of the ith component ion; C_i is the concentration of the ith component in the solution; t_e is the duration of electrolysis; δ_i is the effective width of the diffusion layer for the ith component. This relation does not account for the electrode potential (voltage across the cell) because deposition is assumed to occur at a potential conforming to a steady limiting diffusion current of the ith component ions towards the electrode.

In selective concentration, the electrolysis potential is determined by the discharge potential of the component to be separated [382–386]. When trace impurities are group separated, it is desirable, as a rule, to isolate the maximum possible number of trace elements on the cathode, and to vary the cell voltage in very wide ranges. If the matrix is electrochemically inactive, there would not be any limitations in principle on the values of the potential. However, other conditions being equal, with the rise in cell voltage the degree of isolation first increases and then, at a definite (for each material) voltage, stabilizes or even decreases. For weak electrolytes, the optimum voltage is 30–40 V [387, 388], and for strong electrolytes, it is 5–10 V [389–391]. Liberation of hydrogen shadowing the electrode surface at large voltages has a significant effect on these values. As is shown in [392], the earlier mentioned values of voltage ensure deposition of trace impurities on a graphite electrode, which can be divided into three groups by the chemical composition of the obtained precipitate. Group 1 includes metals that deposit in the elemental state (Ag, Bi, Cd, Cu and Pb); group 2 contains elements depositing as oxides (Co, Cr, Fe, Mn and Ni);

96

elements depositing as oxides, hydroxides and alloys (Ba, Ca, Mg, Mo, Ti and V) form group 3.

Trace elements can be deposited completely or partially. Quantitative separation of trace elements was achieved, for example, in analysis of oxalic, tartaric and citric acid [388] and also of some other substances [387, 393–396]. In most cases, complete separation is achieved only during prolonged electrolysis [381, 397]. For instance, 98–100% deposition of lead and cadmium present in aqueous solutions at a level of $3 \cdot 10^{-6}$% was attained only after 10–15 h [389]. With the decrease in the concentration of trace elements the electrolysis time necessary for attaining complete deposition increases; this is consistent with the theoretical concepts [381]. Zinc was completely electrodeposited from NH_4F solutions on a graphite electrode in 33 h [395]. Owing to the length of the process, partial deposition of trace elements is used in the majority of analytical operations. In this case to obtain correct results it is necessary that the degree of separation of the trace metals should be independent of their concentration in the solution during the selected interval of electrolysis ($\partial Q_i / \partial t_e$ = constant). Under actual conditions this equality is not maintained owing to different side reactions (passivation of the electrode surface by the deposited oxides, liberation of hydrogen, adsorption of surfactants) [392, 398, 399]. The effect of trace elements concentration on the degree of their separation from pure acidified solutions on thin graphite microdiscs has been studied in refs. 392 and 400. When the trace impurities are at a level of $1 \cdot 10^{-7}$ g, 30–40% of the initial amount of group 1 elements, 20–30% of group 2 elements and 10–20% of group 3 elements are deposited (these groups have been mentioned earlier). For the elements of groups 2 and 3, the degree of separation decreased by two to five times when their contents ranged between 10^{-7} and 10^{-5} g. In depositing Bi, Co, Fe and Zn on the inner surface of a graphite tube [395] it was observed that the concentration of trace elements had no effect on the degree of their deposition. In practice, the dependence $\partial Q_i / \partial t_e = f(C_i)$ has to be taken into consideration, otherwise it can lead to incorrect results.

Theoretically, in the absence of other factors affecting the degree of deposition of trace elements, the dependence $C_i = f(t_e)$ should be exponential. But in actual systems the dependence of the degree of deposition on the electrolysis time is described by saturation curves. The time in which saturation is attained is determined by the electrolysis conditions and the composition of the electrolyte and, in the majority of cases, ranges between 15 and 30 min [387, 391, 399]. On the other hand, the trace elements are completely separated from the solution when they are deposited on electrodes that are replaced after short intervals. This enables one to conclude that the appearance of a plateau on the curves showing dependence of the degree of separation on the electrolysis time is exclusively caused by the state of the electrode surface and, in particular, by the deposition of oxides passivating the electrode's working surface [391]. This conclusion is in agreement with the results of the studies of $\partial Q_i / \partial t_e = f(C_i)$ curves.

A rational increase in the cathode surface has a favourable effect on the concentration process [400–402]. Krasil'shchik *et al.* [402] increased the electrode surface by sinking it deep into the solution to be analysed and using, simultaneously, two to four cathodes in one cell. If the concentrate is to be analysed by the AES method, remember that an increase in the electrode surface adversely affects the determination.

The degree of separation of trace elements is also affected by the factors which can cause a change in the rate of mass transfer of elements to the electrode surface, *e.g.* temperature, electric conductivity, viscosity and density of electrolyte [385, 387, 389–391, 399, 401, 403, 404].

For the most part the data published are empirical. The conditions for quantitative and reproducible separation of trace elements to be determined are selected in each individual case. It is difficult to elaborate the general recommendations for selecting concentration conditions, but it is quite possible and rather advisable to estimate and to account for the possible effects of different factors in developing concentration techniques.

Earlier we considered electrodeposition on solid electrodes. Electrolysis on a mercury cathode holds an important place in the concentration of trace elements. The method enables many elements to be separated, including those from acid solutions; this is assisted by high overvoltage of hydrogen on mercury. Usually a mercury pool serves as cathode, as, for instance, in Fig. 2.18. The recent trend is to use solid electrodes preliminarily coated with mercury. Trace elements are very often isolated on a mercury cathode using inverse voltammetry. However, this technique is used in other cases also. Thus, Salin and Habib [405] have described a method of concentrating copper on a mercury–graphite electrode for its subsequent determination by the AES–ICP method. The graphite electrode is dipped into mercury(II) nitrate solution, and mercury is electrolytically deposited on it. The electrode so

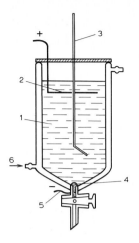

Fig. 2.18. A cell for electrolytic separation of trace elements using a mercury cathode. 1 = Solution to be analyzed; 2 = platinum coil; 3 = mixer; 4 = mercury; 5 = platinum current supply; 6 = cooling water.

obtained is then placed in the solution to be analysed; after deaerating the solution, mercury is deposited by electrolysis. The electrode is washed with water, and mercury is removed by induction heating the electrode. Thereafter, by varying the conditions, copper is determined in an argon plasma.

Electrodeposition of trace elements is combined with various methods of determination. Often, preliminary electrodeposition is combined with subsequent electrochemical transformations of the concentrate deposited on mercury or solid electrodes (by inverse voltammetry or other stripping analysis methods) [382, 397, 406–409] (see elsewhere). Electrolytic decomposition is also combined with photometric, X-ray fluorescence or neutron activation methods [13, 21, 24]. Next to inverse voltammetry and related methods, it is most widely used in combination with AES and AAS [4, 21, 410, 411]. Solid electrodes with trace elements deposited on them can be directly used for determination. For AES or AAS it is not so important in what form the determinable trace element deposits on the electrode. The properties of the obtained precipitate and its distribution on the electrode surface play a smaller role than in inverse voltammetry. This enables the number of determinable elements to be significantly widened. Depending on whether the trace elements deposit completely or partially, various calibration techniques are used in atomic emission analysis. After quantitative separation of the trace elements, use is made of the three-standards method, and after partial separation, the method of standard additions is used. The latter is particularly important if uncontrollable amounts of the matrix deposit on the electrode and affect the excitation and atomization processes in subsequent determination.

Trace elements are deposited in electrolytic cells of various designs [387, 398, 400, 412]. The cells are usually fed from rectifiers ensuring a voltage of 40–50 V and current up to 5–10 A. Sometimes potentiostats are used. The effect of the cell volume, the distance between the electrodes, the mixing rate, and of other parameters on the concentration efficiency has been repeatedly studied [387, 398, 400, 412]. The cathode material and its design are of great importance; they are chosen with due consideration for the requirements of subsequent determination. In atomic absorption determination, a thin filament or a rod of platinum [383, 386], iridium [413, 414], tungsten [384, 400], gold [415] or carbon [416] can serve as cathode. For analysis by the atomic emission method, preference is given to thin disc [417] or cylinder-type carbon electrodes with a device to limit the working surface [389, 393, 403, 415, 418, 419]. One such cell is shown in Fig. 2.19. The electrode can be copper when spark excitation is used [419].

Sometimes, electrodeposition is carried out on the inner surface of graphite tubes under hydrodynamic conditions [395, 420]. In those cases when the electrode is not directly used for determinations, the deposited trace metals are washed from its surface [393] or the surface layer is removed mechanically [393, 401].

Electrodeposition is also used for concentration of trace elements prior to their determination by other instrumental methods. Boslett *et al.* [421] have worked out a procedure for determining Cu, Ni and Zn by

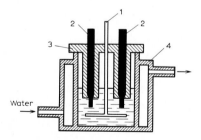

Fig. 2.19. Electrolytic cell for the collection of trace elements on carbon rods. 1 = Platinum anode; 2 = graphite cathodes; 3 = Teflon holder; 4 = quartz cell (the arrows show the direction of water flow).

100

electrodeposition, and for analysis of the obtained precipitate by the X-ray fluorescence method. High-purity graphite rods of diameter 0.6 cm served as working electrodes, electrodeposition being carried out for 6 h. The detection limit of copper was 20 ng. Ando and Tanabe [422] concentrated As, Cd, Fe, Hg, Ni, Pb, Sb, Sr and Zn, and then determined them by the X-ray fluorescence method with a detection limit of $n \cdot 10^{-7}$%.

Electrodeposition of the matrix is less convenient and is not so commonly used as separation of trace components. First, this method is unsuitable for analysis of multicomponent materials. Second, the determination of trace elements with more positive potentials than the potential of the matrix element is not possible. Third, there is always a danger of loss of trace elements which have more negative potentials than the matrix, owing to mechanical adsorption of trace components by the precipitate on the electrode and to the formation of solid solutions or intermetallic compounds – a process associated with variation of discharge potential of trace impurities. To decrease these undesirable effects, separation of the matrix should be carried out with continuous control of the electrode potential.

Separation of the main and alloying constituents of steels (Co, Cr, Fe, Mn and Ni) by electrolysis using a mercury cathode, and subsequent determination of the trace impurities of alkali and rare earth metals are suggested in refs. 423 and 424. The application of the considered variant is described for analysis of high-purity copper [425] and bismuth [426]. Copper bulk was electrodeposited on a platinum cathode at a voltage of 2.5–3 V (current 1.5–2.0 A). To prevent deposition of the trace impurities together with the matrix element, electrolysis was stopped at a copper content of 200 mg. The remaining copper served as collector of impurities in their determination by the atomic emission method. In the analysis of bismuth the matrix was separated on a platinum net at a controlled potential (-0.2 V). 97–99% bismuth present in the solution was separated by electrolysis in 1–3 h depending on the weighed sample being analysed. The remaining bismuth also served as collector for the trace impurities. The potentiostatic conditions of electrolysis enabled only Ag, Au, Hg and Tl to be separated together with the matrix. In the neutron activation determination of lanthanum (10^{-5}–10^{-7}%) in steels, the matrix was separated prior to radiation [427]. The sulphuric acid solution of the sample was transferred to the electrolytic cell with mercury cathode and anode of platinum wire, and electrolysis was carried out for 90 min (voltage 5–7 V, current 2 A);

thereafter, the solution was evaporated and irradiated with a neutron flux of $1 \cdot 10^{14}$ neutrons $cm^{-2}s^{-1}$.

2.5.2 ELECTRODEPOSITION AS PART OF STRIPPING VOLTAMMETRY

As has been mentioned, electrodeposition forms an integral part of the stripping methods of electroanalytical chemistry, of which stripping voltammetry is widely used. In this method, a linear voltage sweep is applied to the electrode after electrolysis; the anode or cathode current of electrodissolution of the electrolysis product is displaced as function of potential or time. We feel that here it is not necessary to dwell on this important field of analytical chemistry as it is discussed in several monographs and reviews. We shall demonstrate only the possibilities of this method by considering an example of inverse voltammetry of solid phases. By the nature of electrochemical transformations of solid phases (preconcentration itself), inverse voltammetry is divided into three main types [381, 406].

Type 1 includes discharge-ionization of elements on the solid electrode surface with the participation of elemental precipitates (inverse voltammetry of metals):

$$Me^{n+} + ne \rightleftharpoons Me \qquad (2.31)$$

This is how the elements, for instance, of the iron subgroup are determined.

Type 2 is the inverse voltammetry of variable valency ions and is based on the interaction of the electrode reaction products and a corresponding component of the solution with the formation of a partially soluble compound on the electrode surface

$$Me^{n+} + me^- \rightarrow Me^{(n \pm m)+} \qquad (2.32)$$

$$Me^{(n \pm m)+} + (n \pm m)R \rightarrow MeR_{n \pm m} \qquad (2.33)$$

$$MeR_{n \pm m} + me^- \rightarrow Me^{n+} + (m \pm n)R \qquad (2.34)$$

This method is used to determine, for example, Au, Cu, Pb, Sn and Tl. The formation of precipitate (the concentration process) proceeds in several stages. Metal ions are first transported to the electrode surface from the solution volume, then migration of electrons takes place, and, finally, a barely soluble compound forms on the electrode surface. The necessary condition for preconcentration of metals is that the chemical reaction of formation of the barely soluble compound should proceed faster than the removal of ions entering into the reaction from the electrode surface. The precipitate in this case should adhere well to the electrode surface. The amount of the compound formed in the electrodeposition process is a linear function of the product of concentration of the electro-active ions in solution and the duration of electrolysis.

Stripping voltammetry of variable valency ions enables multicomponent systems to be analysed with the use of hydroxyl ions or organic ions as reagents. The selectivity of determination is attained through correct selection of reagent precipitant and the use of electrochemical properties of the determined element. If, for example, two oxidation levels, Me^{n+} and $Me^{(n+m)+}$, are typical of element Me, the latter forming a barely soluble hydroxide, then the following process may proceed on the electrode:

$$Me^{n+} \rightarrow Me^{(n+m)+} + me^- \tag{2.35}$$

$$Me^{(n+m)+} + (n+m)OH^- \rightarrow Me(OH)_{n+m} \downarrow \tag{2.36}$$

In this case, the determinable element is concentrated on the electrode surface. Thus, thallium is concentrated on an electrode by oxidation of Tl(I) and by hydrolysis of the electrode reaction product. The precipitate formed is composed mainly of $Tl_2O_3 \cdot nH_2O$ [428].

On using organic reagents the determinable element forms on the electrode surface a barely soluble compound with the reagent. The solution containing the reagent is electrolysed at a potential ensuring oxidation or reduction of the determinable ions; the process on the electrode is followed by the formation of less soluble compounds. The following reactions are typical for the elements to be deposited at a higher oxidation state:

$$Me^{n+} \rightarrow Me^{(n+m)+} + me^- \tag{2.37}$$

$$Me^{(n+m)+} + (n+m)RH \rightarrow MeR_{n+m} \downarrow + (n+m)H^+ \tag{2.38}$$

The following reactions proceed when the element in a low oxidation state binds with a reagent of the same type.

$$Me^{(n+m)+} + me^- \rightarrow Me^{n+} \tag{2.39}$$

$$Me^{n+} + nRH \rightarrow MeR_n \downarrow + nH^+ \tag{2.40}$$

As organic reagent, use is made of 8-hydroxyquinoline, dimethylglyoxime, dithizone, 1-nitroso-2-naphthol and others which form chelates with the cations to be determined. The anion forms of elements are conveniently precipitated with cations of basic dyes, *e.g.* brilliant green and methyl violet. Thus, cobalt(II) is electrochemically concentrated on a graphite electrode as barely soluble cobalt(III) nitrosonaphtholate [429]; antimony is deposited as $SbCl_6^-$ with Rhodamine B cation or triphenylmethane dyes [430].

Type 3 (stripping voltammetry of anions) is based on the reaction of determinable ions with the ionization products of the electrode material, when an insoluble product is formed on the electrode surface. The following reactions describe the principle of the method:

$$Me - ne^- \rightarrow Me^{n+} \qquad\qquad\qquad (2.41)$$
$$\left.\begin{array}{l}\\\end{array}\right\} \text{Preconcentration}$$
$$xMe^{n+} + yR^{m-} \rightarrow Me_xR_y \qquad\qquad (2.42)$$
$$Me_xR_y + ne^- \rightarrow xMe + yR^- \qquad \text{Determination} \qquad (2.43)$$

In electrolysis of solutions, quite complex multicomponent precipitates are usually formed on the surface of electrodes, which, together with individual components, can contain their reaction products, solid solutions and intermetallic and chemical compounds.

The form or the state of the precipitate and its distribution on the electrode surface are of great significance for stripping methods, because the analytical signal is caused by the dissolution of substance only from a definite phase state. Neiman *et al.* [431] have studied the peculiarities of the electrochemical behaviour of lead using graphite and mercury–graphite electrodes. In the case of electrodeposition of lead traces on the surface of pure graphite, the active centres of the electrode material get filled. As the discharge ionization cycles are repeated, the active centres completely saturate and the residue of the metal participates in the formation of the crystal phase. Introduction of mercury(II) into the solution (change over from a graphite to a mercury–graphite electrode) ensures rapid filling of active centres of the electrode material. When such an electrode is used, traces of lead precipitate exclusively on mercury drops, and this ensures a low detection limit and better precision of determination compared to a graphite electrode. A similar effect was noticed in determination of Cu, Sb and Sn [432].

Let us now consider examples of electrochemical deposition used in determination of trace elements by inverse voltammetry of solid phases.

Lipchinskaya *et al.* [433] concentrated nickel and cobalt on a glassy carbon electrode, which has several advantages: chemical stability, large mechanical strength and a wide range of working temperatures.

There are other works also which demonstrate the analytical importance of glassy carbon electrodes [434]. The reaction of manganese(IV) with periodates was employed by Brainina et al. [435] for concentration of manganese and its determination by stripping voltammetry of solid phases using a disc-type graphite electrode impregnated with a mixture of paraffin and polyethylene. The compound formed on the electrode was assumed to have a composition of $MnHIO_6$. This method enabled manganese to be determined in nickel. The detection limit was $1 \cdot 10^{-6}\%$. For determining anions by inverse voltammetry, Parubochnaya et al. [436] concentrated on a silver electrode a film of partially soluble compound of the electrode material with determinable trace components, e.g. sulphide, thiocyanate, ferrocyanide, arsenate and arsenite ions.

Stripping analysis methods including the preconcentration stage can be automated. Lithium in water (10^{-6}–$10^{-7}\%$ mass) was automatically determined with a background of 10^{-2}–10^{-3}-fold amounts of sodium and potassium using the polarographic measuring complex PIK-A-01 [437]. The complex consists of sample selection and measuring aids, a block for supplying the analysed solution into the polarograph sensor, and registration and control devices. The circulating electrolyzer [438], which has several advantages over other electrolytic cells, was used as sensor of an automated system for analyzing natural and sewage waters for Cd, Cu, Pb and Zn. Gunasingham et al. [439] have considered the possibility of automating the determination of elements in anodic inverse voltammetry and have described the determination of Cd, Cu, Pb and Zn for different methods of polarization of solid indicator electrodes. Electrodes are convenient for determining elements in a flow. It is proposed to employ the given scheme for automatic analysis of natural waters including sea waters. Miniaturization of analytical equipment enables analysis to be carried out close to the sites where samples are taken. Thus, a device has been designed for group inverse voltammetric determination of In, Ti and Zn; Bi and Sb; Ag, As, Au, Fe and Hg in various natural substances both under laboratory and field conditions [440]. The content of metals in natural and sewage waters can be continuously controlled by using a flow injection analysis system, detection being done as per the principle of inverse electrochemical methods [441]. Such in-line detectors are convenient for effective control of the composition of environmental samples.

Japanese analysts [442, 443] have developed a minicomputer-controlled automatic Coulomb-potentiographic analyzer for determining down to $1 \cdot 10^{-8} M$ of metals in solutions including natural waters. The background electrolyte was cleaned, trace elements hindering determination were removed from the sample solution, and the trace elements were concentrated and determined; these operations were performed in four electrolytic cells. The analyzer has also been used for determining lead in solutions and lead and zinc in natural waters.

2.5.3 OTHER ELECTROCHEMICAL METHODS

2.5.3.1 Cementation

This is one of the most simple and accessible methods of electrochemical preconcentration. The difference in actual oxidation–reduction potentials of the electrode material and the presence of the metal ions in the solution are the motive forces of the spontaneous process of cementation. As an electrode or, as it is called, cementator, use is often made of metals in the powder form with sufficiently negative oxidation–reduction potentials (aluminium, magnesium and zinc) which make it possible to carry out group separation of more electropositive elements. The rate at which the trace elements are deposited in dilute solutions is basically governed by the diffusion rate of the ions of the element. Therefore, the completeness with which a trace constituent is isolated in a prescribed time increases with the intensity of mixing and temperature.

Ten'kovtsev [444] used cementation for concentration of Bi, Cu and Sb in the analysis of lead salts. Sponge lead with a highly developed active surface was utilized as metal reducer. It was obtained in the analysed solution by the reaction

$$Pb^{2+} + Zn \rightarrow Zn^{2+} + Pb \downarrow \qquad (2.44)$$

Quantitative separation of these elements was carried out from $0.2 M$ HNO_3 at a temperature of about 100°C; separation was completed in 6–10 min by continuously mixing the solution. Copper was concentrated on fine metallic zinc powder [445]; after dissolving the powder in nitric acid, mercury was determined by the AAS method. Copper, lead and noble metals were concentrated on aluminium powder by column chromatography [446]; the concentration coefficient was more than $1 \cdot 10^3$. The concentrated elements were determined by the spark source mass spectrometry method. The application of cementation separation of trace elements from waters and cadmium solution using metals as collectors (Al, Mg, Zn and others) having simple emission spectra is described in refs. 447 and 448. The process ends with atomic emission analysis.

Cementation of trace metals with amalgams of electronegative metals can compete with electrodeposition on a mercury cathode [449]. Generally, cementation is a combination of two electrochemical processes: cathode (separation of the metal to be cementated) and anode (dissolution of the cementating metal).

Two methods of cementation with amalgams are described in ref. 450.

(i) The amalgam of the analysed metal is brought into contact with the matrix metal salt solution; in so doing, the electronegative trace elements transfer to the solution from the amalgam.

(ii) The solution of the analysed metal is treated with the matrix metal amalgam; the electronegative trace elements are concentrated in the amalgam. If the discharge-ionization potentials of trace elements and the matrix are nearly similar, then they can be moved apart by adding complex-forming agents.

Cementation by amalgams was used, for example, for radiochemical separation and concentration of Ag, Au, Bi, Cd, Cu, In, Ir, Pb, Pd, Pt, Se, Sn, Te, Tl and Zn [451].

In the studies of electrochemical behaviour of metals on amalgam electrodes, in electrolytes capable of forming complex ions or slightly soluble compounds with mercury, it has been ascertained that mercury can cementate several trace elements [452]. As, Bi, Cu and Ge (iodide solutions), As and Bi (bromide solutions), Sb and Sn (sulphide–alkali solutions) were concentrated on a mercury drop. Toropova *et al.* [453] studied the possibilities of using electrolytic precipitation on a stationary amalgam microelectrode for concentration and subsequent determination of thallium by the anode current of dissolution of its amalgam. First, the amalgam of the cementating metal (lead) was obtained by electrolysis of the solution containing its salts. Then, the lead amalgam was disconnected from the voltage source, washed with a suitable background electrolyte and placed in a Teflon cell holding the solution to be analysed. Concentration was effected without external polarization under stationary mixing conditions. As the potential of the amalgam electrode abruptly increased (termination of preconcentration), mixing was stopped and the required initial voltage was applied, and the anodic voltage–current curve was recorded. This is a quite selective method of concentration; determination of thallium is not hindered even by 1000-fold amounts of Pb(II) and Sb(III), 100-fold amounts of Cu(II), and 50-fold amounts of Bi(III).

Mizuike *et al.* [454] have developed a rapid method of isolating trace amounts of silver from solutions containing large amounts of copper, iron and lead. The method involves addition of mercury and stirring of the mixture in an ultrasonic field until a water–mercury emulsion is formed. The silver amalgam formed is separated, washed with acetone, and mercury is volatilized at 350°C; the residue is dissolved and analysed by the atomic absorption method. This procedure has been used for determining $n \cdot 10^{-4}\%$ silver in metallic lead.

2.5.3.2 Preconcentration using mercury as chemical reagent

The methods of non-electrolytic concentration of metals on a stationary dropping mercury electrode (SDME) are of interest. These methods enable metal amalgams to be obtained separately. Thus, use has been made of the reaction where metallic mercury reduces bismuth in the presence of iodides with the formation of amalgams of the metal to be concentrated [455]

$$\text{MeL}_m^{(3-my)} + x\text{Hg} \rightleftharpoons \text{Me(Hg)} + \text{HgL}_n^{(2-ny)} + (m-n)\text{L}^{y-}$$

The equilibrium of the reaction appreciably shifts towards the right if the difference in actual potentials of the $\text{MeL}_m^{(3-my)}/\text{Me(Hg)}$ and $\text{HgL}_n^{(2-ny)}/\text{Hg}$ pairs increases as a result of the formation of stable mercury(II) complexes. Such reactions make it possible to concentrate certain trace elements on a stationary mercury drop without its external polarization. The trace elements may be determined from anode peaks of amalgam dissolution. Conditions have been found under which bismuth gathers in the presence of iodides, bromides, thiocyanates and thioureas, and antimony in the presence of iodides in strong acids. Solutions containing 10^{-6}–10^{-7} M metal salts, 1 M perchloric or sulphuric acid, and different amounts of complex-forming substances are placed in a Teflon cell to be thermostatted. The cell is closed with a cover fitted with SDME, a reference electrode, and a gas-supply tube. The electrodes come into contact with each other. An amalgam is obtained on intensively mixing the solution with a magnetic stirrer and by passing the hydrogen formed by electrolysis of alkali solution. After the accumulation is over, the electrodes are placed in a cell containing 1 M $HClO_4$, and anodic dissolution is carried out. In the presence of 1 M

$HClO_4$, 0.01 M KI and 0.1 M thiourea it is possible to determine up to 0.02 $\mu g/ml$ bismuth from the anodic current of dissolution of its amalgam containing antimony (in amounts exceeding up to 100 times) and copper (1000-fold amounts).

For concentration of Se(IV) and Te(IV), Toropova et al. [456] also employed oxidation–reduction reactions between the compounds of determined substances and metallic mercury, which proceeded on a mercury drop surface without external polarization. In this case, the following reactions take place

$$H_2SeO_3 + 4Hg + 4HCl \rightarrow Se + 2Hg_2Cl_2 + 3H_2O \qquad (2.45)$$

$$H_2TeO_3 + 4Hg + 4HCl \rightarrow Te + 2Hg_2Cl_2 + 3H_2O \qquad (2.46)$$

The selectivity of determination can be appreciably increased by choosing the acidity and composition of the electrolyte and also the complex-forming substances. As an analytical signal, Toropova et al. [456] used the value of the current of dissolution of reaction products during subsequent cathodic polarization of a mercury microelectrode. Iron, bismuth, copper, zinc, lead and other accompanying elements do not hinder the determination. Zinc concentrates containing selenium and tellurium up to $1 \cdot 10^{-4}\%$ were analysed.

Toropova et al. [457] have studied the possibility of concentrating amalgam of bismuth and antimony on SDME by reacting mercury and complex-forming reagents directly in the extract. Thiourera served as complex-forming reagent, as it forms much stronger complexes with mercury than with bismuth and antimony. Bismuth and antimony(III) were extracted with ethyl acetate from 1 M solutions of hypochloric acid containing $8 \cdot 10^{-2} - 1 \cdot 10^{-1} M$ thiourea. The thiourea complexes of bismuth and antimony react with mercury in an organic phase to form amalgams, and this confirms the appearance of anode peaks of these metals when amalgam is electrodissolved in the presence of 0.2–1.0 M hydrochloric acid. This technique raises the selectivity of the inverse voltammetry method, and has been used for determination of traces of bismuth and antimony in pure tin.

2.5.3.3 Electrodissolution

Electrodeposition may cause not only precipitation of matrix or trace elements, but also dissolution of the electrochemically active matrix. For example, for determining trace impurities in mercury [458] a mercury anode was anodically polarized at a current density of

$6{-}10\,\text{A/dm}^2$ and at 40–70°C in 20% HCl; the latter was continuously stirred. Trace impurities of Al, Cu, Fe, Mn, Ni, Pb and Sn were transferred from the matrix into solution where they were determined. In other works [459, 460], the anode was made from analysed steel, and the base (iron) was dissolved at a controlled potential. Different inclusions (carbides, nitrides, etc.) remained in the residue, which was then analysed.

Jackwerth devised an original method of preconcentration with the use of mercury. The method involves partial dissolution of the matrix when the impurities do not go into solution, but remain in the residue of the matrix element. Initially, a method based on the difference in the potentials of the matrix and the contaminating metals was developed to concentrate trace amounts of gold and palladium in analysis of mercury [461]. Preconcentration is effected owing to partial dissolution of mercury in nitric acid. For this, 10 g metallic mercury is treated with 3 ml of water and 7 ml of 65% HNO_3 and the mixture is heated slightly. When about 100 mg of undissolved mercury is left, dissolution is stopped. The residue contains gold and silver quantitatively. The solution obtained is decanted and the residue is washed with water. Thereafter, the residue is dissolved in nitric acid, the solution obtained is further treated, and gold with rhodamine B and palladium as iodide complex are determined.

This technique has been utilized for concentration of traces in other high-purity metals. For this, the latter were wetted with pure metallic mercury and the sample obtained was partially dissolved in a suitable medium.

For determining trace amounts of Ag, Au, Bi, Cd, Co, Cu, In, Ni, Pb, Pd, Sn and Tl in high-purity metallic zinc [462], a thin layer of mercury (1–5% of the sample mass) was applied on the surface of a precompressed (into a briquette) sample of mass 5–25 g. This sample was dissolved in concentrated HCl on heating moderately. As the concentration of the acid decreased, more acid was added until 100–200 mg zinc was left. For separating mercury, the residue was washed with water and dissolved in 5 ml concentrated nitric (sulphuric) acid or *aqua regia*. The solution obtained was evaporated to 1 ml, diluted with water up to 30 ml, 5 ml concentrated nitric acid was added, and the solution was heated; mercury was separated as metal. The solution was decanted and, after suitable treatment, Au, Bi, Cu, Ni and Pd were determined spectrophotometrically; Cd and Pd were determined by the voltammetric method; Ag, Cd, Bi, Co, In, Ni, Sn and Tl were determined by the atomic emission method. The AAS method enabled the nitric acid solution to be directly analysed for Cd, Cu, Ni and Pb without separating mercury. Depending on the mass of the sample and of the residue obtained after dissolving, concentration coefficients of 10^3–10^4 were attained. The recovery of Ag, Au, Bi, Cd, Co, In, Ni, Pb, Pd, Sn and Tl exceeded 95%, of copper, 90% and of Ga, 80%.

In atomic absorption determination of traces of Bi, Cd, Ga, In, Tl and Zn in aluminium [463], the detection limits being 10^{-6}–10^{-4}%, use was also made of the fact that the matrix and the contaminating elements had different electrochemical properties. Because of this, the impurities remained in the mercury layer, with which the sample was covered.

In a similar manner the impurities in high-purity gallium [464, 465], cadmium [466], manganese [467], indium and tin [468] were determined using partial dissolution of the matrix; the concentrates obtained were analysed by the atomic absorption method.

2.5.3.4 Electrodialysis

This is an effective method of separating impurities of ionized compounds from electrolyte solutions by means of an electric field [469–471]. This is done as follows. The solution to be analysed is placed in the middle chamber of a three-chamber electrodialyzer, the chambers being separated by ion-passing membranes. A voltage is applied on the electrodes placed in the lateral chambers. As current passes through membrane packets, the cations and anions concentrate respectively in the cathode and anode chambers; similar charge ions separate out due to the difference in the velocities at which they pass through membranes. The separation and concentration efficiency of ions is determined by the composition and temperature of electrolyte, membrane material, applied potential, number of chambers, and some other factors. The efficiency of electrodialysis, as a method of concentration, increased abruptly after ion-exchange resins were used as membrane material. These resins made it possible to significantly lower the background diffusion of ions into the middle chamber from the electrode chambers, improved the distribution of total voltage in the electrodialyzer chambers, and increased the determination selectivity.

Let us illustrate this by considering an example. Electrodialysis in combination with atomic emission analysis was used in analysis of neodymium oxalate for Ca, Co, Cu, Fe and Zn [472]. Fig. 2.20 shows a

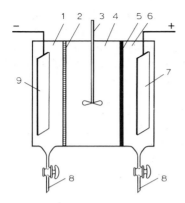

Fig. 2.20. Schematic diagram of a three-chamber electrodialyser. 1 = Cathode chamber; 2 = filtering membrane; 3 = mixer; 4 = middle chamber for the sample suspension; 5 = anion exchange membrane; 6 = anode chamber; 7 and 9 = electrodes; 8 = tubes with taps for discharging solution.

schematic diagram of a three-chamber electrodialyser made from acrylic plastic and used in this study. The chambers have the following parameters: internal diameter, 4.5 cm; volume of the middle chamber, 64 cm^3; volume of the side chamber, 15 cm^3. Electrodes were made from 25 × 25 mm plates of platinum or platinized titanium and spaced at about 60 cm. A high-voltage rectifier was used as a source of direct current, voltage being regulated in the range of 0 to 2 kV. Maximum current was 0.5 A. The middle chamber was separated from one of the lateral chambers with cation or anion exchanger membranes; the other chamber was separated with a high polymer material of high filtration capacity. The membranes were located relative to the lateral chambers so that the direction of electrolytic transfer of impurity ions coincided with that of the flow of solutions through filtration membranes (in concentrating iron, the middle chamber was separated from the anode chamber, and in the case of other elements, from the cathode chamber). The formed suspension of neodymium oxalate was placed in the middle chamber and the lateral chambers were filled with distilled water. The suspension was mixed with a mixer at 500 rpm. The removal of electrolyte from the middle chamber was controlled by the current and the pH of solutions near the electrodes, and was completed at a current strength of about 10 mA and a cell voltage of about 200 V. With the use of radioactive indicators, it was shown that 10^{-2}–$10^{-4}\%$ impurities concentrated with a yield of more than 90%. The relative detection limits of impurities determined by the atomic emission method were lowered by 1–1.5 orders of magnitude. These authors [473] concentrated trace impurities of Ca, Co, Cu and Zn from neodymium hydroxide also.

Moskvin *et al.* [474] studied preconcentration of Pd and Pt from chloride solutions using a liquid membrane which represented porous Teflon impregnated with tributylphosphate. The anode space of the cell was filled with tributylphosphate and the cathode space with HCl solutions (0.1–$10\,M$) containing the element to be concentrated. By the action of an electric field the negatively charged chloride complexes of the type $HMeCl_x^-$ transferred from the catholyte to the anolyte through the membrane. Cox and Carlson [475] have shown that high concentration coefficients can be attained in potentiostatic electrodialysis; a galvanostatic regime does not offer large advantages.

2.5.3.5 Electroosmosis

The process of transferring solvent using an electric field but without the application of hydrostatic pressure on the membrane system is known as electroosmosis [469–471]. This contrasts with electrodialysis where ions undergo transfer. In practice, these two processes (electrodialysis and electroosmosis) can proceed simultaneously to some extent.

A technique involving the use of electroosmosis has been developed for concentration of positively charged ions and colloidal forms of elements with the aid of synthetic ion-exchanger KU-2-8 in the H-form, the latter being used as collector of trace impurities [476–479]. Positively charged porous membranes of thickness 2.5–3.0 mm were prepared by fusing a mixture (3:2) of chromatographic aluminium oxide and Teflon powder at 380°C.

Fig. 2.21 shows a concentration cell. Prior to concentration the membrane was impregnated with deionised water which was also added to the anode chamber, and aqueous solution containing the trace elements to be concentrated was supplied to the cathode chamber from a Mariotte's vessel. Thereafter, a constant voltage was applied to the electrodes and the electroosmotically transferred water was driven off into the storage tank. Iron and cobalt concentrated quantitatively in the absence of cation exchangers. However, significant sorption of trace elements on the surface of the membrane, and also their electrodeposition on the cathode were observed. These undesirable effects were eliminated by adding the cation exchanger KU-2-8 in the H^+-form.

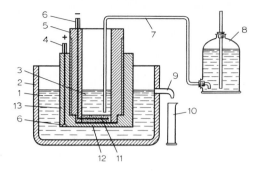

Fig. 2.21. Cell for electroosmotic concentration of trace impurities. 1 = Anode chamber; 2 = beaker made from acrylic plastic; 3 = cathode chamber; 4 = anode; 5 = inner cylinder; 6 = cathode; 7 = PVC hose (inlet); 8 = Mariotte's vessel; 9 = outlet pipe; 10 = measuring cylinder; 11 = ion-exchange resin; 12 = membrane; 13 = outer cylinder.

No less than 99% trace elements (Co, Cu, Fe and Ni) were concentrated in the ion exchanger layer under such conditions. After concentration, the ion exchanger was washed, dried, mixed with charcoal and analysed by the atomic emission method. This technique has been used for determining Co, Cu, Fe and Ni and other elements in corrosion products of constructional materials. The concentration coefficient was about $1 \cdot 10^3$. The proposed membranes have high electroosmotic permeability in respect to aqueous solutions of boric acid, hydrogen peroxide, ethylene glycol and glucose. Quantitative separation of the ionic and colloidal forms of trace impurities to be determined is achieved.

2.5.3.6 Electrophoresis

This method is based on different mobilities of ions, solid particles, gas bubbles, liquid drops and colloidal particles in a liquid or solid phase by the action of an external electric field. Basically it is used for separation and not for preconcentration [480–484]. It is mainly applied in organic analysis, principally in biochemistry. Low-molecular-weight compounds like amino acids, amines, alkaloids, acids, phenols, sugars, hormones and vitamins are separated by this method. In inorganic analysis, electrophoresis assists in solving some problems of concentration and particularly of fine separation of trace elements. Examples of separation by this method are given in refs. 184 and 482–485, electrophoresis being discussed mainly in combination with paper chromato-

114

graphy and thin layer chromatography. Such a combination is called electrochromatography: using a direct current the components of the mixture migrate selectively from the initial narrow zone to the permeable solution of the main electrolyte through a porous medium.

Trace impurities of Eu, Lu, Sm, Tb and Tm were concentrated and separated by electrophoresis in analysis of spectral-pure gadolinium [486]. Thin layers of acetyl cellulose (0.24–0.30 mm) were impregnated with 0.2–0.4 M α-oxyisobutyric acid. Electrophoresis was carried out at a current strength of 8.2–8.7 mA and a potential gradient of 40–80 V/cm. The zones formed on the plate were irradiated with thermal neutrons and then discrete rare earth elements were identified by typical half-lives using a multichannel γ-spectrometer.

2.5.3.7 Electrodiffusion

The phenomenon of electrodiffusion has been used [487] for concentrating trace impurities with sufficiently high electric mobility in liquid fusible metals. A gallium sample containing $n \cdot 10^{-3}\%$ bismuth was used for the purpose. Difficulties associated not with the process of concentration, but with the calculation of initial concentration of a trace component from the data on its content in the enriched sample were faced by the authors of this work [487]. They have developed two techniques for determining the initial content.

Concentration was carried out in a cell comprising a 19–37 cm long capillary (internal diameter 0.6–0.7 mm) which widens to form a reservoir for taking the initial sample of liquid metal. The cell was housed in a water jacket to maintain a constant temperature. Tungsten electrodes were placed on and under the cell. The current density was varied between 1800 and 3700 A/cm^2 in different experiments. On passing current through liquid gallium, bismuth concentrated in the anode space. The authors [487] succeeded in raising the concentration of bismuth in an enriched sample by 10–200 times compared to the initial concentration in 100–400 h.

Mikhailov et al. [488] have worked out a procedure for electrodiffusion concentration of small amounts of Bi, Pb and Cu (10^{-4}–$10^{-6}\%$) in liquid gallium (120°C). For a weighed sample of 0.1–0.3 g, the following concentration coefficients were attained after 200 h: Pb, 7–21; Cu, 5–14; Bi, 20–50. The theory of electrodiffusion and examples of its use as a method of analytical concentration are given in ref. 489. This method is hardly suitable for practical analytical purposes because of its duration.

2.6. Evaporation methods

These methods are based on the use of the difference in distribution coefficients of macro and trace elements in liquid–vapour or solid–vapour systems. They are applied in the analysis of diverse samples. Wide scope, simplicity and rapidity are some important advantages of these methods. Numerous reviews are dedicated to these methods [4, 13, 21, 24, 490–492].

Probably, for analytical practice no convenient classification of the considered methods of concentration is yet available. Nevertheless, it is possible to discern simple distillation (evaporation), rectification, molecular distillation (distillation under vacuum) and sublimation. Dry and wet ashing of organic and biological samples, distillation of inorganic substances, *e.g.* arsenic and phosphorus as arsine and phosphine (after reduction), and silicon as tetrahalide (after oxidation), and other methods are important. Specialists in electrochemical analysis make use of the so-called fractional evaporation method.

2.6.1 EVAPORATION

These methods are based on the difference in liquid and vapour compositions; they include simple distillation, rectification and molecular distillation [403].

The difference in liquid–vapour composition is expressed in terms of the distribution coefficient

$$D = \frac{x}{1 - x} \bigg/ \frac{y}{1 - y} \tag{2.47}$$

where x and y are the molar fractions of the trace component respectively in liquid and gas. This expression is set up in such a manner that D is more than unity.

The distribution coefficient is a function of the nature and composition of the mixtures being separated, and also of temperature. The total content of trace components is, as a rule, much less than the content of a macrocomponent; therefore, it can be taken that $x \ll 1$ and $y \ll 1$. Now it can be assumed that the distribution coefficient of a matrix trace component mixture is independent of the mixture composition, and is constant at a given temperature:

$$D = x/y \tag{2.48}$$

This relationship applies to the case when a trace component is concentrated in a liquid phase. For the opposite case

$$D = y/x \qquad (2.49)$$

Unlike technological processes and preparative methods, the implementation of distillation in an analytical laboratory is simplified because it is not always necessary to retain the fraction, stripped of the trace components, quantitatively.

Simple distillation (evaporation) is a one-stage process. It is applicable for separation and concentration of trace components when the enrichment coefficient is comparatively large. Evaporation is extensively used for concentration of traces in analysis of waters, acids, organic solvents, etc. By taking necessary precautions, the analysed sample is evaporated to a small finite volume or to dryness (often in the presence of a collector) and the obtained concentrate is analysed by a suitable method. Evaporation is often applied in combination with other methods of concentration: extraction, ion exchange, electroosmosis, etc. It does not require reagents.

Rectification is a multistage separation-and-concentration process which is carried out in packed, platelike or filmy columns. With this method one succeeds in separating components with very close properties, *e.g.* isotopes of an element, when $D = 1.001-1.005$. Rectification is extensively used for technological or preparative separation of various organic and inorganic mixtures. It is employed for obtaining different isotopes, high-purity substances, petroleum products, etc. In inorganic analysis this method is only occasionally used because substances with macro- and trace components that appreciably differ in volatility are mostly analysed.

Successful use of evaporation depends on how chemically stable are the macro- and trace components during prolonged heating. In the case of substances having low thermal stability, evaporation is effected at a low pressure; this lowers the boiling point of the solution. If separation is carried out at a pressure of 1.3–1.8 kPa, then such a process is known as molecular distillation or distillation in high vacuum. The process can also be accomplished in non-equilibrium conditions. The expression for the distribution coefficient can be written as

$$D = \frac{x_{\text{cond}}}{1 - x_{\text{cond}}} \bigg/ \frac{x}{1 - x} \qquad (2.50)$$

where x_{cond} is the molar fraction of the trace component in the conden-

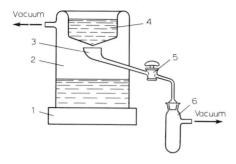

Fig. 2.22. Cell for molecular distillation. 1 = Heater; 2 = distillation vessel; 3 = collector; 4 = cooling agent; 5 = tap; 6 = receiver.

sate; x is the molar fraction in the evaporating liquid at the given instant of time.

Fig. 2.22 shows a cell for carrying out concentration by molecular distillation.

In the case of molecular distillation, the evaporation and condensation surfaces are spaced at a distance equal to the free path length of the molecules of the substance being separated (20–30 mm). Mass transfer includes diffusion of molecules of a more volatile component from a deep-seated layer of molecules to the evaporation surface, evaporation of molecules from the surface, transfer of molecules from the evaporation to the condensation surface, and finally condensation of molecules on the surface. Diffusion of molecules is usually the limiting stage.

(i) Volatilization of matrix. Simple volatilization of the matrix of an analysed substance is a very important technique of concentration. Evaporation of aqueous solutions of salts is a common process. A method has been developed, for instance, for determining trace quantities of Co, Cr, Cu, Mo, Ni and Pb in drinking and river waters; the method is based on slow evaporation of the sample (1 l in several days) and subsequent analysis of the concentrate by the atomic emission method [493]. To determine Cd, Cu and Pb in different waters the sample was evaporated directly in a cup-shaped graphite atomizer which made it possible to operate with a sample of volume up to 30 ml [494]. An electrothermal atomic absorption method has been developed for determining silver (up to $1 \cdot 10^{-6} \mu g/ml$) in rain water and snow [495]. It involves concentration of the trace element by evaporation of a 1 ml sample (in lots of $50 \mu l$ at 80°C) in a tube atomizer of large dimensions. This device can be used in the analysis of various liquid samples.

Automatic AA determination of Ag, Al, Cd, Co, Cr, Cu, Fe, Mn, Ni, Pb, Sb and V in water circulating in radioactive areas of nuclear power reactors is based on evaporating the sample directly in a graphite crucible and atomizing the dry residue [496]. A microprocessor controls the system. Torok et al. [497] have described a device for automatic neutron activation determination of about twenty trace elements in natural and sewage waters. Water is first evaporated and then the dry residue is irradiated; the trace elements are radiochemically separated into groups convenient for gamma-spectrometric measurements and determined.

Evaporation is an indispensable method of preconcentration of impurities in analysis of high-purity acids. The method of determining traces of Co, Cr, Mn, Mo, Pb and V in hydroiodic acid involves evaporation of the sample on a graphite collector and analysis of the obtained concentrate by the atomic emission method [498]. The detection limits of elements are 10^{-6}–$10^{-8}\%$. A procedure has been developed for determining 33 elements in different acids [499]. It involves evaporation of a 100 g sample to dryness in a Teflon vessel in the presence of 10 mg of graphite powder and two drops of concentrated sulphuric acid. Evaporation is carried out with an IR lamp (fitted in a special box) in a weak flow of high-purity nitrogen. The dry residue was also analysed by the AES method. The following detection limits were attained: Be and Mg, $5 \cdot 10^{-8}\%$; Ag, Ba, Ca, Cu, Mn, Al and Si, $(1–5) \cdot 10^{-7}\%$; Li, $5 \cdot 10^{-5}\%$; other elements, $(1–10) \cdot 10^{-6}\%$.

Evaporation is employed in analysis of volatile halides of arsenic, tin, tungsten, platinum, antimony, bismuth, germanium, silicon and other elements. Analysing BBr_3, PCl_3, $AsCl_3$, $POCl_3$ and $SbCl_5$ for Ag, Al, Au, Bi, Cd, Cu, Fe, Mg, Mn, Ni, Pb, Sb, Tl and Zn with detection limits ranging between $1.5 \cdot 10^{-4}\%$ (Zn in $SbCl_5$) and $5 \cdot 10^{-8}\%$ (Mn in $AsCl_3$), Kuzma et al. [500] volatilized the matrix in dry nitrogen in the presence of a graphite collector. The concentrate obtained was analysed by the atomic emission method. Larin et al. [501] used volatilization for concentration of trace impurities in silicon tetrachloride, their subsequent determination by the atomic emission method being carried out in a device made basically of Teflon and molybdenum glass. 20 ml of the sample and 60 mg of spectrally pure graphite powder were taken in a Teflon vessel. The system that was formed was evacuated to a pressure of 0.1 Pa and the matrix was volatilized in a chilled ampoule. After the completion of distillation, the graphite powder was removed and analysed by the atomic emission method. The temperature of the

119

Teflon vessel and of the ampoule was 30 and 20°C, respectively. The detection limits of fifteen trace elements were $2 \cdot 10^{-6} - 6 \cdot 10^{-8}$%.

Rayleigh distillation was employed by Devyatykh et al. [502] for concentration of impurities of non-volatile compounds of Al, Ca, Cr, Fe, Mn and Sb in the atomic emission analysis of titanium tetrachloride. Use was made of the concentrator described in ref. 501 with the difference that the Teflon vessel containing the sample was replaced by a glass vessel in order to observe the process. The sample was heated up to 40°C and the flask receiver was immersed in liquid nitrogen.

Another form of evaporation is drying under vacuum in the frozen state (lyophilic drying). The method is convenient in that it makes it possible to reduce the losses of highly volatile trace elements in analysis, say, of water. The behaviour of 21 elements present in natural waters has been studied with radioactive indicators [503]. Losses of only mercury and iodine occur, their chemical yield being 60–70%. The method has been used in combination with instrumental neutron activation analysis. In neutron activation analysis of natural waters, Lieser et al. [504] also dried the sample (before irradiation) in the frozen state for concentration of Au, Ba, Br, Ca, Ce, Co, Cr, Eu, Fe, K, La, Mo, Na, Sb, Sc, Se, U and Zn.

With the aim of lowering the detection limit of chlorine in metallic sodium, Takahashi et al. [505] separated the matrix by molecular distillation. After a suitable treatment, the chloride ions (ca. $1 \cdot 10^{-4}$%) were determined in the residue by coulometric titration with potentiometric indication of equivalence points.

Matrix evaporation is employed for determining elements/impurities in organic solvents and other volatile organic substances. Erickson et al. [506] have compared a number of techniques of evaporating organic solvents (for the purpose of organic analysis).

Evaporation of matrix may be accompanied by losses of determinable trace elements because of mechanical entrainment of sample from the gas phase, evaporation of volatile forms of trace elements, and sorption on the surface of the vessel used in evaporation. Necessary safety measures must be taken. One of the authors of this book has chosen conditions under which high-purity dioxane and isopropyl alcohol can be evaporated to obtain a concentrate with elements/impurities in the solvents. Two techniques were employed: heat from below using a synthetic quartz water bath; heat from above with an infrared lamp. Experiments were conducted under sterile conditions (acrylic glass box with dust-free air, crucibles made from synthetic quartz, Teflon, platinum) and the solvents were not allowed to boil. Then 100 ml of organic solvent containing trace elements (10^{-4}–10^{-7}% of the mass) was evaporated to dryness on 30 mg spectrally pure graphite powder, and the powder mixed with 10 mg of the same powder containing 10% sodium chloride. The mixture obtained was analysed by the atomic emission method. Parallel to this, to take account of the impurities coming from outside and those present in the solvent, similar solvents were evaporated in the absence of trace elements.

Significant losses (up to 50–70%) of Ag, Al, Bi, Co, Cr, Cu, Fe, Ga, Mg, Mn, Ni, Pb, Ti and Zn were noticed when evaporated on a water bath. No losses were recorded on evaporating with an IR lamp. Apparently, in the latter case, evaporation of the solvent is effected from above, while on a water bath the vapour phase bubbles that are formed in the bulk of the solvent carry away fine droplets of the liquid. Heating from above enabled the systematic and often the random errors to be eliminated. The crucible material had no effect on the results.

In a similar manner, Yudelevich and Shelpakom [490] studied the losses on evaporating solutions of hydrochloric and nitric acids on a graphite collector. It was shown that, excepting impurities with high vapour tension as Ge(IV), Sn(IV) and As(III), the losses of trace elements Bi, Cd, Co, Cr, Ga, In, Ni, Pb, Tl and Zn became significant even at 80°C; at 90–100°C the losses reached a value of 10–15%. This is explained by mechanical drift of fine droplets of the solution with vapours.

(ii) Volatilization of trace components. This technique is less often used than matrix evaporation because the determinable elements are seldom present in the form of readily volatile compounds. Generally, such compounds are first obtained by methods such as chemical transformations. These methods are considered elsewhere.

2.6.2 SUBLIMATION

The equation pertaining to a liquid–vapour system can be used for calculating distribution coefficients for a solid–gas system. Such calculations are, however, very approximate [507] and require a thorough experimental verification. Unlike evaporation methods in which the thermal motion of molecules tends to establish equilibrium between the phases and to equalize concentration within the phase, the degree of establishment of equilibrium in sublimation (volatilization) depends on the homogeneity of trace element distribution in the volume of the particle (crystal); therein the diffusion processes proceed very slowly. Good results are therefore obtained by sublimating the matrix. Alternatively, the sample has to be carefully crushed before analysis.

For determining Ag, Bi, Ni, Pb, Sn and Tl in cadmium, Barinov and Aidarov [508] placed the sample (100 mg) in the crater of a graphite electrode, and then sublimated the matrix at 400°C and at a pressure of 6.6–6.7 Pa. The residue was analysed by the atomic emission method. In the case of atomic emission analysis of iodine for Ag, Al, Au, Ca, Cr, Cu, Fe, Mg, Mn, Ni, Pb, Sb, Sn, Ti, V and Zn the sample was mixed with carbon collector, and then the matrix element was sublimated at 60–65°C [509]. The detection limits were 10^{-5}–10^{-8}% for weighed quantities of 2–12 g. Using sublimation under vacuum, we concentrated trace impurities of Al, Cu, Fe and Mn present in 8-hydroxyquinoline. Determination was completed by the atomic emission method (see ref. 9, p. 152).

2.6.3 VOLATILIZATION AFTER CHEMICAL TRANSFORMATIONS

The concentration method based on conversion of trace components or the matrix into volatile compounds by chemical reactions and subsequent volatilization is widely used. Examples of this are: volatilization of arsenic and antimony as arsine, stibine, trihalides; concentration of boron as boromethyl ether $B(OCH_3)_3$, germanium and silicon as tetrachloride and tetrafluoride, nickel as tetracarbonyl, sulphur as oxides and hydrogen sulphide, and of chlorine, bromine and iodine in the elemental state. Dry and wet mineralization are other methods used for removing the matrix.

2.6.3.1 Mineralization of organic and biological samples

In determining trace elements in organic compounds or inorganic samples containing organic substances, and biological samples, quite often the sample has to be mineralized. Therefore, mineralization is a widely used method in the elemental analysis of soils, plants, animal tissues and human excretion, plastic, rubber goods, petroleum products, etc. In most cases the organic substance of the sample is oxidized.

In dry ashing of organic substances, air (or, rarely, pure oxygen or chlorine) serves as oxidant. CO_2, H_2O and N_2 are the main reaction products when oxidation is carried out in air or in a stream of oxygen.

Sulphur dioxide also appears if the samples to be analysed contain sulphur. Dry ashing is generally carried out at moderate temperatures, *i.e.* at 400–500°C; the sample is placed in a cold muffle furnace or another heater and gradually heated to the given temperature.

The advantages of dry ashing are simplicity and absence of reagents. Complete or partial loss of volatile forms of trace elements or their drift as aerosols, frothing (observed sometimes), "splashing" of sample, and difficulty of dissolving the residue are the main disadvantages of this technique.

Even in 1955, Tirs (cited in ref. 510) listed the disadvantages of this technique. As per his data, various compounds of mercury, halides of As, Ge, S, Sb and Si, chlorides of Al, Cd, Sn, Ti, V, Zn and Zr, fluorides of Mo, Nb, Ta, Ti and W, and some chelates, As, I, S, Sb, Te and Zn are lost partially or completely in the elemental state. These losses are often irrecoverable. Non-volatile trace components can be carried away from the gas phase as aerosols. Certain organic substances are ashed very slowly, stick to the crucible or cup material, and sometimes react with it.

These disadvantages limit the application of the dry ashing technique. But in a number of cases the disadvantages can be overcome by properly selecting the mineralization conditions (see, in particular, refs. 4 and 9): by adding an inert collector or a substance which converts highly volatile compounds into relatively involatile compounds (for example, reduction of chelates into oxalates which upon ashing change into oxides); by carrying out ashing under pressure or, on the contrary, in a low vacuum. Sometimes ashing is carried out in the presence of oxygen.

For determining Ca, Co, Fe and Mn in natural rubber, the sample was ashed and the residue was analysed by the atomic emission method [511]. The concentration coefficients were 40–60. Analysing crude oil and petroleum products for Al, Ca, Cr, Cu, Fe, Mg, Mn, Ni, V and Zn, Farhan and Pazanden [512] calcinated the sample (0.2–0.5 g) at 500°C for 30 min after drying it. The dry residue was analysed by the atomic emission method. The detection limits were 10^{-4}–10^{-5}%.

Dry mineralization is also used for carbon- and sulphur-containing materials. Thus, impurities of twenty elements present in graphite were determined by the AES method after burning the sample to ash [513]. For determination of the trace impurities Al, Ca, Co, Cr, Cu, Mg, Mn, Ni, Pb, Si, Sn and Tl in high-purity carbon materials, Gorbunova *et al.* [514] burnt the sample in air at 700°C; 5 g of sample was mineralized in

6–7 h. The residue after concentration was mixed with sodium chloride and analysed in a dc arc. The detection limits were 10^{-6}–10^{-7}%. A method of determining Al, Ba, Bi, Ca, Cd, Cr, Cu, Fe, Mg, Mn, Ni, Pb, Sn, Ti and V in high-purity sulphur is described in ref. 515. A sample (1 g) was heated in air at 300°C; the matrix element was removed as sulphur dioxide. The residue containing impurities was dissolved in a small volume of 6 M HCl, and 10 mg of a mixture of 90% graphite powder and 10% sodium chloride was added to the solution. The mixture obtained was placed in the crater of a carbon electrode and evaporated to dryness. Thereafter, the sample was excited in a dc arc.

For dry mineralization, use is quite often made of low-temperature oxygen plasma. Patterson [516] has designed a laboratory-scale device for ashing biological materials in oxygen plasma. The device consists of a high-frequency generator for feeding electrode-less lamps used in atomic absorption spectrometry. A vessel containing the sample to be ashed is placed in the inductor of the generator. The ashing time depends on the amount of the material being ashed, its nature, and the power input. The ashed residue can be analysed by any suitable method.

Volodina *et al.* [517] have described a method of decomposing organic substances in low-temperature high-frequency discharge oxygen plasma. A weighed quantity (2–3 mg) taken in a quartz boat was placed in a reactor, and the system was evacuated to a residual pressure of 0.2–0.5 kPa. Oxygen was introduced and a voltage (1100–1200 V) was applied to the system. The sample decomposed in 3–10 min. Under these conditions, Co, Cu and Fe present in organic substances form higher oxides; nickel yields NiO, and noble metals take the elemental form. Decomposition proceeds faster with a mixture of oxygen and fluorine under a radiofrequency field [518]. This technique has been used for determination of Cr, Fe, Pb and Zn in biological samples and food products by AAS. Decomposition was effected in low-temperature pure oxygen plasma for analyzing soft biological tissues [519].

Knapp [520], in the review dedicated to the methods of decomposing organic and inorganic substances, believes that conventional dry composition methods have many advantages. He prefers to carry out decomposition in autoclaves in an atmosphere of oxygen. He has considered arrangements for decomposing organic samples with high-frequency or microwave plasma at a temperature of about 100°C, and also for burning samples in a current of oxygen. Fig. 2.23 shows one such arrangement intended for determination of nano- and picogram

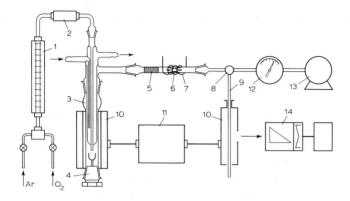

Fig. 2.23. System for the decomposition of biological samples and rocks in microwave induced oxygen plasma and atomic emission determination of mercury in a microwave induced argon plasma. 1 = Rotameter; 2 = a tube with sorbent for gas purification; 3 = quartz vessel for the decomposition of a sample; 4 = sample holder; 5 = quartz-wool; 6 = gold absorber; 7 = heating coil; 8 = three-way tap; 9 = quartz capillary; 10 = microwave cavity; 11 = microwave generator; 12 = manometer; 13 = vacuum pump; 14 = atomic emission spectrometer.

amounts of mercury in biological samples and rocks [521, 522]. Concentration was done in a closed quartz system using microwave oxygen plasma. Mercury is absorbed by gold and an amalgam is formed. The system is then heated in an argon atmosphere, and mercury that isolates is determined by AES using microwave argon plasma. Systematic error is minimized under such conditions.

Special methods of ashing organic samples (rubber, vegetable materials and blood) have been worked out for subsequent analysis of ash by the atomic emission method [523]. The samples are ashed in an arc furnace and the spectrum of the smoke is registered; use is also made of incomplete ashing of sample at 100–400°C without the formation of smoke, but by obtaining semicoke. Ashing continues for 1–20 min. Designs of ashing furnaces are also discussed in this paper. The techniques mentioned above made it possible to avoid the losses of As, Ba, Be, Ca, Cu, Fe, Mn and V. The losses occur at 450–600°C and are associated with the removal of trace elements by the intensively formed smoke and gas.

In wet ashing, concentrated acids and their mixtures (HCl, HNO$_3$, H$_2$SO$_4$), hydrogen peroxide, potassium perchlorate in HCl medium and potassium permanganate in acid and alkali media are used as oxidizers. Generally, wet ashing involves small losses of trace components, but it takes even longer than dry ashing and requires reagents; this can result in the introduction of impurities.

The following is an example of simple decomposition at atmospheric pressure. Procedures have been developed for determining Cu, Fe, Ni, Pb and V in crude petroleum, benzene and some naphtha fractions [524]. They involve ashing of a 100-ml sample with concentrated sulphuric acid on heating, burning of the remaining carbon in an electric furnace at 470°C, and dissolution of the ash in 5 M HCl. Vanadium was determined by a kinetic method using the oxidation reaction of n-phenetidine with BrO_3^-; iron was determined spectrophotometrically with 1,10-phenanthroline; copper, nickel and lead were determined by the extraction spectrophotometric method using bathocuproin, α-furyldioxime and dithizone, respectively. Fig. 2.24 shows the device developed by Bajo et al. [525] for wet mineralization of organic substances (coals, plastics and food products) by sulphuric and nitric acids at atmospheric pressure. Here, ashing (mineralization) is effected almost automatically.

For complete and rapid transfer of slightly soluble substances into solutions, they are dissolved in autoclaves at elevated temperatures and pressures. This method of mineralization is discussed in detail in the review by Jackwerth and Gomišček [526]. Wet ashing under pressure offers a much higher temperature of decomposition than in open vessels. Losses of volatile compounds and contamination of the laboratory air are eliminated. Generally Teflon or glassy carbon vessels are used for ashing. Selection of temperature (170 or 220°C) and pressure, nature and concentration of acids, and ashing features with due consideration of the method for subsequent determination are covered in the review.

Here are a few examples. For determining Be, Co, Cr, Cu, Mn, Ni, Pb, V and Zn in coal [527] the sample was decomposed with a mixture of acids (HNO_3 and HCl) in a closed Teflon vessel. 0.5 g coal decomposes in 7–14 h at 150–160°C; it is then further heated with HF for 2 h. The elements are determined by the flame atomic absorption method. For determining metals in food products, Jackson et al. [528] have developed an automatic system involving wet mineralization of samples in a modified "Technicon" autoclave at 400°C using sulphuric and nitric acids and hydrogen peroxide. The solution obtained was neutralized and the diethyldithiocarbamates of Cd, Co, Fe, Ni, Pb and Zn were extracted with methyl isobutyl ketone, and then the extract was analysed by the AES–ICP method. The process of determination, which

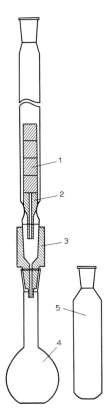

Fig. 2.24. Apparatus for wet-ashing. 1 = Teflon cylinder plugs; 2 = Teflon diffusion funnel; 3 = Teflon stopper with a capillary; 4 = 250 ml Duran volumetric flask; 5 = 70-ml quartz vessel.

is fully automatic, lasts 5–8 h; the output of the system is three samples per hour.

The Beckman sample preparation system [529] for analysis of organic and inorganic substances deserves attention. The system makes it possible to use solid (fusion) and liquid reagents (acidic and basic decomposition, oxidation by concentrated nitric and sulphuric acids, reduction by hydroiodic acid, etc.) and gases (oxidation by oxygen, chlorine, etc. and reduction by hydrogen). Samples are decomposed both under pressure and in ordinary conditions. The device has been utilized in decomposing various minerals, alloys and ores, and enables highly volatile components to be separated at the same time, *e.g.* silicon as tetrafluoride.

Several other devices are available for wet mineralization. Among them are the autoclaves manufactured by Perkin-Elmer and others. Similar autoclaves have been developed at the Giredmet [530] and in the VNII IREA [531] of the USSR.

2.6.3.2 Volatilization of inorganic compounds with preliminary chemical transformation

This is also a widely used method of preconcentration of trace components. With its application, both the matrix and the determinable elements can be volatilized. Simplicity and wide scope are the advantages of this method. The obtained concentrate of trace components can be analysed by different methods, but preference is given to multielemental techniques.

To this method of concentration is dedicated the detailed review by Bächmann [491]; the author has considered concentration of traces present in solid samples. Micro- and macrocomponents are removed by forming inorganic compounds whose boiling points are below 1000°C. The volatile compounds suitable for concentration of trace elements are listed elsewhere. Chlorides: Al, As, Au, Bi, Cd, Ce, Cr, Fe, Ga, Ge, Hf, Hg, In, Mn, Mo, Nb, Os, P, Pb, Po, Re, Ru, S, Sb, Se, Si, Sn, Ta, Tc, Te, Ti, V, W, Zn and Zr; fluorides: As, Bi, Ge, Hf, Hg, Ir, Mo, Nb, Os, P, Re, Rh, Ru, S, Sb, Se, Si, Sn, Ta, Tc, Te, Ti, V, W and Zr; oxides: As, Cd, Hg, Ir, Mo, Os, Po, Re, Ru, S, Se, Tc, Te, W and Zn; elements: noble gases, halogens, As, Bi, Cd, Cs, Fr, Hg, K, Li, N, Na, O, P, Pb, Po, Rb, S, Sb, Se, Sn, Te, Tl and Zn; and hydrides: As, Bi, P, Pb, Sb, Se, Sn and Te.

Besides, sulphur can be removed as H_2S, and tin as SnI_4. As reagents for chemical transformation, use is made of many gaseous, liquid and solid substances, for example, CCl_4, Cl_2, HCl, Br_2, BBr_3, F_2, HF and $AlCl_3$.

Here are a few examples of removing the matrix. Chlorination of metals and volatilization of the macrocomponent chloride so obtained is a technique effectively used in analysis of tin [532], titanium [533], zirconium [534] and vanadium [535]. Yudelevich and co-workers have developed a method for atomic emission analysis of tin [536], antimony [537] and gallium arsenide [538] for up to 21 impurities. The method involves volatilization of the matrix in a special reactor as bromides. A dry inert gas or dried air serves as gas carrier. Bromination in an

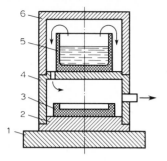

Fig. 2.25. Apparatus for evaporation of matrix in acid vapours. 1 = Electric heater; 2 = stand; 3 = cup with the sample; 4 = reaction chamber; 5 = cup containing acid; 6 = cover of the chamber.

anhydrous medium (bromine solution in carbon tetrachloride) prevented hydrolysis of bromides of the matrix elements and contributed to their more complete removal. Trace impurities were concentrated on a graphite collector and analysed by the atomic emission method. Concentration coefficients of 250–1250 were reached. Dissolution of the sample and volatilization of bromides did not exceed 90 min.

Krasil'shchik [539] has described a device for volatilizing the matrix in mineral acid vapours, in particular as halides in the analysis of boron, antimony and silicon compounds (Fig. 2.25). Cups with samples are placed on a support heated with a closed spiral heater. A cup containing 20–30 ml acid is positioned on the bottom of the "upset" beaker having holes to allow acid vapours and the reaction products to escape into the atmosphere. The temperature in the lower part of the chamber is 180–200°C. Under such conditions the decomposition time for 1 g sample (H_3BO_4, Sb_2O_5) did not exceed 1 h. Other examples are listed in Table 2.11.

Volatilization of the determinable impurities, particularly in determining mercury and the elements that form volatile halides, has still more importance. For determining silicon and arsenic in steels, niobium, tantalum, molybdenum and slags, Kuznetsov et al. [550] volatilized the determinable elements as chlorides. It is convenient to use CCl_4 as chlorinating agent because it ensures complete and rapid transfer of the determinable elements into the gaseous phase. Also, it can be readily purified from detrimental impurities (oxygen, water and compounds of the determinable elements). Chlorides were synthesized at high temperatures (up to 800°C) in a sealed quartz ampoule which contained the material to be chlorinated and carbon tetrachloride.

TABLE 2.11

Concentrating trace elements by forming volatile inorganic compounds of element matrix. Determination by atomic emission method

Matrix	Compound to be removed	Chemical reagent	Determinable trace elements	Ref.
Boron and its compounds	Borate ester	Methyl alcohol	Al, Be, Ca, Cd and 15 other elements	540
Titanium dioxide	TiF_4	HF (gas) at 400°C	Al, Ca, Cr, Cu, Fe, Mg, Mn, Ni, Pb, Si, Sn, V	541
High-purity silicon and its dioxide	SiF_4	HF	Ag, Al, Be, Bi, Ca, Cu, Fe, Ga, In, Mg, Mn, Ni, Pb, Ti, Tl, Zn	542
Germanium and its dioxide	$GeCl_4$	HCl	Ag, Al, Au, Bi, Ca and 15 other elements	543
High-purity tin	$SnBr_4$	Br_2 and HCl (or Br_2 in autoclave)	19 elements	544
Arsenic	$AsBr_3$, $AsCl_3$	Br_2 and HCl at 150°C	Al, Fe, Mg, Mn, Sn	545
High-purity arsenic	As_2O_3	O_2	Cr, Ni, Pb	546
Antimony	$SbCl_3$	HCl (gas) at 300°C	Ag, Al, Bi and 10 other elements	547
Selenium	SeO_2	HNO_3	Ag, Al, Bi and 10 other elements	548
Nickel	$Ni(CO)_4$	CO at 150°C	Bi, Cd, Co and 6 other elements	549

Then the gas phase was directly introduced into the mass spectrometer, and this made it possible to avoid the loss of volatile compounds through condensation and evaporation. Attention may be drawn to the fact that 10^{-4}–$10^{-5}\%$ silicon is removed from various substances by chlorination in quartz glass ampoules. This is because amorphous, crystalline, and vitreous forms of silicon dioxide have very different reactivities towards chlorine.

For determining arsenic and silicon which form volatile compounds, Kuznetsov et al. [551] sealed the samples of copper and steel together with carbon tetrachloride in quartz ampoules and chlorinated them at 650°C. The volatile chlorides formed were introduced into an inductively coupled plasma. The detection limits of silicon and arsenic were found to be respectively $(4–6) \cdot 10^{-5}$ and $(1–2) \cdot 10^{-5}\%$ for a weighed quantity of 200 mg.

In atomic absorption determination of antimony in soils, rocks and ores, the crushed sample (0.25–1 g) was mixed with 0.5 g ammonium iodide and placed in a test tube, the latter was introduced into a gas burner flame [552]. In so doing the sublimate of antimony iodide separating out of the test tube deposited on the cooler part of the test tube. The sublimate was then dissolved in $2\,M$ HCl and the obtained solution was nebulized into an air–acetylene flame. A procedure has been worked out for determining tin in bottom sediments [553]. It involves preliminary separation of tin by volatilizing it as tetraiodide (after adding ammonium iodide and mixing), dissolution of the condensate in a mixture of ascorbic and nitric acids, and determination of tin in the obtained solution by the atomic absorption method in an acetylene–air flame. When determining boron in silicon [554], the sample was dissolved in sodium hydroxide, boron was volatilized as boronmethyl ether, and determined by the photometric method. In a similar manner boron was determined in germanium.

Extensive use is made of the method of evaporating mercury after it is reduced to elementary mercury. Mercury vapours can be collected as amalgams of gold or silver and stored as such; they are analysed (usually by the AAS method) after heating the amalgam. Direct analysis of vapours is also possible, provided the device in which mercury is reduced and volatilized is directly connected to the atomic absorption instrument. Olafsson [555] has used tin dioxide as reducer in determining nanogram amounts of mercury in sea water by the ETAAS method. With a stream of argon, the mercury vapour formed was transferred onto a gold foil (formation of an amalgam); the foil was then heated, and mercury was transferred with argon into the atomization source of the atomic absorption spectrophotometer. Nishimura *et al.* [556] proceeded in a like manner, reducing mercury to metal in analysis of natural waters; thereafter, the mercury vapours were absorbed with porous powder-like silver placed in a glass tube of small dimensions. The tube was then heated in a furnace up to 500°C and mercury was transferred with nitrogen into the atomizer of the device used for the atomic absorption spectrometer.

Evaporation of volatile hydrides is an important method of concentrating arsenic, bismuth, selenium and several other elements, particularly prior to atomic absorption determination. The method of generating hydrides makes it possible to attain high concentration coefficients; it assures "purity" of separation of macro- and microelements and ensures detection limits of $\leqslant 1 \cdot 10^{-7}\%$ for hydride-forming elements. Of no less importance are the possibilities of introducing the gas phase concentrate directly into the ionization or atomization source and simple automatic control. Sometimes it is necessary to resort to preliminary condensation or absorption of the hydrides formed, and only then the sample is introduced into the source. The following volatile hydrides (boiling points in °C are shown in parentheses) have analytical significance: AsH_3 (-62.5), BiH_3 (-22), GeH_4 (-88.5), PbH_4 (-13), SbH_3 (-18.4), H_2Se (-41.3), SnH_4 (-51.8) and H_2Te (-2.3). There is wide use of the hydride-generation method [492].

With hydride generation, concentration and subsequent determination are carried out as follows: the sample is dissolved, a reducing agent is added, the formed hydride is collected and introduced into the excitation or atomization source (often it is transferred by a current of inert gas), and then the determination proper is made. If the analysis ends with atomic emission determination, then an inductively coupled plasma or microwave excited plasma is used as the source of excitation.

For analytical purposes, hydrides are generated mainly by two reactions. The much earlier method – the Gutzeit reaction – involves reduction as per the reaction zinc + HCl (sometimes H_2SO_4); it is suitable mainly for obtaining AsH_3, SbH_3 and H_2Se, and sometimes for BiH_3 and H_2Te. Thus, for determining arsenic in silicon [557] the sample was placed in the test tube of the device shown in Fig. 2.26, dissolved in 20% potassium hydroxide solution, and hydrochloric acid was added to obtain a final concentration of 3 M. Arsenic was then reduced to arsine with granulated zinc and arsine was absorbed with 5% alcohol solution of mercury bromide coated on a disc made of filter paper; the disc was

132

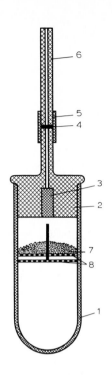

Fig. 2.26. Apparatus for determining arsenic. 1 = Polyethylene test tube; 2 = stopper; 3 = cotton pad; 4 = reaction paper; 5 = connecting tube; 6 = glass tube; 7 = polyethylene chips; 8 = net.

placed in the path of the forming gas. Then the following reaction occurred

$$AsH_3 + 3HgBr_2 \rightleftharpoons As(HgBr)_3 + 3HBr \qquad (2.51)$$

The colouring intensity of the yellow paper depends on the concentration of arsenic. In order to eliminate the interfering effect of hydrogen sulphide, a cotton pad moistened with 1% lead acetate solution was placed in the lower part of the test tube. After the reaction had completed, the paper was taken out and the colour intensity was compared with that of the discs prepared beforehand with arsine under similar conditions from reference samples containing a known content of arsenic. The detection limit of arsenic was $2 \cdot 10^{-6}\%$ for a weighed quantity of 0.5 g.

For determining arsenic in drinking water, use has also been made of the method of reducing it to arsine with metallic zinc in a special flask reactor [558]. The arsine formed was transferred by a helium flow into a trap-freezer immersed in liquid nitrogen. After the concentration was over, the trap was connected to the burner of an atomic absorption spectrometer and heated to room temperature; arsine was transferred by argon into a flame atomizer.

However, as noted by Nakahara [492], the method involving reduction with metallic zinc is not very convenient; the reaction lasts 10 min, and therefore several difficulties are faced in automating the process. The method of obtaining hydrides with the use of sodium tetrahydroborate (boron hydride) is more effective. It is a quick method and enables subsequent multi elemental analysis to be performed. It is very convenient to carry out reduction in a 0.5–10% solution of $NaBH_4$, stabilized by adding sodium (or potassium) hydroxide. Optimal ranges of acidity which are created by adding HCl (or, rarely, sulphuric or nitric acid) depend on the determinable element and are equal to 1–9 M for As, Bi and Sb; 1–3 M for Ge; 0.1–0.2 M for Pb and Sn; 2.5–5 M for Se; and 2.5–3.6 M for Te. Rigin and Verkhoturov [559, 560] have suggested a convenient method of generating hydrides of arsenic and tin, where electrolytic reduction on platinum electrodes is carried out in a 5% solution of potassium hydroxide.

The present-day trend is to use the generated hydrides directly in the excitation or atomization source. The review by Godden and Thomerson [561] considers in detail the ways of obtaining volatile hydrides for subsequent atomic absorption analysis, as well as a method of molecular emission in cavity and an atomic fluorescence method. Therein are discussed different reduction techniques, designs of reaction vessels, and other questions. Information on making partly or fully automated systems, including the flow injection analysis systems, as applied to AES–ICP, AAS and AFS, can be found in the review by Nakahara [492].

Automatic methods for group determination of hydride-forming elements by the AES–ICP technique are also known. Usually, such a system comprises a multichannel scanning spectrometer and a reactor into which the analysed solutions and necessary reagents are introduced by a peristaltic pump [562]. Hydrides and elemental mercury which separate out after the addition of H_2O_2 and $NaBH_4$ are carried over into the plasma by an argon current. The method enables trace elements to be determined at a level of $n \cdot 10^{-7}\%$. A system of obtaining hydrides in a continuous flow is combined with the AES–ICP method for determining arsenic and selenium in tap water [563]. The samples

134

and acid solutions are admitted into the mixing chamber; with the use of a peristaltic pump they are transferred into a reactor where they are mixed with an oxidizing agent (1% solution of $NaBH_4$ in 0.1 M NaOH). The obtained hydrides are transferred by a carrier gas (argon) into a plasma source. Liversage and Van Loon [564] suggested the combination of a flow injection system and the generation of hydrides for semi-automatic determination of arsenic by the AES–ICP method. By this technique, geological samples, ashes and river sediments have been analysed.

Information on a self-contained cell for fully automated determination of mercury and hydride-forming elements (As, Sb, Sn, Te and others) is available in the work of Welz and Wiedeking [565]. Absorption of the formed vapours of mercury and/or hydrides was measured after transferring them into a heated quartz cell. The method allows determination of toxic elements at a nanogram level in water and industrial substances; use of a closed cell gives a high precision of results. Whiteside et al. [566] have worked out an automatic device for atomic absorption determination of toxic trace elements in a flow: mercury (in the elemental state), As, Bi, Sb, Se and Te (as hydrides). Detection limits of 0.5–1.3 ng/ml were attained. For atomic absorption determination of selenium and arsenic [567], use has been made of an MHS-10 automatic generator of hydrides (Perkin-Elmer). A schematic diagram of a fully automated device used for series determination of hydride-forming elements is also given therein. Table 2.12 contains examples of determination of hydride-forming elements in various substances.

2.7 Controlled crystallization

Crystallization is a process of formation and growth of crystals from a solution or melt, sometimes from a gas phase. It is one of the oldest analytical, preparative and technological methods of separation and concentration of substances. Thus, simple crystallization is a technique widely used in analytical laboratories and workshops of various industries. Radium salts have been separated from barium salts and pure preparations of rare earth elements etc. have been achieved by fractional recrystallization. Depending on the volume of work to be done, the desired degree of purity of the compounds to be separated and their properties, crystallization can be carried out in periodic and continuous action apparatuses, in counter-current phase, at different

TABLE 2.12

Volatilization concentration of trace elements as hydrides

Determinable trace element	Substance analysed	Method of determination	Remarks	Ref.
As	Atmospheric particles	AAS	Bi, Sb and Se can also be determined	568
	Seaweeds	AAS	—	569
	Waters	AAS	All soluble forms of arsenic are determined	570
	—	AES with microwave discharge	—	571
	Natural waters	AAS	Arsenite, arsenate, mono- and di-methylarsonic acids can be determined separately	572
As and Sb	Geological material	AAS	—	573, 574
As and Se	Soils and precipitations	AAS	—	575
	Geological materials	AAS	—	576
	River water and biological materials	AAS	—	577
Se	Rocks	AAS	Effect of copper and nickel is eliminated by adding a 1% solution of 1,10-phenanthroline in 0.1 M HCl into the hydride generator	578
Bi and Sb	Natural waters and vegetable materials	AAS	Upon heating, hydrides decompose at the inner surface of hollow graphite rod acting also as atomizer	579
Sn	Steels, alloys and biological materials	AES-ICP	As, Co, Cu, Ni and Sb are a hindrance to determination	580
Bi	Urine	AAS	—	581
	Aqueous solutions	AAS	Output 180 samples per hour	582 583
In and Tl	—	AAS	—	584
Ge	Natural waters	AAS	Methylgermanium compounds can be determined	585
As, Ge, Pb, Sb, Se, Sn	Biological samples	AES with glow discharge	—	586
As, Bi, Sb, Te	High purity phosphorus	AAS		587

temperatures, at normal and increased pressures, or in vacuum. Crystallization methods, particularly zone melting, are also applied for obtaining high-purity substances and growing perfect crystals for various purposes. Of the crystallization methods employed for preconcentration of trace elements, wide use is made of controlled crystallization [4, 21, 588–593].

On cooling a liquid, crystallization starts in it at a definite temperature, and nucleation centres appear throughout the liquid phase when proper measures are not taken. Such a spontaneous process is known as volume (uncontrolled) crystallization. Unlike volume crystallization, controlled crystallization is a solidification process when the rate of crystallization is determined by external conditions and may be controlled. Along the material being crystallized, a temperature gradient is set up which ensures directional motion of the crystallization front, *i.e.* of the interface between the solid and liquid phases. By varying temperature, this front can be moved in the direction either of solidification or melting.

Controlled crystallization can be automated. The process does not demand reagents and can be carried out in a controlled atmosphere; this has a positive effect on the metrological characteristics of analytical procedures including concentration by this method.

According to Kirgintsev *et al.* [592], controlled crystallization can be of two types (Fig. 2.27): directed crystallization (liquid crystallizes in the given direction) and zone melting (the liquid zone moves in a definite direction in the solid body). If there is one interface between the solid body and the liquid (crystallization front) when directed crystallization is carried out, then two such boundaries (crystallization and melting fronts) appear in the case of zone melting.

Blank [593, 594] has reviewed the efficiency of crystallization concentration from the viewpoint of bringing down the lower limit of determinable constituents, decreasing random errors of analysis, or saving the expensive analysed material.

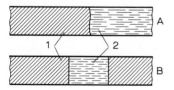

Fig. 2.27. Directed crystallization (A) and zone melting (B). 1 = Solid; 2 = melt (liquid).

Depending on the direction of motion of the crystallization front, directed crystallization can proceed from top to bottom (as per the methods of Chokhral'sky, Stepanov, Tovadze and Kilitaur), from bottom to top (Bridgemen method), and in the horizontal direction (Chalmers or Kapitsa method) [590]. This classification loses its meaning if crystals grow in immediate contact with the walls of a tube-like container, for example, of an ampoule: in this case, directed crystallization can be carried out at different angles of inclination of longitudinal axis of the container. The rate of crystal formation, temperature of the heater and the cooler, and the angle between the container longitudinal axis and the horizontal axis are important factors. Of no less significance are the initial conditions: container's diameter and length and the degree of filling with melt.

During crystallization the components are distributed between the liquid and solid phases. The separation efficiency of components is characterized by the distribution coefficient (D) which is the ratio of equilibrium concentration of a trace component in the solid (C_s) and liquid (C_l) phases in contact with each other.

$$D = C_s/C_l \tag{2.52}$$

It is recommended to use the following relation for calculating the distribution coefficient of a trace component [588].

$$C = C_0 D(1 - g)^{D-1} \tag{2.53}$$

where C_0 is the initial concentration of a uniformly distributed trace component in the sample; C is the concentration of a trace component in the solid phase at the crystallization front; and g is the mass fraction of the crystallized part of the ingot.

Eqn. (2.53) has been derived on the assumption that the distribution coefficient is constant, no diffusion takes place in the solid phase, and the liquid is homogeneous at all times. As a rule, these conditions (Pfann approximations) are fulfilled only at equilibrium and at small rates of directed crystallization.

In a general case, the distribution coefficient can vary depending on the fraction of the crystallized part of the ingot, initial concentration of the trace component and the rate of crystallization. In practice, use is therefore made of an effective distribution coefficient which can be calculated only on the basis of eqn. 2.52 and the condition of mass balance of impurity [595]:

$$D_{\text{eff}} = \frac{\dfrac{C}{C_0}(1 - g)}{1 - \displaystyle\int_0^g \dfrac{C}{C_0}\,dg}$$ (2.54)

However, it is difficult to establish a relationship between the average concentration of a trace impurity in the concentrate (C_c) and in the starting sample (C_0) with the use of D_{eff}. This is the reason why Blank [596] introduced the concept of characteristic distribution coefficient (D_L) which conforms to the following equations irrespective of whether or not the Pfann approximations are fulfilled.

$$C_c = C_0(1 - g_k)^{D_L - 1}, \qquad D_L < 1$$ (2.55)

$$C_c = \frac{C_0(1 - g_k)^{D_L}}{g_k}, \qquad D_L > 1$$ (2.56)

Here, g_k is the mass fraction of the melt crystallized as concentrate. When $D_L > 1$, the initial portion of the ingot acts as concentrate, and when $D_L < 1$, the end portion serves as concentrate. All further discussions will pertain to the usual case, i.e. when $D_L < 1$.

D_L can be determined by using the analytically determined average concentrations $(\bar{C}_i)$ corresponding to the end portions of the ingot [597–599]. The results obtained by analysing the ingot divided into n parts make it possible to determine $(n - 1)$ values of D_L:

$$D_j = \frac{\log\left(1 - \displaystyle\sum_{i=1}^{j} \dfrac{\bar{C}_i}{C_0}\,\Delta g_i\right)}{\log(1 - g)}$$ (2.57)

In this equation, i stands for the number of the portion of the ingot counted from the initial position of the crystallization front; $j < n$ is the total number of analysed portions of the ingot; Δg_i is the relative mass fraction of the ith portion; $g = \Sigma_{i=1}^{j} \Delta g_i$.

The separation of a trace component is quantitative $(R > 0.9)$ only at comparatively small values of distribution coefficients (usually $D_L < 1$). Therefore, use is made of partial concentration of a trace element in a small portion of the ingot. The concentration of a trace component in the analysed sample is calculated by the following equation using the predetermined value of the distribution coefficient [598]:

$$C_0 = \frac{C_c}{(Q_s/Q_c)^{1-D_L}} \qquad (2.58)$$

where Q_s and Q_c are the masses of the analysed sample and the concentrate, respectively.

This method is widely used in analytical practice; it does not lead to significant random errors as it appears at first sight. This is due to the fact that the recovery of a trace component at D = constant and at a constant mass of the ingot depends only on the mass of the concentrate [600–602]. Authentic results are obtained by carrying out concentration under reproducible conditions.

To concentrate impurities in metallic cadmium, Kirgintsev and Gryaznova [603] used the method of partial separation of impurities. The last portion of the ingot – the concentrate of trace elements – was separated mechanically and analysed by the atomic emission method. On analysing the concentrate obtained from a 250 g ingot, the detection limits of Bi, Cu, In, Ni, Pb and Sn were found to be $(1-6) \cdot 10^{-7} \%$. Obviously, the closer the impurity distribution is to the equilibrium, the more effective is the concentration including the partial concentration. This has been confirmed experimentally by Kirgintsev [600, 603]. He and his co-workers [604] have developed a number of techniques for conducting directed crystallization in a horizontal rotating container under equilibrium conditions. These techniques have made it possible to significantly improve preconcentration conditions of Bi, Cu, In, Ni, Pb and Sn in analysis of metallic cadmium.

The form of the obtained concentrate is important. For analysing metallic tin, the impurities were determined by the atomic emission method at the end of the ingot grown in the vertical direction and used as electrode [605]. Concentrating the impurities in salts (from top to bottom), Blank et al. [602] used the narrow portion of an ingot of variable cross section as concentrate to increase the concentration coefficient, the ingot being obtained in a special quartz container.

The condition for effective concentration is that the thickness of the diffusion layer at the liquid–solid body interface should be the minimum possible. This thickness depends on the growth rate and the intensity of natural or forced convection. When the melt is not actively mixed, it is advisable to carry out directed crystallization from top to bottom using natural convection flows caused by a sufficiently large vertical gradient of temperature. Among the methods of active mixing of melt are directed crystallization when the container is in the near vertical position and is rotated along the longitudinal axis, centrifugal crystallization, and others.

A series of studies on preconcentration of trace impurities by directed crystallization has been made by Blank et al. [602]. Use of this method is made to analyse alkali and alkaline earth metal halides which melt at temperatures below 900–1000°C and readily dissolve in water, excepting fluorides of alkaline earth and some alkali metals. We shall discuss some of the results obtained by them.

They recommend the following equation for calculating the decrease in the detection limit of trace components (Π) achieved by crystallization concentration [601]:

$$\Pi = \frac{C_0}{C_0'} = \frac{S_Y}{S_Y'} \left(\frac{q'}{q}\right)^{D_L} \left(\frac{v_0}{v}\right)^{1-D_L} \left(\frac{Q_s}{m_0'}\right)^{1-D_L} \tag{2.59}$$

Here, C_0 is the detection limit of the direct method of analysis; S_Y is the standard deviation of one measurement of a minimum signal for direct analysis (with degree of freedom f); C_0' and S_Y' are the same values for the method with preliminary crystallization concentration; v_0 and v are respectively the volume of the solution from which a useful signal is obtained (a constant value for the given method of analysis) and the total volume of the involved solution; q' and q are the maximum (after preconcentration) and actual concentrations of the preparation in the mixture; Q_s is the weighed amount of the preparation; $m_0' = v_0 q'$ is the optimal mass of the concentrate.

Using this equation, Blank and Afanasiadi [601] computed the necessary parameters of crystallization concentration of Li, K, Rb and Cs for their flame atomic emission determination in sodium iodide. A direct method ensured the determination of trace elements with contents no less than $2 \cdot 10^{-4}$, $5 \cdot 10^{-5}$, $1 \cdot 10^{-4}$, and $2 \cdot 10^{-4}\%$, respectively. After preconcentration under optimal conditions ($S_Y = S_Y'$, $v = v_0$, $q = q'$), a twenty-fold increase in the detection limit of potassium or lithium can be achieved provided $Q_s/Q_c \simeq 100$. The detection limit of rubidium and caesium can be lowered by 55 and 87 times, respectively. With allowance made for flame atomic emission analysis of the concentrate ($v = 10$ ml, $q' = 0.05$ g/ml), the optimum mass of the concentrate ($Q_c = m_0'$) amounts to 0.5 g and of the weighed amount of the sample, Q_s, to 50 g.

Blank et al. [606, 607] chose vertical directed crystallization (from top to bottom) for which the characteristic distribution coefficients of trace components are close to the equilibrium one due to natural convection mixing of melt. The temperature gradient at the crystallization front equalled 80°/cm. They also computed theoretically the shape of an ampoule of variable cross-section, conforming to optimal concentration parameters.

A weighed quantity of sodium iodide (30–50 g) was taken in a weighed quartz ampoule which was placed in a tube furnace. The ampoule was evacuated and the temperature was gradually raised up to 700°C. After the salt had melted, the ampoule was filled with an inert gas and directed crystallization of the substance was started, withdrawing the ampoule from the furnace (in the upward direction) at a rate of 16 mm/h. When the wide part of the ingot had crystallized, the process was stopped, the samples were cooled down to room temperature, and weighed. The narrow portion of the ampoule containing the concentrate was chipped off, sodium iodide was dissolved in water, and, having determined the mass of the narrow portion of the ampoule, the mass of the obtained concentrate was found. The obtained aqueous solution of the concentrate (*ca.* 5% in NaI) was analysed by flame AES. The analysis continued for 6–7 h. The following detection limits were attained: Li, $1 \cdot 10^{-5}\%$; K, $3 \cdot 10^{-6}\%$; Rb, $2 \cdot 10^{-6}\%$; Cs, $2 \cdot 10^{-6}\%$.

Vertical directed crystallization has also been used for concentrating Ag, Al, Bi, Cd, Co, Cr, Cu, Fe, Mn, Mo, Ni, Pb, Sn and Ti in analysis of thallium chloride [606], the melt being mixed by natural convection. The concentrate was analysed by the atomic emission method.

Anions can also be concentrated. For instance, use was made of directed crystallization (from top to bottom in the vertical direction) to concentrate chlorides in analysis of sodium iodide [607]. Along with the quartz ampoules of variable cross-section, knock-down platinum ampoules for multiple use were also used. The chlorides in the concentrate were determined by the phototurbidimetric method involving oxidation of the matrix with sodium iodate and subsequent formation of a suspension of silver chloride in a water-organic solution.

Negative phenomena, like evaporation of volatile trace components and partially of the matrix, their thermal dissociation, reaction of the melt with the container material accompanied by contamination of the sample and losses of impurities due to sorption, can appear if crystallization concentration is carried out in the melt at comparatively high temperatures. They can be eliminated to a large extent with low-temperature crystallization of crystallohydrates and also of water–salt solutions of eutectic composition [608]. Due to small values of the

142

distribution coefficients, it is possible to concentrate impurities, even the isomorphous ones, with the matrix component, *e.g.* rubidium in caesium iodide. Making use of the possibilities of low-temperature crystallization, Blank *et al.* [608] lowered the detection limits of the impurities by 1.5–2 orders in analysis of caesium iodide, potassium chloride, iodate and dehydrate of sodium iodide and tetrahydrate of cadmium nitrate.

This technique is also used for concentrating anions, for example chlorate [609] and nitrate [610] in potassium chloride, and chloride [609], chromate [611] and sulphate [612] in caesium iodide (with subsequent photometric determination). For concentration of sulphates in the analysis of caesium iodide solutions [612] use was made of directed crystallization of the salt solution of cryotectic composition (37.9 g salt in 100 g water; phase transition temperature equalled $-4°C$). The salt solutions were crystallized in glass ampoules placed in a thermostat. Caesium iodide (6.9 g) and water (18.1 ml) were taken in the ampoule. After the sample had dissolved, the test tubes were placed in cassette holes, glass stirrers were introduced into the solution thermostatted at $-13°C$, and the stirrers were turned on. This made it possible to control the temperature gradient at the crystallization front, to create a symmetrical temperature field, and to attack the concentration compaction of impurities in the melt at the crystallization front. Directed crystallization of solutions was then initiated by lowering the cassette with ampoules into the thermostat. After the crystallization was over, boiling water (0.5 ml) was poured into every ampoule in the vertical position. After the dissolution front had moved by 2–2.5 mm, which corresponded to 0.6–0.7 g concentrate, the obtained solutions were prepared for analysis and the sulphates were determined by the phototurbidimetric method. The detection limit was $1 \cdot 10^{-4}\%$.

Blank and Esperiandova [611] have succeeded in bringing down by two orders of magnitude, to $2 \cdot 10^{-6}\%$, the lower limit of determinable contents of chromium in caesium iodide by concentration.

Similarly, lithium, rubidium, copper and iron were concentrated in the analysis of caesium iodide [613, 614].

The disadvantage of vertical directed crystallization is the duration of the process (3–6 h). This period can be reduced to 1 h by carrying out crystallization in a field of centrifugal forces [615]. Blank *et al.* [615] associate the increase in concentration efficiency with the removal of dendrites, containing melt inclusions, from the surface of the growing ingot, this being achieved by forced convection mixing of the liquid phase and the elimination of solid particles present in the melt but differing from it in density. The method was employed to concentrate Co, Cu, Mg, Mn, Ni and Pb out of sodium nitrate melt. Use was made of a specially designed device which made it possible to centrifuge the samples sealed off in glass ampoules and to move them at the same time relative to the heaters towards the axis of rotation. Sharp protrustions were made in the ampoules to ensure the formation of the crystallization front when the ingot started growing. The speed of the centrifugal rotor was 1300 rpm and the centrifugal acceleration at the crystallization front was 340 g. Determination was completed with the atomic emission method.

Considering concentration of chlorides in the analysis of alums, Blank *et al.* [616] revealed that centrifugal directed crystallization was preferred to vertical crystallization: the distribution coefficients were significantly less and the increase in effective distribution coefficients with the ingot growth rate was less abrupt. The centrifugal field in the direction of motion of the crystallization front produces a positive effect on the equilibrium conditions in segregating trace impurities, on the kinetics of the trace component entering the solid phase, the profile of the diffusion layer and the quality of the structure of the interface [617]. Centrifugal directed crystallization was employed in this work to concentrate the trace impurities in sodium nitrate and caesium iodide. Depending on the substance and the amount of the sample, concentration coefficients of 30–300 can be attained (two to five times more compared to normal directed crystallization).

2.7.2 ZONE MELTING

Zone melting has come into the analytical laboratory from engineering. Its use for concentration of impurities is considered in refs. 588–592 and 618–620.

In zone melting, the ingot in the container is gradually moved along one or several heaters. In the heating zone the ingot melts and, on cooling, crystallizes. Besides the existence of two boundaries between the solid and the liquid phases, zone crystallization differs from directed crystallization in that not all the material, but part (zone) of it melts at a time. The first portions of the formed solid phases are enriched with the components having distribution coefficients greater than unity; the higher the distribution coefficient, the faster this substance crystallizes. The end portions of the ingot are enriched with components having distribution coefficients less than unity.

144

The distribution of a component along the ingot length depends on the number of heating zones (n); limiting concentration is attained when $n \to \infty$. As shown by calculations [592], the time spent for quantitative concentration increases abruptly as D and C_c/C_0 increase. Zone concentration has certain limitations: it is difficult to attain concentration coefficients of the order of 100 or more; the efficiency is insufficient with regard to isomorphous impurities; the concentration efficiency decreases with low concentrations of minor component; the technique is unsuitable for analysis of complex-composition samples and substances less stable at melting temperatures; sometimes, evaporation of impurities and a change in their chemical form and composition takes place and the analysed substance also reacts with the container's material. The behaviour of impurities when concentrated by zone melting can be seen from the matrix impurity constitution diagrams which make allowance for the interaction of macro- and trace components in the liquid and solid phases.

Downarowicz [621] employed zone melting for concentration of Ag, Cd, Cu and Ni preliminary to atomic emission or voltammetric analysis of high-purity lead. Konovalov and co-workers [622, 623] have worked out procedures for determining Ag, Al, B, Ca and thirteen other impurities in metallic bismuth (detection limits of 10^{-4}–$10^{-7}\%$) and for silver and copper in metallic lead (10^{-4}–$10^{-5}\%$) [624]. The concentration coefficients of trace impurities can be 20–30. The obtained concentrates are analysed by the atomic emission method. Zone melting in combination with spark-source mass spectrometry and the X-ray fluorescence method was used by Horn [625] for analysing semiconductor silicon. In the analysis of cadmium telluride, Ag, As, Bi, Cr, Cu, Fe, Ga, In, Pb, Sb and Sn were concentrated by vertical zone melting with stirring [626]. With a 90–98% degree of separation, the impurities concentrated in the "dirty" end of the ingot, the length of this end being equal to 0.1 part of the initial length, after one traverse of the zone at the rate of 16 mm/h.

2.8. Pyrometallurgical methods

2.8.1 FIRE ASSAY

This is the main method of preconcentrating noble metals present in ores and rocks and their treatment products. As a part of fire assay, fire melting is used to ascertain the concentration of precious metals in alloys and articles. It is often applied in combination with gravimetric, atomic emission, atomic absorption, or neutron activation analysis. Fire melting combines decomposition of the sample and concentration of trace elements. For preconcentration of noble metals, use is made of the property of molten lead or other collector to readily dissolve these metals with the formation of low-melting-point alloys and to quickly oxidize with the oxygen of the air.

The classic fire assay, as a method of analysis of materials for gold and silver, involves blending (mixing of a ground and weighed amount of the sample with the required quantity of an oxidizer or reducer, flux and collector), crucible or cupel melting (see elsewhere), scorification (if necessary) of the obtained alloy, cupellation (oxidation melting of the alloy), weighing of gold–silver bead, quartation, if necessary, treatment of the bead with nitric acid, washing, drying, calcination of the dry gold residue and its weighing.

Depending on the composition of the material being analysed and of the determinable elements, litharge, sometimes red lead, zinc acetate, plumbite and metallic lead serve as lead suppliers. Besides lead, use is made of lead–silver alloy, metallic silver, copper, nickel, tin, some sulphide alloys and other collectors that dissolve noble metals. Fluxes, which transform infusible compounds into low-melting slags, include quartz, ground glass, borax, sodium carbonate and potash, and include reducers, powder-like charcoal, wheat- or rye-flour, cream of tartar, starch, iron fillings and paper; saltpetre (potassium nitrate), litharge and red lead oxide play the role of oxidisers (besides the oxygen of the air).

Crucible melting is a process of reduction and dissolution: litharge or any other compound of lead is reduced to metal. Lead dissolves noble metals forming alloys with them; the latter settle as a separate phase at the bottom of the crucible. The melting temperature depends on the type of the concentrator used. Thus, on using lead, melting can be carried out at 900–950°C, and on using copper, at 1150–1200°C. The bead obtained is the concentrate of noble metals. In crucible melting, a slag phase is also obtained due to the interaction of rock with fluxes. Crucible melting is used for analysis of ores and their treatment products containing noble metals more than 1 g/ton of the material analysed. For rich ores (containing > 100 g of noble metals in 1 ton of ore) use is made of cupel melting.

Cupel melting is a reduction–oxidation process which proceeds in the presence of granular lead, borax and quartz. The impurities get oxidised and go into slag while the noble metals are collected by metallic lead. An oxidizing atmosphere is created by the air that enters into the furnace and by the litharge obtained from lead. In this case, the charge is melted at 900–950°C in a muffle (with closed doors) and the molten lead is oxidized by allowing the air to enter; lead partially evaporates as oxide, but a large part of it goes into slag. The alloy is poured into a casting mold and is separated from the slag after it has cooled.

In fire assay, the obtained alloy is cupelled, *i.e.* burnt off in an oxidizing atmosphere on a small porous cup – cupel made from bone meal, magnesite or other suitable refractory materials. Cupellation gives rise to lead oxides and other non-noble metal oxides which are absorbed by the cupel. The bead of gold–silver alloy remains on its surface. After weighing, the bead is treated with dilute nitric acid. Silver goes into the solution and the obtained bead of gold is weighed after washing, drying and calcination.

Information on special features of fire melting of various ores and minerals, their treatment products, wastes of non-ferrous metallurgical plants, native metals and alloys can be found in the manuals [627, 628].

Preconcentration by fire melting is characterized by selectivity; it ensures the representativeness of the sample and sufficiently high values of concentration coefficients. Using a microbalance, the following detection limits can be attained in fire assay: gold, 0.1–0.2 g/ton; silver, 5–10 g/ton. At these concentratiions, the separation of elements is difficult, however. This sharply impairs the precision and accuracy of the determination results. The combination of fire melting with modern highly sensitive instrumental techniques of determination has eliminated these limitations and significantly widened the range of its application. It has now become a normal practice to use fire melting with spectrophotometry, atomic emission and atomic absorption analysis. Unfortunately, such combined methods of determining noble metals have the disadvantages of classic fire assay such as difficulty of chemical treatment of the concentrate for subsequent determination of elements. This operation can be appreciably simplified with the use of instrumental neutron activation analysis, thanks to modern semiconductor detectors of high resolving power.

Now we shall illustrate the possibilities of fire melting and the basic trends of development of this special technique of preconcentration by considering examples.

Zdorova and Popova [629] have worked out a procedure for determining gold and silver in sulphide ores and their treatment products after fine grinding and thorough mixing with the charge: litharge, sodium carbonate, borax and potassium nitrate. The mixture is mixed in a paper packet, sodium carbonate and borax are added and placed on a heated fireclay crucible. The crucible process is carried out at 1100–1200°C. The obtained lead alloy (30–40 g), which is the concentrate of gold and silver, and the slag are easily separated mechanically after pouring them in cast iron moulds and cooling. The lead alloy is then melted in the presence of an oxidiser (cupellation) at 850–950°C for 30–40 min until a bead of gold–silver alloy is formed. After weighing, the bead is dissolved in HNO_3 (1:7), converting silver into nitrate. The residue, which is gold, is weighed. The content of silver equals the difference in the mass before and after dissolving in nitric acid.

Recourse has to be made to quite lengthy and sometimes to difficult separations if silver, gold and platinum-group metals are to be determined simultaneously. Some achievements have been made in this direction also, which simplify the analysis of noble metals.

An interesting procedure has been developed for isolating ruthenium and osmium after concentration on lead beads [630]. On treating the beads with a mixture of 70% HCl and ice-cold acetic acid (10:1) at 160–180°C, Pd, Pt and Rh dissolve while only part (26%) of the osmium dissolves and evaporates, and ruthenium does not dissolve at all. The vapours of osmium are absorbed in sodium hydroxide solution. After treating the bead with an acid, Pd, Pt and Rh are determined in the solution using suitable methods. The insoluble residue is fused with sodium peroxide; ruthenium and osmium are leached out with water. The solution obtained is poured into an alkali solution containing osmium, $NaBrO_3$ and H_2SO_4 are added, and RuO_4 and OsO_4 are volatilized. The latter (RuO_4 and OsO_4) are absorbed respectively in $6\,M$ HCl and thiourea solution in $6\,M$ HCl. The elements are then determined spectrophotometrically.

If lead is used as collector of noble metals, then the lead bead obtained on melting usually has a mass of 25–30 g. It is difficult to determine the noble metals in such a bead. Therefore, a large portion of lead is removed by scorification and cupellation in order to obtain a small bead. Partial cupellation was used to increase the concentration coefficient to $1 \cdot 10^3$ [631]. A weighed amount of 10 g was crucible melted without adding silver, and then the lead bead was scorified. Thereafter, cupellation was done and lead beads of mass between 8 and 0.01 g were obtained. With the decrease in the mass of lead beads, the losses of noble metals remain practically constant and do not exceed 2.7, 2.3, 9.4

and 5.6% for platinum, palladium, rhodium and gold, respectively. Thus, Au, Pd, Pt and Rh can be concentrated by incomplete cupellation in a lead bead of mass up to 0.01 g. The determination ends with the AAS method, without separating lead. Fishkova *et al.* [632] have developed a combined method for determining small amounts of gold and silver in mineral raw materials (up to 0.005 g/ton Au and 0.1–0.2 g/ton Ag). The method involves their concentration in a lead alloy which is then decreased to 0.5–0.2 g by incomplete cupellation. The metals are determined by the AAS method.

The assaying experience shows that more than 90–95% losses of noble metals upon cupellation occur at the last moment of this process. Incomplete cupellation eliminates this source of losses and helps to achieve concentration coefficients up to $1 \cdot 10^4$.

Partial cupellation can, however, produce additional errors in the analysis results. Thus, attention was drawn to the inhomogeneity of lead beads obtained by fire melting and partial cupellation of various copper–nickel ores and concentrates [633]. Therefore, to obtain quantitative results, it is recommended to either use the bead completely or to transfer it preliminarily into a solution.

Besides chemical methods of separating a concentrator and noble metals, lead is volatilized under vacuum – under conditions when the operator does not inhale lead vapours. For determining iridium in ores and their enrichment products, Grinzaid *et al.* [634] decreased the mass of lead beads (from 200 to 10 mg) by vacuum-volatilization of lead at 850°C. A lead bead was placed in a graphite crucible heated by a resistance heater; as the lead was volatilized, an iridium-rich phase was concentrated in the upper portion of the melt. Together with iridium, other noble metals, *e.g.* Au, Rh and Ru were also concentrated during volatilization. The graphite crucible with the concentrate left after volatilization was fixed in a special carbon holder and analysed by the AES method.

For analysis of low-grade ores and materials, wide use is made of preconcentration of platinum-group metals in copper–nickel–iron, copper and tin concentrators [635–637]. Such concentrators enable all the platinum-group elements and gold to be isolated quantitatively. It is advantageous to use copper or copper–nickel concentrators which make it possible to utilize copper and nickel from the sample being analysed, to lower the fire melting temperature to 1200°C, and to completely separate iron [638]. Lead–silver (ref. 636, p. 110) and tin [637] concentrators are used.

A procedure has been worked out for atomic emission or atomic absorption determination of platinum-group metals in the copper and nickel electrolysis products [639]. This procedure involves isolation of noble metals on a copper concentrator. A fast and highly sensitive method of neutron activation determination of platinum-group metals and gold in rocks and ores has also been developed [640]. It involves preconcentration (prior to irradiation) of the determinable elements by fire melting at 1000°C; high-purity metallic nickel is used as collector. The bead obtained (nickel sulphide), after grinding, is dissolved in concentrated HCl, and the residue, after washing and drying, is irradiated in a nuclear reactor channel. γ-Spectra of the sample and of the reference samples are recorded after the sample has "cooled".

As a concentrator of noble metals, Kuznetsov et al. [641] also used nickel sulphide which, in their opinion, has the following advantages: high degree of isolation of platinum-group metals ($\geqslant 98\%$); low melting point (about 800°C); the possibility of using large weighed amounts (up to 100 g), the analysed material can contain much nickel and appreciable amounts of copper; the samples containing large amounts of metal sulphides need not be preliminarily treated chemically. Other advantages include the fact that the bead can be easily separated from the slag and readily ground. This concentrator was used [641] to determine noble metals in copper–nickel sulphide ores and their treatment products (Fig. 2.28). The sample was melted with the charge (Na_2CO_3, $Na_2B_4O_7$, glass, NiS). A bead of nickel sulphide was ground and dissolved in concentrated HCl until hydrogen sulphide evolved. The undissolved residue contained Au, Ir, Os, Pd, Pt, Rh and Ru. Silver which had transferred into the solution (92–97%) was determined by the AAS method. This very method was employed for determining Au, Pd, Pt, Rh and Ru (Ir, Os and Ru were determined by the atomic emission method). Maximum value of the concentration coefficient was $1 \cdot 10^3$. The behaviour of noble metals in fire melting with the use of nickel sulphide has been studied in detail [642]. Certain amounts of gold and silver are lost with slag when the charge contains sulphur in excessive amounts as against that required for the formation of nickel sulphide.

Taking into consideration the fact that the interest in copper and copper–nickel concentrators is growing, Danilova et al. [643] have worked out a procedure for their analysis after dissolving in a mixture of HCl and H_2O_2 and sorption of Au, Ir, Pd, Pt, Rh and Ru from solutions using Milton-T fibre. The fibre is then burnt and the residue subjected to atomic emission analysis.

150

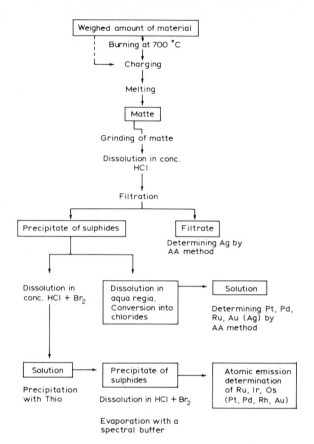

Fig. 2.28. Fire assay atomic emission (and atomic absorption) analysis of copper–nickel sulphide ores.

In the analysis of copper slime and of slime cinder, iridium was preconcentrated using a silver collector [644]. For separating Ir, Rh and Ru, a silver alloy was dissolved in concentrated nitric acid. The residue containing platinum-group metals was analysed by the X-ray fluorescence method. An atomic absorption method has been developed for determining Au, Pd, Pt and Rh in silver beads [645].

As mentioned earlier [637], a method involving preconcentration of noble metals in a tin alloy has been suggested, which, unfortunately, complicates to some extent the sample preparation and the analysis procedure and results in more losses of noble metals. Adjusting the composition of the charge, the temperature and the melting time,

TABLE 2.13

Examples of using fire assay

Elements to be concentrated and determined	Substance analysed	Method of determination	Ref.
Au	Ores	Assay	649
	Lead-production materials	Spectrophotometry	650
	Mineral raw material	NAA	651
	Geological samples	NAA	652
	Natural substances	AES	653
Os	Natural substances	Spectrophotometry and NAA	654, 655
Pt	Ores	Impulse voltammetry	656
Ir	Natural materials	NAA	657–659
	Slimes and platinic concentrates	XRFS	660
	Slags, copper ores and concentrates	AES	661
	Platinized materials	AAS and photometry	662
Ag	Products containing Al, Fe and Mn	AES and AAS	663
Pt	Ores	AAS	664
Ag and Au	Ores	AAS and gravimetry	665
Au, Ir, Os, Pd, Pt, Rh, Ru	Natural substances	AES	666
Au, Ir, Pd, Pt, Rh	Geological and other materials	AES	667
Au, Pd, Pt, Rh	Rocks	AES	668
Ag, Au, Pd, Pt	Minerals	AES	669
Ag and Au	Ores	AAS	670
Au, Pd, Pt	Ores and concentrates	AAS	671
Au and platinic metals	Ores	NAA	672
	Ores and their processing products	NAA	673
Au, Ir, Pt	Natural substances	NAA	674
Pd and Pt	Rocks	AAS	675

Kirillov et al. [646] succeeded in isolating Au, Ir, Pd, Pt and Rh quantitatively in a tin alloy and in using this procedure in analysis of copper-nickel sulphide ores and their treatment products.

Artem'yev and co-workers [647, 648] have revealed that fire melting enables other chalcophilic metals to be preconcentrated along with the noble metals; determination of elements of both groups can be carried out simultaneously. They used fire melting in combination with neutron activation analysis. Preconcentration of selenium and tellurium before irradiation significantly simplified the decomposition of the sample and the separation. After irradiation, selenium and tellurium were determined by measuring the radioactivity of lead beads obtained from the specimens and the reference samples. The following procedure was used: 1 g finely ground specimen was mixed with the charge (5 g litharge, 0.5 g sodium carbonate, 0.5 g sugar and 0.1 g borax), the mixture was transferred into a refractory clay crucible and placed in a muffle furnace. The mixture was melted at 1100°C, molten lead was separated from slag by pouring the melt into a stainless-steel mould. The lead bead obtained and the reference samples prepared in a like manner were irradiated. Thereafter, by γ-activation, tellurium was determined after ten days, and selenium, after twenty days. The following detection limits were attained: tellurium, $3 \cdot 10^{-6}\%$; selenium, $1 \cdot 10^{-6}\%$.

Artem'yev et al. [648] state that fire melting preconcentration is also suitable for determination of uranium (by fission products), bismuth, tin and indium. There are other works devoted to the use of fire melting (see Table 2.13).

2.8.2 OTHER PYROMETALLURGICAL METHODS

In this group are included the methods usually used in pyrometallurgy, but which have been tailored to the needs of analytical laboratories. Among them, pyrometallurgical slag formation is a special method of preconcentration of impurities in metals. It is based on partial oxidation of the matrix by oxygen; the trace components having large, in absolute value, isobaric–isothermal potentials of oxide formation reactions, change into oxides and, by forming slag, concentrate on the surface of the solidifying ingot. Thus, a weighed amount of metallic silver was twice melted in air [676]. The impurities Bi, Cu, Fe, Pb and Tl present in it transferred into the surface layer of the metal bead as a result of oxidation; after separation, the bead was analysed by the atomic emission method. To accelerate slag formation, Karabash [677] passed damp air saturated with HCl vapours through the melt of the analysed metal. The small amounts of oxides, chlorides and oxychlorides of the matrix element formed were separated into a discrete phase and served as collector of trace components that transfer into similar chemical forms. The method was used to concentrate nineteen elements in the analysis of high-purity bismuth and tin.

For concentration purposes, Chalkov [678] has utilized the interaction of melts of metals and salts. Such an interaction is used in nonferrous metallurgy for refining metals. For molten bromides (at 700°C) the electrochemical series takes the following form [679]: Na, Ca, Mg, Al, Mn, Tl, Zn, Ga, In, Cd, Fe, Rb, Cr, Sn, Ag, Cu, Co, Ni, Bi, Sb, As, Te and Au. When the melts of metallic indium and its bromides interact, the metals on the right of indium in the series will concentrate in the latter. Use of this fact was made in the atomic emission analysis of high purity indium. A weighed amount of indium of mass 3.3 g was taken. Part of the indium (1.3 g) was converted into bromide by dissolution in hydrobromic acid and subsequent evaporation of the solution to dryness. The remaining sample was brought into contact with the obtained bromide in a quartz cell at 225–230°C and at a pressure of 13 Pa. In the opinion of the authors, $InBr_2$ and $InBr$ were formed by reduction of $InBr_3$, and the impurities Ag, As, Au, Bi, Cd, Co, Cr, Cu, Fe, Ni, Pb, Sb, Sn and Te in the metallic form concentrated in the indium bead. Then the temperature was raised by 30–50°C and indium bromides were volatilized. The remaining bead (indium) after a suitable treatment was subjected to atomic emission analysis. For a concentration coefficient of 100, the detection limits of $1 \cdot 10^{-5}$–$2 \cdot 10^{-7}\%$ were attained. Chalkov [678] claims that this method helps to reduce the consumption of hydrobromic acid and, therefore, to lower the blank (an advantage of the method).

Preconcentration of trace elements upon interaction of thallium with cadmium chloride has also been studied [678]. Impurities Ag, As, Au, Bi, Co, Cu, Fe, Ga, In, Ni, Pb, Sb and Sn were concentrated in a thallium bead. To lower the detection limits of certain impurities to 10^{-6}–$10^{-8}\%$, Chalkov [678] used a concentration procedure involving distribution of elements between two immiscible melts; the elements in the molten state can interact both by the mechanism of physical dissolution and with the formation of chemical compounds. Thus, on fusing lead or bismuth with metallic sodium, tellurium, selenium and chlorine form Na_2Te, Na_2Se and $NaCl$. A weighed amount (50–500 g) of lead or bismuth was fused at 360–400°C with 0.2–0.5 g metallic sodium and

154

10–15 g sodium hydroxide. The mixture was mixed for 5–10 min and the alloy, after cooling, was treated with water. The obtained alkaline solution was analysed for tellurium, selenium and chlorine using voltammetry, luminescence, potentiometry, nephelometry and other methods.

2.9. Flotation

Flotation is a process by which dispersed solid or liquid particles, precipitations, colloidal particles and water-dissolved substances can be brought to the surface of an aqueous phase with the aid of a rising flow of gas bubbles. The substance to be floated must be hydrophobic; this creates conditions for their contact with gas bubbles. Special agents are used to isolate hydrophilic substances and to make them hydrophobic.

Flotation is widely used as a highly effective method of enriching mineral resources after grinding them. For this, a suspension of ground raw material in water is treated with collectors which are anion- and cation-active and non-ionogeneous surfactants. The collectors are adsorbed on the surface of the separable component particles, and lower their wettability. Air in the form of fine bubbles is passed through the suspension and the separable components are floated. In the presence of frothing agents (compounds which can be adsorbed on the water–air boundary and can moderately stabilize the froth), a sufficiently stable layer of froth enriched with components to be separated is formed on the suspension surface. Other reagents can also be added into the suspension. They are suppressors which prevent sticking of air bubbles to ballast substances and their drift into the concentrate, and activators which increase adsorption of collectors and their hydrophobizing action. If necessary, reagents that help to regulate the acidity of the medium may also be introduced. Flotation at its simplest consists of the following. The gas bubbles and the hydrophobic solid surface come closer; under definite conditions the layer of water that separates them becomes unstable and ruptures spontaneously. Bubbles adhere to the particle surface and drift the particles into the froth over the suspension surface.

Probably Fukuda and Mizuike [680] were the first to apply flotation for concentration of trace elements.

Flotation concentration is carried out in two ways [681]. (1) The inorganic and organic precipitates present in aqueous solution are floated with surface-active substances of opposite charge or without them. Usually, trace elements are coprecipitated with small amounts of a precipitator carrier which is floated. (2) Ionic flotation; here ions (usually complex ions) of the elements present in aqueous solution are floated with surface-active substances having a charge opposite to that of the ion of the determined element.

Flotation is accomplished in special cells ensuring the following: introduction and dispersion of air in the suspended matter, mixing of the suspension with air and keeping the particles in the suspended state, separation of suspension and froth, removal and transportation of enriched products. Typical cells are shown in Fig. 2.29.

Some problems related to analytical application of flotation as a method of preconcentration are dealt with in the reviews [21, 681–684]. Therein, attention is paid, in particular, to the factors that determine the degree of separation of trace components: surface activity of the compound to be isolated, its affinity to associate with a collector, pH of solution, concentration of activators and suppressors, and temperature. A high degree of absolute concentration of trace elements is noted.

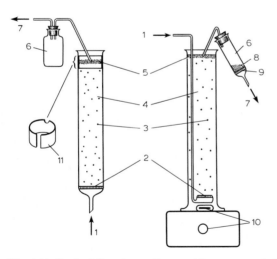

Fig. 2.29. Typical flotation cells. 1 = Nitrogen or air introduction; 2 = porous glass filter; 3 = air bubbles; 4 = sample solution; 5 = foam layer, concentrate of trace elements; 6 = beaker or tube for concentrate sampling; 7 = suction; 8 = foam; 9 = vitreous silica fibre; 10 = magnetic stirrer; 11 = polyethylene insert.

Extensive use is made of flotation for concentration of trace elements after their co-precipitation on suitable collectors. Often hydrides are employed as collectors. For example, a method has been worked out for determination of arsenic in natural waters [685]. The method involves co-precipitation of arsenic on iron hydroxide at pH 8–9 and flotation of the obtained precipitate in the presence of sodium oleate, flotation being carried out by passing air through the solution for 30 s. The concentrate, after dissolving in hydrochloric acid, is carried over to the generator of hydrides in which $NaBH_4$ is already placed. The arsine formed is transported by argon into a flame atomizer (argon–hydrogen). Other examples are listed in Table 2.14.

Organic substances can also be used as collectors. Silver can be separated from aqueous solutions by co-precipitation on adding acetone solution of 2-mercaptobenzthiozole [696]. However, filtration of amorphous precipitate and even its centrifugal separation do not yield positive results. Flotation enables more than 95% silver to be isolated from 3 l sea water (0.1 M in nitric acid). The precipitate that gathers on the liquid surface is separated by passing the sample through a porous glass filter and analysed by the atomic absorption method.

2.9.2 IONIC FLOTATION

This term is not quite accurate because in ionic flotation it is not the ions that are floated, but their compounds using oppositely charged surface-active substances. Surface-active substances should be present in quantities greater than those of the ions to be concentrated; too excessive amounts are undesirable.

A method has been worked out for separation and determination of thorium [697]. The method involves flotation of an ionic associate of zephyramine and of a thorium complex with arsenazo III, which is formed in 0.3 M HCl solution. Nitrogen is passed through the solution and the froth is analysed after a suitable treatment by γ-activity. The method has been used for determining thorium in nuclear decay products. On this very principle is based the procedure of determining uranium traces in sea water [698] (the determination is completed by the neutron activation or the spectrophotometric method) and zirconium in nickel alloys (spectrophotometry) [699]. Other examples are listed in Table 2.15.

TABLE 2.14

Trace elements concentration by flotation of precipitates

Elements to be concentrated	Substance analysed	Collector	Surfactant	Method of determination	Ref.
U	Sea water	Iron(III) hydroxide	Sodium dodecylsulphonate	Spectrophotometry	686
Cu, Zn	Sea water	Iron(III) hydroxide	Dodecylamine	AAS	687
Hg	Sea water	Cadmium sulphide	Octadodecyltrimethyl-ammonium chloride	AAS	688
Se	Sea water	Iron(III) hydroxide	Sodium dodecylsulphonate	Spectrophotometry	689
Sc	—	Iron(III) hydroxide	Sodium oleate	Spectrophotometry, AAS	690
Cd, Pb	Atmospheric precipitations	Hydroxides of Fe(III) and of aluminium	Sodium salts of fatty acids	AAS	691
Cd, Co, Cu	Sea water	Iron(III) hydroxide	Octadecylamine, sodium lauryl sulphate	AAS	692
Co, Cu, Ni	Water	Zirconium hydroxide	Sodium oleate or dodecylsulphonate	AAS	693
As	Natural waters	Iron(III) hydroxide	Sodium dodecylsulphonate	AAS	694
As, Ge, Sb, Se	Acidified aqueous solutions	Iron(III) hydroxide	Sodium lauryl sulphonate	AAS	695

TABLE 2.15

Trace elements concentration by ionic flotation

Elements to be concentrated	Substance analysed	Complex-forming agent	Surfactant	Method of determination	Ref.
Mo	Sea water	–	Cetylpyridine chloride	Photometry	700
20 elements	–	Thiocyanate	Cationic surfactants	–	701
Cd, Cu, Mn, Zn	–	1,10-Phenanthroline	Bromopyrogallol red	AAS	702
Cr	–	Diphenylcarbazide	Sodium lauryl sulphonate	Spectrophotometry	703
Cu	Waters	Potassium butylxanthate	Cetyltrimethyl ammonium bromide	AAS	704
U	Sea water	Arsenazo III	Tetradodecyldimethyl-benzyl ammonium chloride	Spectrophotometry	705
Au, Ir, Pd, Pt, Rh	Dilute solutions	Chloride	Cationic surfactants	NAA and spectrophotometry	706

References pp. 179–203

There is another variant of ionic flotation which does not involve blowing of air or other gases. Here, use is made of an organic solvent which does not mix with water [707]. Sometimes, this is known as ionic sublimation. The elements to be concentrated are collected at the interface without the use of surface-active substances. An example of this is the flotation of less-soluble ionic associates that are formed by multicharge anionic complexes of metals and by hydrophilic basic dyes. More often, this concentration technique is used in combination with spectrophotometric determination. Another example of this is flotation–spectrophotometric determination of osmium(IV) [708]. The associate $OsCl_6^{2-}$ is floated with a cation of Rhodamine 66. Toluene or hexane serves as organic solvent. The ionic associate is collected at the interface; the compound floated by toluene has the following composition: $[(R^+)_2 OsCl_6^{2-}] \cdot 2(R^+ Cl^-)$, where R^+ stands for Rhodamine 66. The residue is dissolved in acetone and determined photometrically. Marczenko [707] has reviewed such methods.

2.10. Filtration and other related membrane methods

Filtration, as it is known, is a process of motion of liquid or gas through a porous medium; it is accompanied by the isolation of suspended solid particles. This method is used mainly for concentration of solid particles present in atmospheric air and in the air of production areas. It can also be employed for separating particles from aerosols and even from colloidal solutions. In the latter case, ultrafiltration is employed, particles being passed through a semipermeable membrane with pore sizes 10^{-2}–10^{-1} μm. In analytical laboratories, filtration is usually carried out in periodic action devices working under normal or increased pressures or under vacuum.

There are reviews [709, 710] dedicated to air sampling techniques used in determination of metals. Therein are discussed the laboratory equipment employed for these purposes (rotary compressors, pumps, rotameters, gas-filled counters, filters for entrapping suspended matter). The possibilities of various instrumental methods of analysis of atmospheric pollution are also discussed, in particular neutron activation, spectrochemical, atomic absorption, X-ray fluorescence methods, inverse voltammetry, electronic and microprobe methods.

Paper [711, 712], graphite [713], porous glass and quartz [714], glass fibre [715] and synthetic materials [716, 717] have been used as filtering-element material in the analysis of air. The procedures for determining trace elements in air do not principally differ from each other. They involve sampling, combined with the concentration stage, by passing a definite volume of air through the filtering element, removal of particles from the filter, and determination.

As an example, we shall mention a method [713] of determining a number of elements in atmospheric dust. To this end, air was passed for 24 h (at the rate of 6–8 l/min) through graphite filters 47 mm in diameter. The concentrate is analysed by the atomic emission method. Aleksandrov et al. [716] have developed an atomic emission method for determining As, Co, Cr, Mn, Mo, Ni, Pb, Sb, Sn, Sr, Ti, V, Zn and Zr in atmospheric dust. The detection limits of this method are 10^{-6}–$10^{-7}\%$. Here air is also passed through a filtering element and the filter is ashed and analysed. For determining the elemental composition of particles present in atmospheric air, air was passed through a silver filter with pore size of $0.8\,\mu m$ [718]. The filters were then pressed into electrodes and analysed by spark-source mass spectrometry.

Sometimes, chemical reactions take place on filtering elements. Thus, to determine mercury in the air of premises, the trace metal was concentrated on filters made of fibre glass paper impregnated with iodine (the paper was immersed in methanol solution of iodine for 24 h) [719]. The air was sucked through two such filters at the rate of 200 l/h. After sampling, the filters were washed with water, sulphuric acid and hydrogen perioxide. The filtrates were analysed by AAS or AES.

An automatic atomic absorption method has been worked out for determining arsenic in atmospheric particles [720]. The air sample is passed through a filtering element made of fibre glass. A disc is cut out from the filter and treated with a mixture of concentrated sulphuric and nitric acid; arsenic is reduced to arsine and carried to an electro-thermal atomizer.

Ultrafiltration is commonly employed for separating high- and low-molecular-weight compounds in a liquid phase [469, 470, 721, 722] with the aid of selective membranes that allow mainly low-molecular-weight compounds to pass through them, the motive force being the difference in pressure between both sides of the membrane. It is used for technological and preparative purposes, and also as a separation and concentration method in organic and biological analysis, and for carrying out analytical studies for medico-biological purposes.

Trace elements may be concentrated by ultrafiltration through a membrane; the technique involves selective bonding of elements into

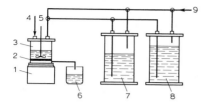

Fig. 2.30. Arrangement for membrane filtration separation of elements with polymeric reagents. 1 = Magnetic stirrer; 2 = membrane; 3 = cell; 4 = sample injection (1st way); 5 = stirrer; 6 = filtrate; 7 = vessel with washing solution; 8 = vessel with sample solution (2nd way of sample introduction); 9 = pressure source.

complexes with water-soluble polymers containing complex-forming groups [723]. Those ions of metals which do not form complexes with water-soluble polymers pass through the membrane, and those bonded in complex compounds concentrate in solution over the membrane. Filtration is carried out under pressure. The diversity of water-soluble complex-forming polymers indicates a useful future for ultrafiltration. Fig. 2.30 shows the arrangement used for carrying out ultrafiltration [723].

Under development is another prospective method: reverse osmosis. It involves filtration of solutions through semipermeable membranes which allow solvent molecules to pass through them, but retain completely or partially the molecules or ions of the dissolved substance [469, 470, 721, 722]. The method utilizes the phenomenon of osmosis – solvent spontaneously goes into the solution through the membrane. The pressure at which equilibrium is attained in the system is called osmotic pressure. When a pressure exceeding osmotic pressure is applied from the solution side, the solvent starts moving in the reverse direction. This method has no practical application in analytical chemistry.

As an example of reverse osmosis, we shall mention a method of separating radium from dilute nitrate, chloride or sulphate solutions [724] by passing them through porous membranes (of cellulose acetate) at 2–2.75 MPa; 88–99% radium separates out.

Dialysis as a method of separating and concentrating dissolved substances that have very different molecular masses is based on the fact that these substances diffuse at different rates through the semipermeable membrane separating the concentrated and dilute solutions. Due to unequal ionic forces of solutions on either side of the membrane, *i.e.* there exists a concentration gradient, the trace com-

162

ponents will diffuse from the solution with the lesser ionic strength to the solution with the greater ionic strength (receiver). The motive force in dialysis is the difference in chemical potentials on both sides of the membrane. Initially, only cellulose, including nitrocellulose and cellulose acetate, membranes were used in dialysis. Later, cation- and anion-exchanging membranes as well as amorphous membranes were shown to exhibit the desired properties. Note that when one electrolyte is separated from other electrolytes the process is known as Donnan dialysis. In conventional dialysis, one non-electrolyte is separated from other non-electrolytes or electrolytes present in the mixture.

Donnan systems [725] have two inherent properties which make them suitable for concentration of trace elements. (1) An ion may spontaneously diffuse through the membrane against its concentration gradient. In other words, an ion may go from the solution where its concentration is small to a solution where it (the ion) is present in a much higher concentration. (2) The concentration of ion in the electrolyte-receiver will be proportional to its initial concentration in the analysed substance and to the duration of dialysis. High values of concentration coefficients can be attained under correctly selected conditions. This method has been used for concentration of trace elements prior to their atomic absorption [726, 727] and voltammetric [728, 729] determination. Other examples can be found in Table 2.16.

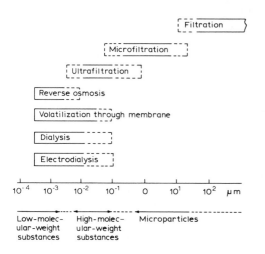

Fig. 2.31. Particle size ranges of membrane methods of separation and preconcentration of trace elements.

TABLE 2.16

Concentration of trace elements by dialysis

Trace elements to be concentrated	Substance analysed	Method of determination	Remarks	Ref.
Co, Ni	Natural waters	AAS	Cation exchange membrane P-1010 is used	730
Ca, Co, Cu	–	AAS	Cation exchange membrane P-1010 is used	731
Cd, Co, Cu, Ni	Waters	Ion chromatography	Concentration coefficient 80–100	732
Co, Cu, Ni, Zn and others	–	–	Membrane is of cellulose acetate; continuous method; EDTA is used	733
Cd, Cu, Pb, Zr	–	Ion-selective electrodes	Cation exchange membrane Nafion 811 was used	734
Cd, Eu, Ni, La, Lu	–	AAS	–	725

Fig. 2.31 shows the range of application of membrane methods of separation and concentration which are of interest from the viewpoint of analytical use. It is based on the figure given in the monographs [470, 722].

2.11. Chemical transport reactions

Transport reactions are widely used for obtaining high-purity compounds and also epitaxial layers and monocrystals. Besides the ease of controlling the process and the possibility of its automation, these reactions have no waste, *i.e.* processes can be carried out in a closed cycle and the reagent gas or carrier gas always returns to the cycle. Chemical transport reactions are convenient for comparatively simple substances which can form intermediate gaseous products capable of transforming again into the solid state when the conditions (temperature, pressure) vary. We shall illustrate this by considering the following examples:

$$3Cu_2O(s) + 6HCl(g) \xrightarrow{600°C} 2Cu_3Cl_3(g) + 3H_2O$$

$$\longrightarrow 3Cu_2O(s) + 6HCl(g) \tag{2.60}$$

$$Si(s) + SiBr_4(g) \xrightarrow[Ar]{1300°C} 2SiBr_2(g) \xrightarrow{850°C} Si(s) + SiBr_4(g) \tag{2.61}$$

$$2Ge(s) + 2I_2(g) \xrightarrow[H_2]{600°C} 2GeI_2(g) \xrightarrow{350-400°C} Ge(s) + GeI_4(g) \tag{2.62}$$

$$2CdS(s) + 2I_2(g) \underset{400°C}{\overset{1000°C}{\rightleftharpoons}} 2CdI_2(g) + S_2(g) \tag{2.63}$$

Transport reactions resemble sublimation; here not the vapours of the substance, but the vapours of a volatile intermediate compound serve as carrier of the substance. The substance is carried from one zone to another in three ways: by the flow of reagent gas with reaction products; by molecular diffusion of reagent gas and of the reaction products; by convective diffusion of reagent gas and of the reaction products.

Fig. 2.32 shows a schematic diagram of the device used for carrying out transport reactions with the participation of nickel tetracarbonyl. Nickel is placed at one end of the device. Atmospheric air is replaced by carbon monoxide. Nickel tetracarbonyl is formed at the cold end of the tube:

$$Ni + 4CO \xrightarrow{45-50°C} Ni(CO)_4 \tag{2.64}$$

References pp. 179–203

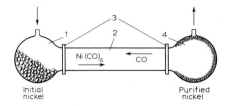

Fig. 2.32. Reactor for carrying out transport reactions with nickel tetracarbonyl. 1 = Cold zone; 2 = tube; 3 = flange joint; 4 = hot zone.

Diffusing to the hot end, this intermediate product decomposes

$$Ni(CO)_4 \xrightarrow{180-200°C} Ni + 4CO \qquad (2.65)$$

The isolated nickel settles on the walls at the hot end and carbon monoxide returns by diffusion to the cold end to transport a new portion of nickel.

Such a process could be used for both purification of nickel and, perhaps, for concentration of trace impurities in it. We have not found examples of using this method for analytical preconcentration of trace elements. The reader is, however, referred to the papers dedicated to purification of substances from inorganic trace impurities. The content of aluminium in silicon, on purifying it by transferring silicon tetrachloride, was lowered by about two orders, of iron by more than one order and of zinc by about an order [736]. Antimony is almost completely removed from silicon on transporting silicon tetraiodide [737]. The study of the transport reaction of germanium with iodine (intermediate compound, GeI_2) has revealed [738] that the matrix element is practically completely purified from indium and gallium, the content of which does not exceed $10^{-6}-10^{-7}\%$.

2.12. Thermodiffusion

Thermodiffusion, as a method of separation, is based on the appearance of a concentration gradient in a gas or liquid mixture in a non-uniform temperature field [735]. It can be carried out in an ordinary glass tube sealed at both ends, one end of which is cooled while the other is heated. This is a quite effective method for separating gas mixtures, in particular for solving a complex problem of separation of uranium isotopes after transferring the matrix into gaseous hexafluoride. In analytical practice it is not yet used.

2.13. Combination of preconcentration methods

Concentration methods can be combined in two ways. In one of them, on the basis of known techniques, a new method referred to as "exsorption" has appeared [739]; it combines extraction and sorption. The suspension of a sorbent in an organic solvent immiscible with water is used. Other examples are the methods of concentrating solutions prior to their flame atomic absorption determination [740, 741]. Aerosol from a normal pneumatic atomizer is dried in a heating chamber and collected on a filament electrode at high voltage [740]. Dust is collected by electrostatic precipitation. When a sufficient amount of the substance settles on the electrode, the latter is heated to evaporate the deposited substance; the vapours are sent to a chamber. The substance is then carried to a flame by the gas oxidizer. In so doing, the concentrate does not get contaminated, and a very high coefficient of concentration may be attained. This technique has some disadvantages, too. The filament electrode intensively heats up in the flow of active gas; the electrode has a limited melting point and a limited stability to a gas oxidizer. Therefore, only salts with a low volatilization temperature (Cd, Pb and Zn) evaporate from it. This difficulty has been overcome by using the so called liquid (water) electrode [741].

In the second way, various preconcentration methods are used in succession. Often, one of these methods is an integral part of the sample preparation technique. Any additional stage complicates the analysis and entails an increase in the consumption of reagents and time. Therefore, such a combination is most often used as a constrained measure.

The analysis of geological samples, soils, plants, tissues and secretions of living organisms, and synthetic organic substances and materials often involves two-stage preconcentration of trace components: (1) the sample is decomposed by dry or wet ashing with partial removal of the matrix; (2) concentration is carried out using another suitable method. For example, in analysis of geological samples silicon is usually removed with hydrofluoric acid. Thus, for determining Cd, Cu, Ni and Zn in silicate rocks, Japanese researchers [742] decomposed the sample with hydrofluoric acid and *aqua regia* in a closed Teflon vessel. After the silicon was removed as tetrafluoride, the trace elements were sorbed on the chelate-forming resin Chelex-100. The column was washed with water, the elements to be determined were eluted with $2\,M$ HNO_3, and the eluate was analysed by the atomic absorption method.

Small amounts of selenium (up to $6 \cdot 10^{-6}\%$) in rocks were determined by the same method [743] using an acetylene–air flame. The sample was treated with a mixture of hydrofluoric and nitric acids to separate selenium. After a suitable treatment of the solution, selenium was extracted with 1% phenol in benzene from bromide solutions. The extract was evaporated to dryness, the residue was dissolved in water, and the solution obtained was evaporated in a tantalum boat placed in a flame.

The analysis of organic, biological and carbon-containing materials requires that they should be ashed irrespective of the method of subsequent preconcentration of trace elements. Thus, for determining nanogram amounts of mercury in carbon, leaves of fruit trees and liver [744], the samples were mineralized with a mixture of HNO_3 + H_2SO_4 + $HClO_4$ in a flask fitted with Raschig rings and a reflux condenser. Mercury is not lost in this case. After the organic base was removed, the sample was transferred into a reaction vessel, a reducing mixture ($NaCl$ + $SnCl_2 \cdot 2H_2O$ + H_2SO_4) was added, and the vessel was immediately connected to the adsorption cuvette via a Teflon tube. A carrier gas (argon) was passed, mercury was transported to the atomizer, and the analytical signal was recorded. This method enabled mercury to be determined down to $1 \cdot 10^{-5}\%$.

In the analysis of plants for Be, Hf, Nb, Ta and Zr [745], the samples were calcinated at 500°C; the dry residue was dissolved. Thereafter, at pH 4 (Hf, Nb, Ta and Zr) and 8.5–8.7 (Be) the trace elements were sorbed with pyrogallol formaldehyde resin under static conditions. The concentrate was mixed with graphite collector and analysed by the atomic emission method. In the analysis of high-purity graphite [746], graphite was burnt, the residue was dissolved, and six determinable elements were concentrated by sorbing them on complex-forming sorbents. Determination was completed by the AES–ICP method.

Other combinations of methods are also used. The inverse voltammetric determination of cadmium in aerosols [747] involves its concentration by combining filtration and sublimation. The fine particles present in air are first separated on a glass-fibre filter placed in a quartz tube. The tube is introduced into the device for sublimation. The sample is then mineralized for 20–25 min in a current of oxygen at 400–450°C. After this, cadmium is sublimated in a current of hydrogen at 650–700°C. The metallic mirror formed is dissolved and cadmium is electrodeposited on a mercury electrode or on an amalgamated glassy carbon electrode. Cadmium is then dissolved anodically. This method enables 0.15–0.25 ng cadmium to be determined.

In stripping voltammetry, use is sometimes made of two-stage concentration. Solvent extraction is combined with electrochemical concentration from the extract mixed with a suitable background electrolyte. This method, known as extraction stripping voltammetry, will be considered in the following chapter.

Several procedures have been developed, which involve successive application of extraction and sorption or distillation and other methods. Such combinations are discussed in the literature (see, in particular, ref. 9). Other examples illustrating approaches to the problem of combined techniques for concentration of trace elements are given elsewhere.

Jackwerth et al. [748] concentrated impurities present in thallium by precipitating the matrix as iodide. EDTA was added to eliminate the observed coprecipitation of Bi, Cd, Cu and Pb. After centrifugal separation of the residue, the traces (Bi, Cd, Co, Cu, In, Ni, Pb and Th) left in the solution were extracted (at pH 4) with chloroform as diethyldithiocarbamates. The extract was evaporated and analysed by the atomic emission and photometric methods.

In atomic emission analysis of high-purity indium arsenide for fourteen impurities [749], arsenic tribromide was volatilized after dissolving the sample in hydrobromic acid in the presence of bromine. Indium was then extracted with 2,2'-dichloroethyl ether from $8\,M$ hydrobromic acid. The aqueous phase was evaporated on carbon powder and analysed spectrographically using a dc arc. Nedler et al. [750] determined 10^{-5}–$10^{-6}\%$ Al, Cd, Mg, Mn, Cu, Ni, Pb and Zn in indium phosphide; one of the matrix elements (indium) was extracted from bromide solutions with diethyl ether, the other (phosphorus) was volatilized as phosphine. After the removal of the matrix, the solution was evaporated on carbon powder and analysed by the atomic emission method.

Two-stage concentration of trace elements in analysis of different waters (fresh water, sea water, industrial wastes) is often used. The sample is first evaporated to a small volume and then concentrated by a suitable method. For example, in atomic absorption determination of boron in sea water [751], the sample was evaporated to 25% of the initial volume, the trace element was then extracted with 20% solution of 2-ethyl-1,3-hexanediol in methyl isobutyl ketone, and the extract was nebulized into an acetylene–nitrous oxide flame.

The following technique is also interesting. In analysis of slightly mineralized waters, Ba, Co, Cs, Na and Sr were concentrated by electroosmosis [752]. For fast removal of cations from the cathode space, they were sorbed during concentration on the cation exchanger KU-2-8 placed together with a platinum cathode in a nylon net bag.

Now we shall dwell on successive application of other concentration techniques in one analytical cycle.

A combined method has been developed for concentration of noble metals for their subsequent determination in natural and industrial materials using instrumental methods: atomic emission, atomic absorption and neutron activation [753]. Highly dispersed sulphide acts as a selective collector in co-precipitating platinum-group metals from highly acidic solutions [754]; on the other hand, chelate-forming sorbents have good selectivity with reference to these elements [755]. The authors utilized the merits of these methods in one analytical cycle. Taking into consideration that, in concentration of noble metals on copper sulphide, iridium and rhodium are not completely separated from the solutions obtained after dissolving the sample, the authors used a mixed collector: copper sulphide and 2-mercaptobenzothiazole. Separation of noble metals from copper, which is one of the macrocomponents of many natural materials, was attained by sorbing them with the chelate-forming sorbent PVB-MP-12T. After separation, the sorbent was analysed by the methods mentioned earlier. The recovery in the concentrate amounted to 90–100%.

A method has been developed by Yoshida et al. [756] for isolation of Fe, Hg and Zn from zirconium; it is based on precipitation of tetrakis(di-n-propionylmethanate)zirconium. The chelates of trace components are co-precipitated with the matrix element. Zone melting is applied for additional concentration of trace elements.

For determining Bi, Cu, In, Pb, Ni and Sn in metallic cadmium, Kirgintsev et al. [604] combined controlled crystallization with matrix evaporation. They succeeded in lowering the detection limits of trace impurities by 2.5–3 orders compared to direct atomic emission analysis. Controlled crystallization was effected in such a manner that the ingot did not completely solidify. A small amount of the melt (about 2 g) enriched with impurities was removed. Cadmium was then volatilized at 450°C for 3 h and a ten-fold decrease in the melt mass was obtained. In such a combination of preconcentration methods, no reagents are required. It is not surprising that detection limits of $1.5 \cdot 10^{-8}\%$ were achieved.

As has been mentioned earlier, fire melting is quite often employed for concentration of gold, silver and platinum metals in analysis of natural substances. The metallic bead obtained (concentrate of noble

170

metals) is subjected to additional treatment for separating the metal collector and transferring the determinable elements into a form convenient for analysis. This part of the procedure also often includes concentration. For instance, an atomic absorption method has been worked out for determining Au, Pd, Pt and Rh in natural substances [757]. This method involves combination of crucible melting (lead as concentrator; 1st stage of concentration) with subsequent precipitation (2nd stage) of noble metals from the solutions obtained after

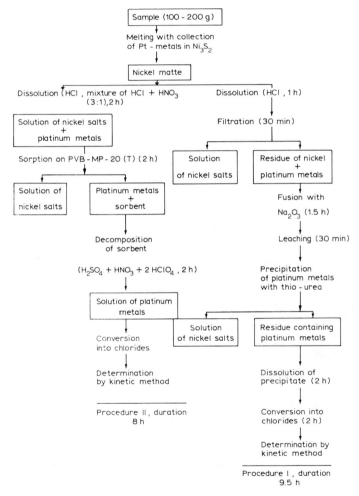

Fig. 2.33. Combined concentration of platinum-group metals in analysis of rocks.

dissolving the lead bead. The method has proved to be more effective than classic cupellation. Davydov *et al.* [758] concentrated platinum metals using fire melting (nickel sulphide as concentrator) and precipitation from the acid solutions, obtained after dissolving the sulphide, with thiourea or selective separation with the chelate-forming sorbent PVB-MP-20T. They succeeded in concentrating noble metals from solutions containing large amounts of nickel (up to 100 mg/ml). Figure 2.33 shows how trace elements are concentrated by the considered methods. These methods made it possible to analyse rocks containing 10^{-6}–10^{-7}% platinum metals.

2.14. Comparison of preconcentration methods

The selection of criteria for unbiased comparison of preconcentration methods and the comparison itself are difficult problems. A clear approach to the comparison has not yet been found, although the properties (parameters) that can be compared are known. Such a comparison can hardly be done without reference to the type of material analysed and the method of analysis of the concentrate. Before selecting a preconcentration method, the requirements for the whole analytical procedure must be clearly defined: elements to be determined, metrological parameters, duration of analysis, number of analyses required, etc. This allows informed decisions to be made.

What characteristics must be taken into consideration in comparing the methods of concentration? These are quantitative characteristics of concentration (concentration coefficient, recovery, trace components distribution coefficient, precision), the nature of the analysed substance and of the determinable trace elements, number of elements that can concentrate in one stage, scope of the method, duration, simplicity, availability and cost of equipment and reagents, adherence of the analyst to any method and his professional skill, safety measures, possibility of automating the process, requirements for working areas, and consumption of sample if expensive substances are analysed.

How to compare such a large number of factors? The factors can be compared in turn and a method, better in all characteristics, can be found. However, only certain characteristics of a method may be superior to the corresponding characteristics of the other method. Subjectivity of estimation, which usually does not go beyond the "worst–better" concept, and the relative significance of parameters cannot be left out of the reckoning. Some factors may merely fall out of the estimation.

On the whole, as has been correctly remarked in ref. 759, a work devoted to the comparison of determination methods, the task of selecting the best method rests on the solution of the problem of comparing multi-parameter systems, certain parameters of which are not compared quantitatively but only "qualitatively". The authors of ref. 760 recommend using the mathematical theory for solving similar problems. However, such suggestions have not yet been practically realized.

One of the main characteristics of preconcentration methods is the concentration coefficient. By this factor, Zolotov [761] tried to compare the methods (Fig. 2.34). However, this, as stressed by the author, is only one factor: often other parameters are of no less importance.

Most often, the methods are compared by one or two characteristics which are probably of maximum interest to the authors of publications.

Various methods of matrix separation prior to inverse voltammetric determination of trace amounts of antimony in metallic iron were compared in terms of time spent for determination [762]. Solvent extraction, ion exchange and constant potential electrolysis were used. The duration of analysis was found to be 60, 120 and 90 min for solvent extraction, ion exchange and electrolysis, respectively. Determining Cu, Fe, Li, Mn, Ni, Sr and Zn in atmospheric precipitations by stripping voltammetry, Kuz'mina et al. [763] used evaporation, ion exchange and extraction. As far as rapidity and recovery of determinable elements

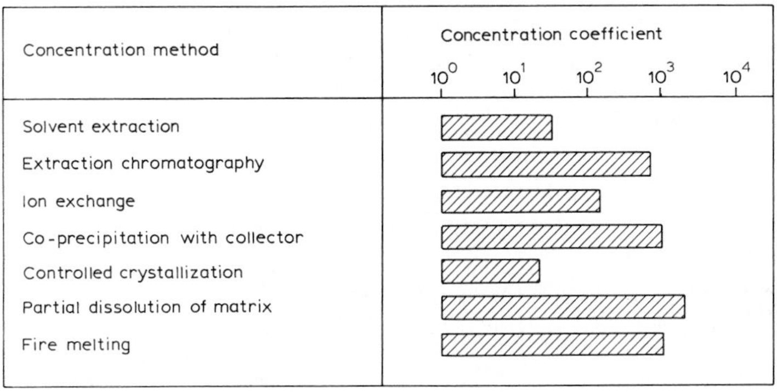

Concentration method	Concentration coefficient 10^0 10^1 10^2 10^3 10^4
Solvent extraction	
Extraction chromatography	
Ion exchange	
Co-precipitation with collector	
Controlled crystallization	
Partial dissolution of matrix	
Fire melting	

Fig. 2.34. Values of concentration coefficients attainable with various methods of preconcentration.

TABLE 2.17

Comparison of different preconcentration methods used for atomic emission analysis of high-purity antimony

Preconcentration method	Trace elements to be concentrated	Maximum concentration coefficient	Standard deviation (%)	Duration of analysis
Extraction of matrix	Ag, Al, As, Bi, Ca, Cd, Co, Cr, Cu, In, Mg, Mn, Ni, Pb, Te, Zn	100	13–18 (for Ca 40)	9.3 h
Volatilization of matrix	Ag, Al, Au, Ba, Be, Ca, Cd, Co, Cr, Cu, Fe, Ga, In, La, Mg, Mn, Ni, Pb, Pt, Tl, Zn	250	11–40 (for Ca 43)	40 min
Ion-exchange concentration	Bi, Ca, Cd, Co, Cr, Cu, Ga, Fe, In, Mg, Mn, Ni, Pb, Tl, Zn	250	13–27	19 h
Controlled crystallization	Ag, Al, Bi, Cd, Co, Cr, Cu, Fe, In, Mg, Mn, Ni, Pb, Sn, Zn			Only crystallization concentration
(a) One-stage		10	7–18	10 h
(b) Two-stage + extraction		1000		20 h
(c) Two-stage + volatilization		2500		11 h

are concerned, the method of partial evaporation of sample proved to be an optimum technique (losses of elements did not exceed 10%). Different methods of separation (mainly from molybdenum) and concentration of rhenium were compared [764], e.g. evaporation, sublimation and others, in analysis of copper–molybdenum ores and their treatment products. Evaporation has been recommended for determination of rhenium in solutions and solid products containing 5–100 μg Re/g, and sublimation for determination in a wide range of concentrations starting from 1 μg Re/g, the latter being characterized by better precision of results compared to other methods used for analysis of solid samples.

Leyden and co-workers [765, 766] have made a critical comparison of five methods of concentrating ten trace elements (Ag, As, Co, Cr, Cu, Fe, Hg, Mn, Pb and Zn) which are of importance in the analysis of environmental samples. Subsequent to concentration, the enumerated elements were determined by the X-ray fluorescence spectrometry method. The following methods were compared: (1) sorption on paper coated with a cation exchanger; (2) co-precipitation with sodium diethyldithiocarbamate; (3) co-precipitation with ammonium pyrrolidinedithiocarbamate; (4) reaction with 8-hydroxyquinoline and subsequent adsorption of the formed chelates on activated charcoal; (5) sorption on glass beads (with controllable pore size) to which diethyldithiocarbamate groups were inoculated. Detection limit, sensitivity, accuracy, range in which the calibration curve is linear, and the dependence of the degree of separation on concentration of metals were used as criteria for comparison. The methods yield good results (practically by all characteristics) when use is made of pure reference solutions. Interfering effects and other difficulties, say, small reaction rate of chromium(III) with dithiocarbamates, were noted. The study has revealed that comparison of preconcentration methods is a difficult task.

The preconcentration techniques employed in analysis of high-purity antimony were compared [767]; preconcentration of trace elements was carried out by extracting the matrix with 2,2'-dichlorodiethyl ether [768] or by volatilization of the matrix [769], separating trace impurities by ion exchange [770] and controlled crystallization [771]. Dispersion analysis was performed for determining errors at each stage of analysis. Random errors caused by extraction, ion exchange and controlled crystallization were negligible; concentration by matrix evaporation introduced significant errors. The authors then compared different methods of concentration by the number of determinable impurities, maximum concentration coefficient, relative standard deviation, and the time taken for the complete analytical cycle (Table 2.17). It is difficult to comment on the results of the study. Concentration coefficients shall have maximum values in the case of successive concentration of two-stage directed crystallization and extraction of the matrix element using 2,2'-dichlorodiethyl ether. The procedure is, however, characterized by the duration and the possibility of determining only fifteen elements. In each case, the analyst chooses a procedure conforming in several characteristics to the requirements of the product analysed.

For comparing two procedures of analysis of high-purity substances, Blank [772] regards the lower limit of determinable amounts of the given trace component to be the most important parameter.

One can judge about the extent to which one or the other method of preconcentration is used, its significance, and the field of application by knowing how much this technique has been covered in the reference literature. It is known that different types of analytical analysis handbooks contain information on the methods that are applied in practice by virtue of their good metrological parameters, stability to disturbances, rapidity of analysis, etc. Williams [773] was guided by these criteria in selecting methods of determination of anions for their inclusion in the *Handbook of Anion Determination.*

The following methods have been chosen for preconcentration of widespread anions:

carbonate	adsorption, precipitation;
nitrate	precipitation, ion exchange, volatilization with preliminary chemical transformation;
silicate	precipitation, extraction, volatilization with preliminary chemical transformation;
fluoride	volatilization with preliminary chemical transformation;
chloride	precipitation and co-precipitation, sorption including ion exchange, volatilization with preliminary chemical transformation;
bromide	ion exchange, precipitation, volatilization with preliminary chemical transformation;

iodide	precipitation, ion exchange, extraction;
thiocyanate	precipitation;
perchlorate	precipitation, extraction;
sulphate	sorption methods including ion exchange, precipitation, volatilization with preliminary chemical transformation;
sulphide	volatilization with preliminary chemical transformation, ion exchange, diffusion;
selenate and selenite	volatilization with preliminary chemical transformation, ion exchange, extraction, precipitation and co-precipitation;
thiosulphate	precipitation, ion exchange, extraction;
orthophosphate	precipitation and co-precipitation, extraction, ion exchange, electroprecipitation;
arsenate	volatilization with preliminary chemical transformation, extraction, sorption methods including ion exchange, precipitation;
tungstate	precipitation and co-precipitation, ion exchange, extraction;
molybdate	precipitation and co-precipitation, ion exchange, extraction, flotation;
chromate and bichromate	precipitation and co-precipitation, extraction, ion exchange;
permanganate	extraction, co-precipitation;
borate	volatilization with preliminary chemical transformation, ion exchange, extraction.

Thus, precipitation and co-precipitation, sorption methods, particularly ion exchange and volatilization with preliminary chemical transformation are most often recommended.

We now compare the widespread methods of preconcentration by the following important, in our opinion, criteria: (1) possibility of group concentration; (2) possibility of individual concentration; (3) stability to hindrances (effect of matrix, etc.); (4) value of the concentration coefficient; (5) precision of concentration results; (6) risk of losing materials; (7) possibility of application to various analysed substances; (8) possibility of application to trace elements of different nature; (9) combination with other methods of separation and concentration; (10) possibility of developing hybrid methods; (11) possibility of automation; (12) rapidity; (13) simplicity; (14) availability; (15) application to

TABLE 2.18

Comparison of various preconcentration methods by twenty criteria

Preconcentration method	Criteria																				Total
	1	2	3	4	5	6	7	8	9	10	11	12	13	14	15	16	17	18	19	20	
Extraction	1	1	1	1	1	1	1	1	1	1	1	1	1	1	1	1	1	1	1	1	20
Sorption methods	1	1	1	1	1	1	1	1	1	1	1	1	1	1	1	1	1	1	1	1	20
Electrodeposition (precipitation and dissolution)	0	1	1	1	1	1	1	0	1	1	1	1	1	1	0	1	1	1	1	1	17
Volatilization after chemical transformations	1	1	0	1	1	0	1	1	1	1	1	1	0	1	1	1	1	1	1	1	17
Precipitation and co-precipitation	1	1	0	1	1	1	0	1	1	1	0	0	1	1	0	1	1	1	1	1	15
Evaporation and sublimation	1	0	0	1	1	0	0	0	1	1	1	1	1	1	0	1	1	1	1	1	14
Flotation	0	1	0	0	1	1	0	0	0	0	0	0	1	1	0	1	1	0	1	1	9
Controlled crystallization	1	0	0	0	1	1	0	0	0	0	0	0	0	1	0	1	0	0	1	1	7

a wide range of problems; (16) requirements as to the qualifications of working personnel; (17) availability of instruments and devices; (18) prevalence; (19) availability of working procedures; (20) cheapness of the technique.

We shall assign 1 (for "good" properties) and 0 (for "bad" properties) to every criterion. We sum up the values of criteria for every technique. Table 2.18 contains the results of such an estimation. The methods are arranged in the following sequence: extraction and sorption techniques > electrodeposition (precipitation and dissolution) and volatilization after chemical transformations > precipitation and co-precipitation > evaporation and sublimation > flotation > controlled crystallization. These conclusions agree with common sense and analytical practice.

References

1 E.B. Sandell and H. Onishi, *Photometric Determination of Traces of Metals*, Wiley, New York, 1978.
2 W.E. Pickering, *Modern Analytical Chemistry*, Marcel Dekker, New York, 1972.
3 *Methoden der Analytishen Chemie und Ihre Theoretishen Grundlagen*, VEB Deutsher Verlag für Grundstoffindustrie, Leipzig, 1971.
4 M.P. Semov, in I. Zil'bershtein (Editor). *Spectrochemical Analysis of Pure Substances*, Adam Hilger, Bristol, 1977.
5 Yu.A. Zolotov, *Pure Appl. Chem.*, 50 (1978) 129.
6 T.P. Yakovleva, V.P. Zhivopistsev, K.I. Mochalov and B.I. Petrov, *Uchen. zap. Permsk. gos. Univ.*, No. 324 (1974) 46.
7 V.P. Zhivopistsev, Yu.A. Makhnev and I.S. Kalmykova, *Zavod. Lab.*, 38 (1972) 145.
8 M. Pinta, *Pure Appl. Chem.*, 37 (1974) 483.
9 Yu.A. Zolotov and N.M. Kuz'min, *Extraction Preconcentration*, Khimiya, Moscow, 1971 (in Russian).
10 A.K. De, S.M. Khopkar and R.A. Chalmers, *Solvent Extraction of Metals*, Van Nostrand, Reinhold, London, 1970.
11 N.M. Kuz'min and Yu.A. Zolotov, *Zavod. Lab.*, 38 (1972) 897.
12 I.P. Alimarin and V.V. Bagreev (Editors). *Theory and Practice of Extraction Methods*, Nauka, Moscow, 1985 (in Russian).
13 J. Minczewski, J. Chwastowska and R. Dybczyński, *Separation and Preconcentration Methods in Inorganic Trace Analysis*, Ellis Horwood, Chichester, 1982.
14 Y. Marcus, *Talanta*, 23 (1976) 203.
15 Yu.A. Zolotov, B.Z. Iofa and L.K. Chuchalin, *Extraction of Halide Complexes of Metals*, Nauka, Moscow, 1973 (in Russian).
16 T. Brana and G. Ghersini (Editors). *Extraction Chromatography*, Akadémiai Kiadó, Budapest, 1975.
17 Yu.A. Zolotov, N.M. Kuz'min, O.M. Petrukhin and B.Ya. Spivakov, *Anal. Chim. Acta*, 180 (1986) 137.
18 N.M. Kuz'min, V.S. Vlasov, V.Z. Krasil'shchik and V.G. Lambrev, *Zavod. Lab.*, 43 (1977) 1.

19 Yu.A. Zolotov and N.M. Kuz'min, *Solvent Extraction of Metals with Acylpyrazolones*, Nauka, Moscow, 1977 (in Russian).

20 *Second All-Union Conference on Concentration Methods Used in Analytical Chemistry, Moscow, December 6–9, 1977*, Abstracts of Reports, Nauka, Moscow, 1977 (in Russian).

21 A. Mizuike, *Enrichment Techniques for Inorganic Trace Analysis*, Springer-Verlag, Berlin, Heidelberg, New York, 1983.

22 Yu.A. Zolotov, V.A. Bodnya and A.N. Zagruzina, *CRC Crit. Rev. Anal. Chem.*, 14 (1982) 93.

23 H. Freiser, *Bunseki Kagaku*, 30 (1981) S47.

24 E. Jercan, *Metode de separaro în chimia analitică*, Technică, Bucuresti, 1983.

25 N.G. Vanifatova, I.V. Seryakova and Yu.A. Zolotov, *Extraction of Metals with Sulphur-Containing Compounds*, Nauka, Moscow, 1980 (in Russian).

26 D.L. Tsalev, I.P. Alimarin and S.I. Neiman, *Zh. Analt. Khim.*, 27 (1972) 1223.

27 D.L. Tsalev, *God. Sofii. Univ. Kl. Okhridski, Khim. Fak.*, No. 66 [1975(1971/1972)] 225.

28 N.I. Tarasevich, G.V. Kozyreva and Z.P. Portugalskaya, *Vestn. Mosk. Gos. Univ., Khim.*, No. 2 (1975) 241.

29 A. Janper, K.-H. Willman and F.J. Simon, *Fresenius' Z. Anal. Chem.*, 320 (1985) 137.

30 N.T. Sizonenko, E.S. Zolotovitskaya, E.I. Yakovenko and L.E. Belenko, in *Methods of Analysis of High-Purity Halides of Alkali and Alkaline Earth Metals*, Vol. 2, VNII Monokristallov Publishers, Kharkov, 1971, p. 50 (in Russian).

31 N.M. Kuz'min, G.I. Zhuravlev, I.A. Kuzovlev, A.N. Galaktionova and T.I. Zakharova, *Zh. Anal. Khim.*, 24 (1969) 429.

32 N.M. Kuz'min, G.D. Popova, I.A. Kuzovlev and V.S. Solomatin, *Zh. Anal. Khim.*, 24 (1969) 899.

33 N.M. Kuz'min, T.P. Dubrovina and O.M. Shemshuk, *Zavod. Lab.*, 35 (1969) 784.

34 N.M. Kuz'min, T.P. Dubrovina and O.M. Shemshuk, *Zh. Anal. Khim.*, 28 (1973) 364.

35 S.V. Freger, A.S. Lozovik, M.I. Ovrutsky and N.B. Tovbis, in *IVth All-Union Conference on Methods of Obtaining and Analysing High-Purity Substances, Gorky, May 30–June 1, 1972*, Abstracts of Reports, Gorky State University Publishers, Gorky, p. 113, 1972 (in Russian).

36 N.T. Sizonenko and Yu.A. Zolotov, *Zh. Anal. Khim.*, 24 (1969) 1341.

37 Sun I-Lyan, Tan Jan-Huan, Ha Yue-Lyan, Chjan Tzen-Pu, Chen Chzen-Tsi and Ha Shou-Chun, *Yuantszynyen*, No. 12 (1965) 1038.

38 A.P. Golovina, I.P. Efimov and V.K. Runov, in *Basic Dyes in Analytical Chemistry, All-Union Symposium, June 12–14, 1974*, Abstracts of Reports, Erevan State University Publishers, Erevan, p. 29, 1974 (in Russian).

39 E.T. Karaseva, *Low-Temperature Extraction-Fluorimetric Method of Determining Rare-Earth Elements with 1-Phenyl-3-Methyl-4-Benzoylpyrazolone-5 (PMBP)*, Institute of Chemistry of DVNTs AN SSSR; deposited in VINITI, No. 1249–74.

40 V.G. Revenko, V.V. Bagreev, Yu.A. Zolotov and L.S. Kopanskaya, *Zh. Anal. Khim.*, 27 (1972) 187.

41 E. Ivanova, S. Mareva and N. Iordanov, in *2nd Nat. Conf. Anal. Chem. Int. Participation, Golden Sands-Varna, 1976*, Summaries of Papers, Bulgarian Academy of Sciences, p. 19.

42 L.I. Pavlenko, O.M. Petrukhin, Yu.A. Zolotov, A.V. Karyakin, G.N. Gavrilina and I.E. Tumanova, *Zh. Anal. Khim.*, 29 (1974) 933.

43 Yu. A. Zolotov, O.M. Petrukhin, L.I. Izosenkova and E.B. Krasil'nikova, *Zh. Neorg. Khim.*, 16 (1971) 2563.

44 L.A. Demina, O.M. Petrukhin and Yu.A. Zolotov, *Zh. Anal. Khim.*, 27 (1972) 593.
45 N.G. Vanifatova, in *Theory and Practice of Extraction Methods*, Nauka, Moscow, p. 95, 1985 (in Russian).
46 G.A. Vorobyeva, Yu.A. Zolotov, L.A. Izosenkova, A.V. Karyakin, L.I. Pavlenko, O.M. Petrukhin, I.V. Seryakova, L.V. Simonova and V.N. Shevchenko, *Zh. Anal. Khim.*, 29 (1974) 497.
47 O.M. Petrukhin, V.N. Zhevchenko, I.A. Zakharov and V.A. Prokhorov, *Zh. Anal. Khim.*, 32 (1977) 897.
48 Yu.A. Zolotov, G.A. Vorobyeva, I.V. Seryakova and A.V. Glembovsky, *Dokl. Akad. Nauk SSSR*, 209 (1973) 909.
49 I.V. Seryakova, G.A. Vorobyeva and Yu.A. Zolotov, *Zh. Anal. Khim.*, 27 (1972) 1840.
50 G.A. Vall, M.V. Ysol'tseva, I.G. Yudelevich, I.V. Seryakova and Yu.A. Zolotov, *Zh. Anal. Khim.*, 31 (1976) 27.
51 V.P. Shaburova, I.G. Yudelevich, I.V. Seryakova and Yu.A. Zolotov, *Zh. Anal. Khim.*, 31 (1976) 255.
52 E.E. Rakovsky, N.V. Shvedova and L.D. Berliner, *Zh. Anal. Khim.*, 29 (1974) 2250.
53 E.E. Rakovsky, N.V. Shvedova and L.D. Berliner, *Zh. Anal. Khim.*, 30 (1975) 1775.
54 I.G. Yudelevich, G.A. Vall, V.G. Torgov and T.M. Korda, *Zh. Anal. Khim.*, 25 (1970) 80.
55 A.V. Nikolaev, V.G. Torgov, V.A. Mikhailov and E.N. Gilbert, *Izv. Sib. Otd. Akad. Nauk SSSR, Ser. Khim. Nauk*, 14 (6) (1967) 120.
56 I.G. Yudelevich, G.A. Vall, V.G. Torgov and T.M. Korda, *Zh. Anal. Khim.*, 26 (1971) 1550.
57 V.G. Torgov, V.N. Andrievskaya and E.N. Gilbert, *Izv. Sib. Otd. Akad. Nauk SSSR, Ser. Khim. Nauk*, 12 (5) (1969) 148.
58 E.N. Gilbert, G.V. Gluhova, G.G. Gluhov, V.A. Mikhailov and V.G. Torgov, *J. Radioanal. Chem.*, 8 (1971) 39.
59 E.N. Gilbert, G.V. Verevkin and V.A. Mikhailov, *Izv. Sib. Otd. Akad. Nauk SSSR, Ser. Khim. Nauk*, 15 (3) (1975) 102.
60 T.M. Korda, V.G. Torgov, T.A. Chanysheva and I.G. Yudelevich, *Zh. Anal. Khim.*, 32 (1977) 1767.
61 A.V. Nikolaev, I.G. Yudelevich, G.A. Vall and V.G. Torgov, *Izv. Sib. Otd. Akad. Nauk SSSR, Ser. Khim. Nauk*, 14 (5) (1974) 93.
62 G.A. Vall, S.S. Shatskaya, I.G. Yudelevich and V.G. Torgov, *Izv. Sib. Otd. Akad. Nauk SSSR, Ser. Khim. Nauk*, 12 (5) (1973) 88.
63 E.E. Rakovsky and M.I. Starozhitskaya, *Zh. Anal. Khim.*, 29 (1974) 2094.
64 O.M. Petrukhin, Yu.A. Zolotov and L.A. Izosenkova, *Zh. Anal. Khim.*, 16 (1971) 3285.
65 N.L. Fishkova and O.M. Petrukhin, *Zh. Anal. Khim.*, 28 (1973) 645.
66 B.L. Serebryany, N.L. Fishkova, O.M. Petrukhin and E.E. Rakovsky, *Zh. Anal. Khim.*, 28 (1973) 2333.
67 P. Senise and T. Levi, *Anal. Chim. Acta*, 30 (1964) 422.
68 M. Mojski, *Talanta*, 27 (1980) 7.
69 R.J.H. Seeverens, E.J.M. Klassen and F.J.M.J. Maessen, *Spectrochim. Acta, Part B*, 38B (1983) 727.
70 R.S. Shulman, L.I. Gindin and A.A. Vasilyeva, *Izv. Sib. Otd. Akad. Nauk SSSR, Ser. Khim. Nauk*, 1 (1) (1972) 142.
71 M.L. Rakhlina, A.S. Lomekhov, I.G. Yudelevich, L.I. Gindin, A.A. Vasilyeva, N.A. Tkacheva, V.N. Andrievsky and G.A. Chernousova, *Izv. Sib. Otd. Akad. Nauk SSSR, Ser. Khim. Nauk*, 4 (2) (1977) 71.

181

72 A.A. Vasilyeva, L.I. Gindin, R.S. Shulman, I.G. Yudelevich, T.V. Lanbina, N.A. Chainikova, G.I. Smirnova, L.V. Dubetskaya and V.N. Andrievsky, *Izv. Sib. Otd. Akad. Nauk SSSR, Ser. Khim. Nauk*, 4 (2) (1977) 65.

73 N.A. Borshch, in *Theory and Practice of Extraction Methods*, Nauka, Moscow, 1985, p. 111 (in Russian).

74 N.A. Borshch, O.M. Petrukhin, Yu.A. Zolotov, E.G. Zhumadin, V.I. Nefedov, A.B. Sokolov and I.N. Marov, *Koord. Khima*, 7 (1981) 1242.

75 N.A. Borshch, O.M. Petrukhin and Yu.A. Zolotov, in *Vth All-Union Conference on Extraction Chemistry, July 11–13, 1978*, Abstracts of Reports, Published by *Sib. Otd. Akad. Nauk SSSR*, Novosibirsk, p. 152, 1978 (in Russian).

76 I.F. Seregina, O.M. Petrukhin, A.A. Formanovsky and Yu.A. Zolotov, *Dokl. Akad. Nauk SSSR*, 275 (1984) 385.

77 B.Ya. Spivakov, V.I. Lebedev, V.M. Shkinev, N.P. Krivenkova, T.S. Plotnikova, I.P. Kharlamov and Yu.A. Zolotov, *Zh. Anal. Khim.*, 31 (1976) 757.

78 B.I. Petrov, *Zh. Anal. Khim.*, 38 (1983) 2051.

79 I.G. Yudelevich, L.M. Buyanova and I.P. Shelpakova, *Chemical Atomic Emission Analysis of High-Purity Substances*, Nauka, Novosibirsk, 1980 (in Russian).

80 V.P. Shaburova, T.A. Chanysheva, I.G. Yudelevich and E.A. Fedyashina, *Izv. Sib. Otd. Akad. Nauk SSSR, Ser. Khim. Nauk*, 14 (6) (1983) 121.

81 V.G. Torgov, E.N. Gilbert, V.A. Mikhailov and V.P. Shaburova, *Izv. Sib. Otd. Akad. Nauk SSSR, Ser. Khim. Nauk*, 9 (4) (1976) 69.

82 I.G. Yudelevich, V.P. Shaburova and V.G. Torgov, *Izv. Sib. Otd. Akad. Nauk SSSR, Ser. Khim. Nauk*, 4 (2) (1973) 79.

83 V.P. Shaburova, I.G. Yudelevich, V.G. Torgov and I.L. Kotlyarevsky, *Zh. Anal. Khim.*, 26 (1971) 930.

84 I.G. Yudelevich, V.P. Shaburova, V.G. Torgov and O.I. Shcherbakova, *Zh. Anal. Khim.*, 28 (1973) 1049.

85 B.Ya. Spivakov, V.M. Shkinev and Yu.A. Zolotov, *Zh. Anal. Khim.*, 30 (1975) 2182.

86 Yu.A. Zolotov, B.Ya. Spivakov and V.M. Shkinev, in *2nd Nat. Conf. Anal. Chem. Int. Participation, Golden Sands-Varna, 1976*, Summaries of Papers, Bulgarian Academy of Sciences, p. 10.

87 B.Ya. Spivakov, V.M. Shkinev and Yu.A. Zolotov, in Proceedings of the *Int. Solvent Extraction Conference ISEC '80, Liège, 6–8 September, 1980*, Vol. 2, Paper No. 80–85.

88 V.M. Shkinev, I. Khavezov, B.Ya. Spivakov, S. Mareva, E. Ruseva, Yu.A. Zolotov and N. Iordanov, *Zh. Anal. Khim.*, 36 (1981) 896.

89 V.M. Shkinev, B.Ya. Spivakov, V.A. Orlova, T.M. Malyutina, T.I. Kirillova and Yu.A. Zolotov, *Zh. Anal. Khim.*, 33 (1978) 922.

90 Yu.V. Yakovlev and V.P. Kolotov, *Zh. Anal. Khim.*, 36 (1981) 1534.

91 V.M. Shkinev, B.Ya. Spivakov, G.A. Vorob'eva and Yu.A. Zolotov, *Anal. Chim. Acta*, 167 (1985) 145.

92 I.A. Blyum and Yu.A. Zolotov, *Zh. Anal. Khim.*, 31 (1976) 159.

93 T.A. Chanysheva, I.G. Yudelevich and E.A. Fedyanina, *Izv. Sib. Otd. Akad. Nauk SSSR, Ser. Khim. Nauk*, 2 (1) (1984) 92.

94 L.V. Zelentsova, I.G. Yudelevich and T.A. Chanysheva, *Izv. Sib. Otd. Akad. Nauk SSSR, Ser. Khim. Nauk*, 2 (1) (1982) 132.

95 A.B. Volynsky, A.I. Subochev, B.Ya. Spivakov, V.A. Slavny and Yu.A. Zolotov, *Zh. Anal. Khim.*, 36 (1981) 98.

96 T.R. Bangia, K.N.K. Kartha, M. Varghese, B.A. Dhawale and B.D. Joshi, *Fresenius' Z. Anal. Chem.*, 310 (1982) 410.

97 Yu.A. Zolotov, N.G. Vanifatova, T.A. Chanysheva and I.G. Yudelevich, *Zh. Anal. Khim.*, 32 (1977) 317.
98 G.A. Vall, L.P. Poddubnaya, N.G. Vanifatova, I.G. Yudelevich and Yu.A. Zolotov, *Zh. Anal. Khim.*, 35 (1980) 260.
99 I.V. Seryakova, G.A. Vorob'eva, A.V. Glembotsky and Yu.A. Zolotov, *Anal. Chim. Acta*, 77 (1975) 183.
100 G.A. Vall, *Zh. Anal. Khim.*, 40 (1985) 1049.
101 Yu.A. Zolotov, *Dokl. Akad. Nauk SSSR*, 180 (1968) 1367.
102 Yu.A. Zolotov and V.I. Golovanov, *Dokl. Akad. Nauk SSSR*, 191 (1970) 92.
103 Yu.A. Zolotov and V.I. Golovanov, *Dokl. Akad. Nauk SSSR*, 193 (1970) 626.
104 Yu.A. Zolotov and A.A. Prokoshev, *Zh. Anal. Khim.*, 28 (1973) 629.
105 Yu.A. Zolotov and Z.Kh. Sultanova, *Zh. Inorg. Khim.*, 18 (1973) 1055.
106 A.B. Sokolov and Yu.A. Zolotov, *Zh. Inorg. Khim.*, 17 (1972) 1123.
107 Yu.A. Zolotov and A.A. Prokoshev, *Zh. Anal. Khim.*, 26 (1971) 2307.
108 V.V. Bagreev, Yu.A. Zolotov, Yu.S. Tseryuta, L.M. Yudushkina, I.M. Kutuzov, C. Fischer and P. Mühl, in *Proceedings of the International Solvent Extraction Conference, Lyon, 8–14 September, 1974*, Vol. 2, Soc. Chem. Ind., London, 1974, p. 1883.
109 Yu.I. Popandopulo, V.V. Bagreev and Yu.A. Zolotov, *J. Inorg. Nucl. Chem.*, 39 (1977) 2257.
110 V.V. Bagreev, C. Fischer, L.M. Yudushkina and Yu.A. Zolotov, *J. Inorg. Nucl. Chem.*, 40 (1978) 553.
111 V.V. Bagreev, C. Fischer, L.M. Kardivarenko and Yu.A. Zolotov, *Polyhedron*, 1 (1982) 623.
112 Yu.I. Popandopulo and V.V. Bagreev, in *Theory and Practice of Extraction Methods*, Nauka, Moscow, 1985, p. 62 (in Russian).
113 N.M. Kuz'min and V.S. Vlasov, *Zh. Anal. Khim.*, 30 (1975) 1924.
114 Yu.A. Zolotov, V.V. Bagreev, P.P. Kish and I.I. Pogoida, *Zh. Anal. Khim.*, 30 (1975) 1692.
115 V.V. Bagreev, Yu.I. Popandopulo and Yu.A. Zolotov, *Zh. Anal. Khim.*, 39 (1984) 1349.
116 F.I. Lobanov, *Extraction of Inorganic Compounds with Melts, Results of Science and Engineering* (Inorganic Chemistry, Vol. 7), VINITI, Moscow, 1980 (in Russian).
117 F.I. Lobanov, *Zavod. Lab.*, 47 (1981) 1.
118 F.I. Lobanov, I.M. Yanovskaya and N.V. Makarov, *Usp. Khim.*, 52 (1983) 854.
119 T. Fujinaga, K. Kuwamoto and E. Nakayama, *Talanta*, 16 (1969) 1225.
120 T. Fujinaga, K. Kuwamoto, T. Yonekubo and M. Satake, *Jpn. Analyst*, 18 (1969) 1113.
121 T. Fujinaga, M. Satake and T. Yonekubo, *Jpn. Anal.*, 19 (1970) 216.
122 T. Fujinaga, M. Satake and T. Yonekubo, *Bull. Chem. Soc. Jpn.*, 48 (1975) 899.
123 T. Fujinaga, M. Satake and M. Shimizu, *Jpn. Anal.*, 25 (1976) 313.
124 T. Fujinaga and B.K. Puri, *Talanta*, 22 (1975) 71.
125 T. Fujinaga and B.K. Puri, *Indian J. Chem., Sec. A*, 14A (1976) 72.
126 F.I. Lobanov, A.G. Buyanovskaya and I.M. Gibalo, *Zh. Anal. Khim.*, 26 (1971) 1655.
127 F.I. Lobanov, V.A. Leonov and I.M. Gibalo, *Radiochem. Radioanal. Lett.*, 16 (1974) 201.
128 A.C. Gillet, *Fresenius' Z. anal. Chem.*, 247 (1969) 163.
129 A.C. Gillet, *Fresenius' Z. anal. Chem.*, 254 (1971) 35.
130 F.I. Lobanov, E.A. Terentyeva, I.M. Yanovskaya and N.V. Makarov, *Zavod. Lab.*, 49 (1983) 11.

131 T. Nagahiro, K. Uesugi, M. Satake and B.K. Puri, *Bull. Chem. Soc. Jpn.*, 58 (1985) 1115.
132 A. Kawase, S. Nakamura and N. Fudagawa, *Bunseki Kagaku*, 30 (1981) 229.
133 V.P. Zhivopistsev, I.N. Ponosov and I.A. Selezneva, *Zh. Anal. Khim.*, 18 (1963) 1432.
134 V.P. Zhivopistsev, K.I. Mochalov, T.P. Yakovleva and E.A. Selezneva, *Uchen. Zap. Permsk. Gos. Univ.*, No. 207 (1974) 17.
135 V.P. Zhivopistsev, B.I. Petrov, Yu.I. Makhnev and I.N. Ponosov, *Uchen. Zap. Permsk. Gos. Univ.*, No. 324 (1974) 230.
136 V.P. Zhivopistsev, Yu.A. Makhnev and M.I. Degtev, in *Pyrazolone Derivatives as Analytical Reagents, Physico-Chemical Analysis Methods*, Perm State University Press, Perm, p. 95, 1976 (in Russian).
137 B.I. Petrov and K.G. Galinova, *Zh. Anal. Khim.*, 33 (1978) 1481.
138 B.I. Petrov, K.G. Galinova and V.B. Margulis, *Zh. Anal. Khim.*, 40 (1985) 1781.
139 K. Murata and S. Ikeda, *Jpn. Anal.*, 18 (1969) 1137.
140 A. Massoumi and C.E. Hedrick, *J. Chem. Eng. Data*, 14 (1969) 52.
141 H. Kawamoto and H. Akaiwa, *Chem. Lett.*, No. 3 (1973) 259.
142 C.E. Matkovich and G.D. Christian, *Anal. Chem.*, 46 (1974) 102.
143 T.I. Zvarova, V.M. Shkinev, B.Ya. Spivakov and Yu.A. Zolotov, *Dokl. Akad. Nauk SSSR*, 273 (1983) 107.
144 T.I. Zvarova, V.M. Shkinev, G.A. Vorob'eva, B.Ya. Spivakov and Yu.A. Zolotov, *Mikrochim. Acta*, III (1984) 449.
145 V.M. Shkinev, N.P. Molochnikova, T.I. Zvarova, B.Ya. Spivakov, B.F. Myasoedov and Yu.A. Zolotov, *J. Radioanal. Nucl. Chem.*, 88 (1985) 115.
146 S.Yu. Ivakhno, A.V. Afanasyev and G.A. Yagodin, *Membrane Extraction of Inorganic Substances (Itogi Nauki i Tekhniki, Ser. Neorganicheskaya Khimiya)*, Vol. 13, VINITI, Moscow, 1985 (in Russian).
147 T. Braun and A.B. Farag, *Anal. Chim. Acta*, 99 (1978) 1.
148 G.J. Moody and J.D.R. Thomas, *Analyst (London)*, 104 (1979) 1.
149 T. Braun, *Zh. Anal. Khim.*, 39 (1984) 736.
150 T. Braun, J.D. Navratil and A.B. Farag, *Polyurethane Foam Sorbents in Separation Science and Technology*, CRC Press, Boca Raton, FL, 1986.
151 T. Braun, *Cellular Polymers*, 3 (1984) 81.
152 E.G. Korobeinikova, V.A. Luginin and I.A. Tserkovnitskaya, in *Problems of Modern Chemistry of Coordination Compounds*, Issue 7, Leningrad State University Press, Leningrad, p. 162, 1983 (in Russian).
153 T. Khondze, *Bunseki Kagaku*, 37 (1982) 203.
154 I.B. Pierce, *Anal. Chim. Acta*, 24 (1961) 146.
155 G.S. Katykhin, *Zh. Anal. Khim.*, 20 (1965) 615.
156 A.J.P. Martin and R.L.J. Synge, *Biochem. J.*, 35 (1941) 1358.
157 S.M. Mayer and E.R.J. Tomkins, *J. Am. Chem. Soc.*, 69 (1947) 2866.
158 E. Gluckauf, *Trans. Faraday Soc.*, 51 (1955) 34.
159 V.K. Markov, *Tr. Kom. Anal. Khim. Akad. Nauk SSSR*, 14 (1963) 99.
160 T.A. Kulbeda and N.M. Kuz'min, *Zh. Anal. Khim.*, 31 (1976) 1678.
161 N.M. Kuz'min, V.S. Vlasov and T.A. Bokova, *Zh. Anal. Khim.*, 27 (1972) 1807.
162 N.M. Zhukovsky, G.S. Katykhin, A.L. Martynov and M.K. Nikitin, *Radiokhimiya*, 10 (1968) 252.
163 B.K. Preobrazhensky, L.N. Moskvin, A.V. Kalyavin, O.M. Mieva and B.S. Usikov, *Radiokhimiya*, 10 (1968) 377.

164 G.S. Katykhin and M.K. Nikitin, *Radiokhimiya*, 10 (1968) 474.
165 T.A. Bolshova and E.N. Shapovalova, *Zh. Anal. Khim.*, 38 (1983) 1489.
166 G.G. Glukhov, L.A. Larionova and E.N. Gilbert, *Izv. Sib. Otd. Akad. Nauk SSSR, Ser. Khim. Nauk*, 7 (3) (1973) 87.
167 I.P. Alimarin, T.A. Bolshova, G.V. Prokhorova and E.N. Shapovalova, *Zh. Anal. Khim.*, 32 (1977) 650.
168 G.I. Shmanenkova, V.I. Shelkova and M.G. Zemskova, *Nauchn. Tr., Gos. Nauchno-Issled. Proektn. Inst. Redkomet. Promsti.*, 111 (1982) 68.
169 S.S. Grazhulene, B.K. Karandashev and Yu.V. Yakovlev, *Radiochem. Radioanal. Lett.*, 57 (1983) 273.
170 E.N. Gilbert, G.V. Verevkin, V.A. Yakhina and E.S. Gureeva, *Zh. Anal. Khim.*, 35 (1980) 656.
171 G.V. Veriovkin, E.N. Gilbert, A.M. Nemirovsky and E.G. Obrazovsky, *Zh. Anal. Khim.*, 36 (1981) 1073.
172 E.N. Gilbert, G.V. Veriovkin, B.N. Botchkaryov, A.A. Godovikov, V.Ya. Zhavoronkov and V.A. Mikhailov, *J. Radioanal. Chem.*, 26 (1975) 253.
173 E.N. Gilbert, G.V. Veriovkin, V.J. Semenov and V.A. Mikhailov, *J. Radioanal. Chem.*, 38 (1977) 229.
174 L.R.P. Butler, J.A. Brink and S.A. Engelrecht, *Trans. Inst. Min. Metall.*, 76 (1967) 188.
175 J. Rydberg, *Acta Chem. Scand.*, 23 (1969) 647.
176 F.D. Pierce, M.J. Gortatowski, H.D. Mecham and R.S. Fraser, *Anal. Chem.*, 47 (1975) 1132.
177 M. Jones, G.F. Kirbright, L. Ranson and T.S. West, *Anal. Chim. Acta*, 63 (1973) 210.
178 J.C. Kraak, *Trends Anal. Chem.*, 2 (1983) 183.
179 M. Gallego, C.M.D. de Luque and M. Valcárcel, *At. Spectrosc.*, 6 (1985) 15.
180 L. Nord and B. Karlberg, *Anal. Chim. Acta*, 125 (1981) 199.
181 M. Bengtsson and G. Johansson, *Anal. Chim. Acta*, 158 (1984) 147.
182 M. Vatanabe, T. Vatanabe, Yu. Horaoku and K. Hosiha, *Jap. Pat.* KL G01 No. 35/02, G01 No. 21/31, No. 56-37505, presented in 1976, No. 51-23782, published in 1981.
183 W. Rieman and H.F. Walton, *Ion Exchange in Analytical Chemistry*, Pergamon Press, Oxford, New York, Toronto, Sydney, Braunschweig, 1970.
184 M.P. Volynets, *Thin-Layer Chromatography in Inorganic Analysis*, Nauka, Moscow, 1974 (in Russian).
185 G.I. Malofeeva, G.V. Myasoedova and M.P. Volynets, *Mikrochim. Acta (Wien)*, No. 1 (1978) 391.
186 M.M. Senyavin, *Ion Exchange in Production Engineering and Analysis of Inorganic Substances*, Khimiya, Moscow, 1980 (in Russian).
187 O.V. Chashchina, *Ion-Exchange Concentration in Analysis of Trace Impurities*, Tomsk State University Press, Tomsk, 1980 (in Russian).
188 V.E. Bukhtmarov, *Ion-Exchange Methods in Analysis of Metals*, Metallurgiya, Moscow, 1982 (in Russian).
189 G. Schwedt, *Chromatographic Methods in Inorganic Analysis*, Dr. Alfred Hüthig Verlag, Heidelberg, Basel, New York, 1981.
190 M. Marhol, *Ion Exchangers in Analytical Chemistry, Their Properties and use in Inorganic Chemistry*, Academia, Prague, 1982.
191 G.V. Myasoedova and S.B. Savvin, *Chelate-Forming Sorbents*, Nauka, Moscow, 1984 (in Russian).

192 J. Inczedy, *Ion-Exchangers and Their Application*, Ellis Horwood, Chichester, 1986.
193 M.M. Senyavin, *Tr. Kom. Anal. Khim. Akad. Nauk SSSR*, 15 (1965) 311.
194 L.K. Liepin, *Usp. Khim.*, 9 (1940) 533.
195 M. Kimura, *Bunseki*, No. 77 (1981) 297.
196 E.V. Shugurova, *Cand. Sc. Dissertation*, Kazakh State University, Alma Ata, 1980 (in Russian).
197 J.M. Robinson, L. Rhodes and D.K. Wolcott, *Anal. Chim. Acta*, 78 (1975) 474.
198 J.L. Moyers and R.A. Duce, *Anal. Chim. Acta*, 69 (1974) 117.
199 M. Kimura, *J. Chem. Soc. Jpn., Chem. Ind. Chem.*, No. 1 (1981) 1.
200 I.A. Tarkovskaya, *Oxidized Coal*, Naukova Dumka, Kiev, 1981 (in Russian).
201 I.A. Tarkovskaya, K.I. Lazebnaya, M.I. Ovrutsky and A.N. Tomashevskaya, *Ukr. Khim. Zh.*, 42 (1976) 798.
202 E. Jackwerth, J. Lohmar and G. Wittlet, *Fresenius' Z. Anal. Chem.*, 270 (1974) 6.
203 E. Beinrohr and H. Berndt, *Mikrochim. Acta*, I (1985) 199.
204 H. Berndt and E. Jackwerth, *Fresenius' Z. Anal. Chem.*, 290 (1978) 369.
205 H. Berndt, *Arch. Eisenhuettenwes.*, 54 (1983) 503.
206 H. Berndt and J. Messerschmidt, *Fresenius' Z. Anal. Chem.*, 308 (1981) 104.
207 M. Doǧan and L. Elci, *Spectrochim. Acta, Part B*, 39B (1984) 1189.
208 B. Vanderborght, J. Verbuck and R. van Grieken, *Bull. Soc. Chim. Belg.*, 86 (1977) 23.
209 B.W. Vanderborght and R. van Grieken, *Anal. Chem.*, 49 (1977) 311.
210 M. Kimura, *Talanta*, 24 (1977) 194.
211 H. Berndt, E. Jackwerth and M. Kimura, *Anal. Chim. Acta,*. 93 (1977) 45.
212 E. Jackwerth, *Fresenius' Z. Anal. Chem.*, 271 (1974) 120.
213 H. Hashitani, M. Okumura and K. Fujinaga, *Fresenius' Z. Anal. Chem.*, 320 (1985) 774.
214 Cheng Jai-Kai, Lin Jin-Chum and Jiang Zu-Cheng, *Inorg. Chim. Acta*, 94 (1984) 154.
215 I. Sanemasa, E. Takagi, T. Deguchi and H. Nagai, *Anal. Chim. Acta*, 130 (1981) 149.
216 Z.I. Otmakhova, O.V. Chashchina and N.I. Slezko, *Zavod. Lab.*, 35 (1969) 685.
217 Z.I. Otmakhova, O.V. Chashchina, N.I. Slezko and G.A. Kataev, *Tr. Tomsk. Univ.*, No. 237 (1973) 75.
218 O.K. Tikhonova, Z.I. Otmakhova and O.V. Chashchina, *Zh. Anal. Khim.*, 28 (1973) 1288.
219 H. Zavanzka, D. Barałkiewicz and H. Elbanowska, *Chem. Anal. (PRL)*, 22 (1977) 913.
220 A.T. Kashuba and C.R. Hines, *Anal. Chem.*, 43 (1971) 1758.
221 K. Sarzanini, E. Mentasti, M.C. Gennaro and C. Baioechi, *Ann. Chim.*, 73 (1983) 385.
222 I.A. Shevchuk, N.P. Dovzhenko and Z.N. Kravtsova, *Ukr. Khim. Zh.*, 47 (1981) 773.
223 N.I. Slezko, O.V. Chashchina and A.G. Synkova, *Bull. Bismuth Inst. (Brussels)*, No. 18 (1977) 1.
224 I.P. Alimarin, V.K. Karandashev, Yu.V. Yakovlev and N.N. Dogadkin, *Zh. Anal. Khim.*, 36 (1981) 1510.
225 G. Küppers and G. Erdtmann, *Fresenius' Z. Anal. Chem.*, 307 (1981) 369.
226 Yu.A. Zolotov, O.A. Shpigun, L.A. Bubchikova and E.A. Sedel'nikova, *Dokl. Akad. Nauk SSSR*, 263 (1982) 889.
227 J. Slanina, F.P. Bakker, P.A.C. Jongejan, L. van Lamoen and J.J. Möls, *Anal. Chim. Acta*, 130 (1981) 1.
228 H.R. Hering, *Chelatbildende Ionenaustauscher*, Akademie-Verlag, Berlin, 1967.
229 G.V. Myasoedova and S.B. Savvin, *Zh. Anal. Khim.*, 37 (1982) 499.

230 P. Pakarinen, J. Pallon and R. Akkelsson, *Nucl. Instrum. Meth. Phys. Res.*, B231 (1984) 168.
231 J.W. Jones and Th.C. O'Haver, *Spectrochim. Acta, Part B*, 40B (1985) 263.
232 L.R. Hathaway and G.W. James, *Adv. X-Ray Anal.*, 20 (1977) 453.
233 R.E. van Grieken, C.M. Bresselurs and B.M. Vanderborght, *Anal. Chem.*, 49 (1977) 1326.
234 M. Grote and A. Kettrup, *Anal. Chim. Acta*, 172 (1985) 223.
235 I.I. Antokol'skaya, G.V. Myasoedova, L.I. Bolshakova, M.G. Ezernitskaya, M.P. Volynets, A.V. Karyakin and S.B. Savvin, *Zh. Anal. Khim.*, 31 (1976) 742.
236 G.V. Myasoedova, I.I. Antokol'skaya, O.P. Shvoeva, L.I. Bol'shakova and S.B. Savvin, *Talanta*, 23 (1976) 866.
237 F.I. Danilova, I.A. Fedotova and R.M. Nazarenko, *Zavod. Lab.*, 48 (1982) 9.
238 G.V. Myasoedova, I.I. Antokol'skaya, F.I. Danilova, I.A. Fedotova, V.P. Grin'kov, N.V. Ustinova, E.A. Naumenko, E.Ya. Danilova and S.B. Savvin, *Zh. Anal. Khim.*, 37 (1982) 1578.
239 V.A. Orobinskaya, N.E. Chulkova, R.M. Nazarenko, G.A. Dmitrieva, V.P. Ivanova, I.G. Shepet'ko, Z.A. Rogovin and M.O. Lishevskaya, in *Analysis and Technology of Noble Metals*, Metallurgiya, Moscow, 1971, p. 90 (in Russian).
240 M.S. Shakurova, I.I. Litvinskaya, S.E. Kashlinskaya and A.F. Karpova, *Zavod. Lab.*, 40 (1974) 1088.
241 F.I. Danilova, V.A. Orobinskaya, V.S. Parfenova, R.M. Nazarenko, V.G. Khitrov and G.E. Belousov, *Zh. Anal. Khim.*, 29 (1974) 2142.
242 Zs. Horváth, K. Falb and M. Varju, *Atom. Absorpt. Newsl.*, 16 (1977) 152.
243 Zs. Horváth, A. Lástity and O. Szakács, *Anal. Chim. Acta*, 173 (1985) 273.
244 Zs. Horváth, R.M. Barnes and P.S. Murty, *Anal. Chim. Acta*, 173 (1985) 305.
245 K.H. Lieser, H.-M. Röber and P. Burba, *Fresenius' Z. Anal. Chem.*, 284 (1977) 361.
246 K.H. Lieser, E. Breitwieser, P. Burba, M. Röber and R. Spatz, *Mikrochim. Acta*, I (1978) 363.
247 S. Imai, M. Muroi, A. Hamaguchi, R. Matsushita and M. Koyama, *Anal. Chim. Acta*, 113 (1980) 139.
248 M. Muroi, S. Imai and A. Hamaguchi, *Analyst (London)*, 110 (1985) 1083.
249 J. Smits and R.V. Grieken, *Anal. Chim. Acta*, 123 (1981) 9.
250 G.V. Myasoedova, M.P. Volynets, T.A. Koveshnikova and Yu.I. Belyaev, *Zh. Anal. Khim.*, 29 (1974) 2252.
251 M. Hiraide, S.P. Tillekeratne, K. Otsuka and A. Mizuike, *Anal. Chim. Acta*, 172 (1985) 215.
252 P. Minkkinen, *Finn. Chem. Lett.*, No. 4–5 (1977) 134.
253 C.J. Toussaint, G. Aina and F. Bo, *Anal. Chim. Acta*, 88 (1977) 193.
254 D. E. Leyden, G.H. Luttrell, W.K. Nonidez and D.B. Werho, *Anal. Chem.*, 48 (1976) 67.
255 D.E. Leyden, G.H. Luttrell, A.E. Sloan and N.J. De Angelis, *Anal. Chim. Acta*, 84 (1976) 97.
256 G.V. Lisichkin, G.V. Kudryavtsev and P.N. Nesterenko, *Zh. Anal. Khim.*, 38 (1983) 1684.
257 P. Sutthivaiyakit and A. Kettrup, *Anal. Chim. Acta*, 169 (1985) 331.
258 T. Seshardi and A. Kettrup, *Fresenius' Z. Anal. Chem.*, 310 (1982) 1.
259 T. Seshardi and A. Kettrup, *Fresenius' Z. Anal. Chem.*, 296 (1979) 247.
260 M. Grote, A. Schwalk and A. Kettrup, *Fresenius' Z. Anal. Chem.*, 313 (1982) 297.

187

261 F. Malamas, M. Bengtsson and G. Johansson, *Anal. Chim. Acta,* 160 (1984) 1.
262 Yu.A. Zolotov, O.M. Petrukhin, G.I. Malofeeva, E.V. Marcheva, O.A. Shiryaeva, V.A. Shestakov, V.G. Miskar'yants, V.I. Nefedov, Yu.I. Murinov and Yu.E. Nikitin, *Anal. Chim. Acta,* 148 (1983) 135.
263 V.A. Shestakov, G.I. Malofeeva, O.M. Petrukhin, E.V. Marcheva, N.K. Esenova, Yu.I. Murinov, Yu.E. Nikitin and Yu.A. Zolotov, *Zh. Anal. Khim.,* 38 (1983) 2131.
264 I.I. Nazarenko, I.V. Kislova, G.I. Malofeeva, O.M. Petrukhin, Yu.I. Murinov and Yu.A. Zolotov, *Zh. Anal. Khim.,* 40 (1985) 2129.
265 V.A. Shestakov, G.I. Malofeeva, O.M. Petrukhin, O.A. Shiryaeva, L.N. Kolonina, E.V. Marcheva, Yu.I. Murinov, G.G. Bikbaeva, Yu.E. Nikitin and Yu.A. Zolotov, *Zh. Anal. Khim.,* 36 (1981) 1784.
266 V.A. Shestakov, G.I. Malofeeva, O.M. Petrukhin, Yu.A. Zolotov, O.A. Shiryaeva, L.N. Kolonina, E.V. Marcheva, Yu.E. Nikitin and Yu.I. Murinov, in *Methods of Isolation and Determination of Noble Metals*; GEOKhI AN SSSR, Moscow, 1981, p. 47 (in Russian).
267 O.A. Shiryaeva, L.N. Kolonina, I.N. Vladimirskaya, G.I. Malofeeva, O.M. Petrukhin, Yu.A. Zolotov, E.V. Marcheva, Yu.I. Murinov and Yu.E. Nikitin, in *Methods of Isolation and Determination of Noble Metals*, GEOKhI AN SSSR, Moscow, 1981, p. 90 (in Russian).
268 V.P. Baluda, V.G. Miskaryants, L.A. Nikitina, O.A. Shiryaeva, L.N. Kolonina, I.N. Vladimirskaya, V.A. Shestakov, G.I. Malofeeva, O.M. Petrukhin, E.V. Marcheva, Yu.I. Murinov, Yu.E. Nikitin and Yu.A. Zolotov, *Zh. Anal. Khim.,* 37 (1982) 1569.
269 O.A. Shiryaeva, L.N. Kolonina, I.N. Vladimirskaya, V.A. Shestakov, G.I. Malofeeva, O.M. Petrukhin, E.V. Marcheva, Yu.I. Murinov, Yu.E. Nikitin and Yu.A. Zolotov, *Zh. Anal. Khim.,* 37 (1982) 281.
270 V.A. Shestakov, G.I. Malofeeva, O.M. Petrukhin, O.A. Shiryaeva, E.V. Marcheva, Yu.A. Murinov, Yu.E. Nikitin and Yu.A. Zolotov, *Zh. Anal. Khim.,* 39 (1984) 311.
271 E. Blasius, K.-P. Jansen, M. Keller, H. Lander, T. Nguyen-Tien and G. Scholten, *Talanta,* 27 (1980) 107.
272 E. Blasius, K.-P. Jansen, W. Adrian, W. Klein, H. Klotz, H. Luxenburger, E. Mernke, V.B. Nguyen, T. Nguyen-Tien, R. Rausch, J. Stockemer and A. Toussaint, *Talanta,* 27 (1980) 127.
273 E. Blasius and K.-P. Jansen, *Pure Appl. Chem.,* 54 (1982) 2115.
274 G. Knapp, B. Schreiber and R.W. Frei, *Anal. Chim. Acta,* 77 (1975) 293.
275 T. Yano, S. Ide, Y. Tobeta, H. Kobayashi and K. Ueno, *Talanta,* 23 (1976) 457.
276 K.M. Ol'shanova, V.P. Vasyutin and F.P. Paderina, in *2nd All-Union Conference on Concentration Methods Used in Analytical Chemistry*, Nauka, Moscow, 1977, p. 104 (in Russian).
277 T. Tanaka, K. Hiiro and A. Kawahara, *Jpn. Anal.,* 24 (1975) 460.
278 M.E. Vdovenko and M.L. Kaplan, *Zavod. Lab.,* 35 (1969) 179.
279 A. Chow and D. Buksak, *Can. J. Chem.,* 53 (1975) 1373.
280 T. Braun and E. Bujdoso, *CRC Crit. Rev. Anal. Chem.,* 13 (1982) 223.
281 T. Braun and A.B. Farag, *Anal. Chim. Acta,* 153 (1983) 319.
282 T. Braun, M.N. Abbas, A. Elek and L. Banos, *J. Radioanal. Chem.,* 67 (1981) 359.
283 A. Chow and D. Buksak, *Can. J. Chem.,* 53 (1975) 1373.
284 T. Braun and A.B. Farag, *Anal. Chim. Acta,* 65 (1973) 115.
285 H.D. Gesser and G.A. Horsfall, in *Recents progr. connais progr. phys. et chim. gallium et composés, Colloq., Marseille, 1976,* B8/1-B8/14.

286 T. Braun, A.B. Farag and M.P. Maloney, *Anal. Chim. Acta*, 93 (1977) 191.
287 V.P. Koryukova, L.I. Koval'chuk, E.V. Shabanov and A.M. Andrianov, in *2nd All-Union Conference on Concentration Methods Used in Analytical Chemistry*, Nauka, Moscow, 1977, p. 120 (in Russian).
288 Y. Chigetomi and T. Kojima, *J. Chem. Soc. Jpn., Chem. Ind. Chem.*, No. 8 (1977) 1242.
289 E.S. Solovyeva, A.P. Fedosov, G.N. Voronskaya, I.Ya. Nikolishin, T.N. Sopova and S.S. Korovin, in *2nd All-Union Conference on Concentration Methods Used in Analytical Chemistry*, Nauka, Moscow, 1977, p. 121 (in Russian).
290 G. Grancini, M.B. Stievano, F. Girardi, G. Guzzi and R. Pietra, *J. Radioanal. Chem.*, 34 (1976) 65.
291 K.H. Lieser, S. Quandt and B. Gleitsmann, *Fresenius' Z. Anal. Chem.*, 298 (1979) 378.
292 S. Aleksandrov, *Fresenius' Z. Anal. Chem.*, 321 (1985) 578.
293 F.D. Pierce and H.R. Broun, *Anal. Lett.*, 10 (1977) 685.
294 M.M. Guedes da Mota, F.D. Römer and B. Griepink, *Fresenius' Z. Anal. Chem.*, 277 (1977) 19.
295 I.H. El-Hag and A. Townshend, in *Abstr. Pap. Pittsburgh Conf. Anal. Chem. Appl. Spectrosc., Atlantic City, NJ, March 9–13, 1981*, Monroeville, PA, p. 622.
296 S.J. Parry, *Analyst (London)*, 105 (1980) 816.
297 Fang Zhaolun, Xu Shukun and Zhang Suchun, *Anal. Chim. Acta*, 164 (1984) 41.
298 Fang Zhaolun, Xu Shukun and Zhang Suchun, *Anal. Chem. (China)*, 12 (1984) 997.
299 P. Hernandes, L. Hernandes, J. Vicente and M.T. Sevilla, *An. Quim.*, B81 (1985) 117.
300 M.A. Marshall and H.A. Mottola, *Anal. Chem.*, 57 (1985) 729.
301 Fang Zhaolun, J. Růžička and E.H. Hansen, *Anal. Chim. Acta*, 164 (1984) 23.
302 S. Olsen, L.C.R. Pessenda, J. Růžička and E.H. Hansen, *Analyst (London)*, 108 (1983) 905.
303 S.D. Hartenstein, J. Růžička and G.D. Christian, *Anal. Chem.*, 57 (1985) 21.
304 C.W. McLeod, I.G. Cook, P.J. Worsfold, J.E. Davie and J. Queay, *Spectrochim. Acta, Part B*, 40B (1984) 57–62.
305 C.B. Ranger, *Am. Lab. (Fairfield, Conn.)*, 17 (1985) 92.
306 A.K. Babko, *Concise Encyclopedia of Chemistry*, Vol. 3, Sovetskaya entsiklopediya Press, Moscow, 1964, p. 784 (in Russian).
307 E. Jackwerth, in B. Sansoni (Editor), *Chemishe Multi-Elementahreicherung – Probleme der Anpassuhg an Problenmaterial und Bestimmungsmethode*, Instrumentelle Multielementanalyse, Verlag Chemie, Weinheim, 1985.
308 A.M. Ustinov and N.Ya. Chalkov, *Zavod. Lab.*, 37 (1971) 931.
309 L.L. Baranova, B.Ya. Kaplan, M.G. Nazarova and L.S. Razumova, *Zavod. Lab.*, 51 (1985) 25.
310 E. Jackwerth, *Pure Appl. Chem.*, 51 (1979) 1149.
311 S. Szwabski and J. Dobrowolski, *Chem. Anal. (PRL)*, 18 (1973) 937.
312 V.G. Tiptsova-Yakovleva, A.G. Dvortsan and I.B. Semenova, *Zh. Anal. Khim.*, 30 (1975) 1577.
313 E. Jackwerth, J. Lohmar and G. Schwark, *Fresenius' Z. Anal. Chem.*, 260 (1972) 101.
314 K. Hirokawa, in *Colloq. Spectroscopic. Int. 16, Heidelberg, 1971, Vorabdr., Bd. 1*, London, 1971, p. 369.
315 E. Jackwerth and P.G. Willmer, *Talanta*, 23 (1976) 197.
316 Y. Yamamoto, H. Yamagushi and S. Ueda, *J. Chem. Soc. Jpn, Chem. Ind. Chem.*, No. 1 (1975) 78.
317 B.Ya. Kaplan, Z.P. Kirillova, Yu.I. Merisov, M.G. Nazarova, E.I. Petrova and G.S. Skripkin, *Zavod. Lab.*, 40 (1974) 256.

189

318 I.N. Bykova and T.G. Manova, *Zavod. Lab.*, 38 (1972) 178.
319 I.N. Bykova, T.G. Manova, V.G. Silakova and G.P. Boznyakova, *Zh. Anal. Khim.*, 28 (1973) 1481.
320 R.V. Moore, *Anal. Chem.*, 54 (1982) 895.
321 P. Marijanović, J. Makjanić and V. Valković, *J. Radioanal. Nucl. Chem., Art.*, 81 (1984) 353.
322 G.L. Everett, *Anal. Proc.*, 19 (1982) 88.
323 I.V. Melikhov and M.S. Merkulova, *Sokristallizatsiya (Co-crystallization)*, Khimiya, Moscow, 1975 (in Russian).
324 N.V. Kovaleva and V.T. Chuiko, *Zh. Anal. Khim.*, 28 (1973) 1985.
325 V.T. Chuiko, A.A. Kravtsova and N.V. Kovaleva, *Zh. Anal. Khim.*, 27 (1972) 805.
326 E. Jackwerth, E. Doring and J. Lohmar, *Fresenius' Z. Anal. Chem.*, 253 (1971) 195.
327 V.T. Chuiko, A.A. Kravtsova and N.V. Kovaleva, *Zh. Anal. Khim.*, 27 (1972) 703.
328 N.A. Rudnev and G.I. Malofeeva, *Tr. Kom. Anal. Khim., Akad. Nauk SSSR*, 15 (1965) 224.
329 V.T. Chuiko, *Main Methods of Concentrating Traces of Metals by Coprecipitation, Their Theoretical Foundations and Application in Chemical Analysis, D.Sc. degree thesis*, Saratov State University, 1962 (in Russian).
330 N.A. Rudnev, G.I. Malofeeva, A.M. Tuzova, V.N. Pavlova, I.M. Rodionova and G.S. Pavlova, in *2nd All-Union Conference on Concentration Methods Used in Analytical Chemistry, Abstracts of Reports*, Nauka, Moscow, 1977, p. 182 (in Russian).
331 A.I. Novikov, in *2nd All-Union Conference on Concentration Methods Used in Analytical Chemistry, Abstracts of Reports*, Nauka, Moscow, 1977, p. 185 (in Russian).
332 V.I. Plotnikov and I.I. Safonov, in *2nd All-Union Conference on Concentration Methods Used in Analytical Chemistry, Abstracts of Reports*, Nauka, Moscow, 1977, p. 187 (in Russian).
333 N.A. Rudnev, L.I. Pavlenko, G.I. Malofeeva and L.V. Simonova, *Zh. Anal. Khim.*, 24 (1969) 1223.
334 M.P. Lebedinskaya and V.T. Chuiko, *Zh. Anal. Khim.*, 28 (1973) 863.
335 H. Uchida, K. Iwasaki, K. Tanaka and Ch. Iida, *Anal. Chim. Acta.* 134 (1982) 375.
336 J.D. Mullen, *Talanta*, 23 (1976) 846.
337 G.M. Kolesov, *Chem. anal. (PRL)*, 20 (1975) 497.
338 V.G. Tiptsova-Yakovleva, A.G. Dvortsan and I.B. Semenova, *Zh. Anal. Khim.*, 25 (1970) 686.
339 V.M. Ivanov, A.I. Busev, L.I. Smirnova and Zh. I. Nemtseva, *Zh. Anal. Khim.*, 25 (1970) 1149.
340 W. Yoshimura, *Jpn. Anal*, 25 (1976) 726.
341 Y. Harada, *Bunseki Kagaku*, 31 (1982) 130.
342 A.S. Buchanan and P. Hannaker, *Anal. Chem.*, 56 (1984) 1379.
343 T.M. Reymont and R.J. Dubois, *Anal. Chim. Acta*, 56 (1971) 1.
344 L.I. Pavlenko, G.I. Malofeeva, L.V. Simonova and O.Yu. Andreichenko, *Zh. Anal. Khim.*, 29 (1974) 1122.
345 R. Iesuè and C. Taroli, *Rass. Chim.*, 30 (1978) 75.
346 E.S. Zolotovitskaya, A.B. Blank, L.E. Belenko and L.V. Glushkova, *Zh. Anal. Khim.*, 36 (1981) 1518.
347 B. Magyar and G. Kauffmann, *Talanta*, 22 (1975) 267.
348 F.A. Chmilenko, V.S. Smirnaya and V.T. Chuiko, *Zh. Anal. Khim.*, 28 (1973) 2176.
349 A. Hirose and D. Ishii, *J. Chem. Soc. Jpn., Chem. Ind. Chem.*, No. 10 (1972) 1996.

350 A. Hirose and D. Ishii, *J. Chem. Soc. Jpn., Chem. Ind. Chem.*, 20 (1974) 17.
351 B.Ya. Kaplan and O.A. Shiryaeva, *Nauchn. Tr. Gos. Nauchno-Issled. Proektn. Inst. Redkomet. Promsti.*, 71 (1976) 49.
352 H. Berndt and E. Jackwerth, *Fresenius' Z. Anal. Chem.*, 283 (1977) 15.
353 E. Jackwerth, G. Graffmann and J. Lohmar, *Fresenius' Z. Anal. Chem.*, 247 (1969) 149.
354 V.I. Kuznetsov, *Zh. Anal. Khim.*, 9 (1954) 199.
355 V.I. Kuznetsov and G.V. Myasoedova, in: *Application of Labelled Atoms in Analytical Chemistry*, GEOKhI AN SSSR, Moscow, 1955, p. 24 (in Russian).
356 V.I. Kuznetsov, *Tr. Kom. Anal. Khim. Akad. Nauk SSSR*, 6(9) (1955) 249.
357 V.I. Kuznetsov and G.V. Myasoedova, *Tr. Kom. Anal. Khim. Akad. Nauk SSSR*, 9(12) (1958) 76.
358 V.I. Kuznetsov, *Tr. Kom. Anal. Khim. Akad. Nauk SSSR*, 15 (1965) 279.
359 G.V. Myasoedova, *Zh. Anal. Khim.*, 21 (1966) 598.
360 V.I. Kuznetsov and T.G. Akimova, *Coprecipitation Concentration of Actinides with Organic Coprecipitants*, Atomizdat, Moscow, 1968 (in Russian).
361 V.I. Kuznetsov and G.V. Myasoedova, in *Materials of the Conference on Problems Concerning Manufacture and Application of Indium, Gallium and Thallium*, Part I, GIREDMET, Moscow, 1960, p. 165 (in Russian).
362 V.I. Kuznetsov and V.V. Gorshkov, *Radiokhimiya*, 5 (1963) 93.
363 W.P. Tappmeyer and E.E. Pickett, *Anal. Chem.*, 34 (1962) 1709.
364 V.I. Kuznetsov and G.V. Myasoedova, in *Application of Photometric Methods of Analysis in Materials Quality Control, Materials of the Seminar*, Mosk. Dom Nauchno-Tekhnich. Propagandy im. F.E. Dzerzhinskogo, Moscow, issue 1, 1963, p. 16 (in Russian).
365 V.I. Kuznetsov, V.V. Gorshkov, T.G. Akimova and I.V. Nikavskaya, *Tr. Kom. Anal. Khim. Akad.Nauk SSSR*, 15 (1965) 296.
366 V.I. Shlenskaya, V.P. Khvostova and V.I. Bulgakova, *Zh. Anal. Khim.*, 29 (1974) 314.
367 T.M. Moroshkina and G.G. Savinova, *Zh. Anal. Khim.*, 24 (1969) 1165.
368 M.V. Virtsavs, O.E. Veveris and Yu.A. Bankovsky, in *2nd All-Union Conference on Concentration Methods Used in Analytical Chemistry, Abstracts of Reports*, Nauka, Moscow, 1977, p. 202 (in Russian).
369 E. Scheubeck, *Mikrochim. Acta*, 2 (1980) 283.
370 K.V. Krishnamurty and M.M. Reddy, *Anal. Chem.*, 49 (1977) 222.
371 Y. Yamamoto, M. Sugita and K. Ueda, *Bull. Chem. Soc. Jpn.*, 55 (1982) 742.
372 G.S. Caravajal, K.I. Mahan and D.E. Leyden, *Anal. Chim. Acta*, 135 (1982) 205.
373 C.L. Smith, M. Motooka and R. Willson, *Anal. Lett.*, A15 (1984) 1715.
374 B.K. Puri, A.K. Gupta, M. Katyal and M. Satake, *Int. Lab.*, Nov.–Dec. (1985) 60.
375 M.F. Hussain, R.K. Bansal and B.K. Puri, *Analyst (London)*, 110 (1985) 1131.
376 M.C. Mehra, B.K. Puri, K. Iwasaka and M. Satake, *Analyst (London)*, 110 (1985) 791.
377 V.A. Nazarenko and I.M. Grekova, *Zh. Anal. Khim.*, 31 (1976) 2137.
378 H. Bem and D.E. Ryan, *Anal. Chim. Acta*, 166 (1984) 189.
379 V.V. Gorshkov, L.I. Mekhryusheva, L.A. Smakhtin, I.A. Frolov and V.S. Zaburdyaev, *Tr. Kom. Khim. Khim. Tekhnol.*, No. 1 (40) (1975) 100.
380 M.N. Chelnokova, A.I. Busev and T.M. Kosareva, *Zh. Anal. Khim.*, 36 (1981) 230.
381 Kh.Z. Brainina, *Stripping Voltammetry of Solid Phases*, Khimiya, Moscow, 1972 (in Russian).
382 K. Hiiro and A. Kawahara, *Jpn. Anal.*, 23 (1974) 1075.

383 W. Lund, Y. Thomassen and P. Dule, *Anal. Chim. Acta*, 93 (1977) 53.
384 W. Lund and B. Larsen, *Anal. Chim. Acta*, 70 (1974) 299.
385 G. Batley and J. Matcusec, *Anal. Chem.*, 49 (1977) 2031.
386 F.O. Jensen, J. Dolezal and F.J. Langmyhr, *Anal. Chim. Acta*, 72 (1974) 245.
387 V.Z. Krasil'shchik and T.G. Manova, *Zh. Anal. Khim.*, 30 (1975) 971.
388 V.Z. Krasil'shchik and G.A. Shteinberg, *Zh. Anal. Khim.*, 26 (1971) 1903.
389 V.Z. Krasil'shchik, T.G. Manova, L.K. Raginskaya and O.A. Kuznetsova, *Zavod. Lab.*, 39 (1973) 935.
390 V.Z. Krasil'shchik and G.A. Shteinberg, *Tr. Vses. Nauchno-Issled. Inst. Khim. Reakt. Osobo Chist. Khim. Veshchestv*, No. 40 (1978) 160.
391 V.Z. Krasil'shchik and A.F. Yakovleva, in *Obtaining and Analysing High-Purity Substances*, Gorky State University Press, Gorky, 1974, p. 147 (in Russian).
392 V.Z. Krasil'shchik and A.F. Yakovleva, *Tr. Vses. Nauchno-Issled. Inst. Khim. Reakt. Osobo Chist. Khim. Veshchestv*, No. 37 (1975) 183.
393 Y. Takata, K. Hirota and K. Sakai, *Jpn Anal.*, 24 (1975) 703.
394 Y. Thomassen, B. Larsen, F. Langmyhr and W. Lund, *Anal. Chim. Acta*, 83 (1976) 103.
395 G. Voland, P. Tschöpel and G. Tölg, *Anal. Chim. Acta*, 90 (1977) 15.
396 A. Mizuike, N. Mitsuya and K. Yamagai, *Bull. Chem. Soc. Jpn.*, 1 (1969) 253.
397 R. Neeb, *Inverse Polarographie und Voltammetrie*, Akademie-Verlag, Berlin, 1969.
398 V.Z. Krasil'shchik, A.F. Yakovleva and G.A. Shteinberg, *Tr. Vses. Nauchno-Issled. Inst. Khim. Reakt. Osobo Chist. Khim. Veshchestv*, No. 34 (1972) 156.
399 V.Z. Krasil'shchik, T.G. Manova and A.F. Yakovleva, *Tr. Vses. Nauchno-Issled. Inst. Khim. Reakt. Osobo Chist. Khim. Veshchestv*, No. 35 (1973) 171.
400 Y. Hoshino, T. Utsunomija and K. Fukui, *Jpn. Chem. Ind. Chem.*, 6 (1977) 808.
401 V.Z. Krasil'shchik, T.G. Manova and N.I. Kuznetsova, *Tr. Vses. Nauchno-Issled. Inst. Khim. Reakt. Osobo Chist. Khim. Veshchestv*, No. 39 (1974) 154.
402 V.Z. Krasil'shchik, T.G. Manova and T.I. Markovskaya, *Tr. Vses. Nauchno-Issled. Inst. Khim. Reakt. Osobo Chist. Khim. Veshchestv*, No. 37 (1975) 194.
403 I.G. Yudelevich, N.F. Zakharchuk, G.A. Kalambet and L.S. Kletskina, *Izv. Sib. Otd. Akad. Nauk SSSR, Ser. Khim. Nauk*, No. 14, issue 2 (1972) 97.
404 V.Z. Krasil'shchik and A.F. Yakovleva, *Metody Anal. Khim. Reakt. Prep.*, No. 21 (1973) 102.
405 E.D. Salin and M. Habib, *Anal. Chem.*, 56 (1984) 1186.
406 Kh. Z. Brainina and E.Ya. Neiman, *Solid-Phase Reactions in Electroanalytical Chemistry*, Khimiya, Moscow, 1982 (in Russian).
407 F. Vidra, K. Stulik and E. Julakova, *Spectrochemical Stripping Analysis*, Ellis Horwood, Chichester, 1976.
408 A.G. Stromberg, A.A. Kaplin and N.P. Pikula, *Zavod. Lab.*, 43 (1977) 385.
409 B.Ya. Kaplan, R.G. Pats and R.M.-F. Salikhdzhanova, *AC Voltammetry*, Khimiya, Moscow, 1985 (in Russian).
410 V.Z. Krasil'shchik and A.F. Yakovleva, *Tr. Vses. Nauchno-Issled. Inst. Khim. Reakt. Osobo Chist. Khim. Veshchestv*, No. 33 (1971) 152.
411 V.Z. Krasil'shchik, N.M. Kuz'min and E.Ya. Neiman, *Zh. Anal. Khim.*, 34 (1979) 2045.
412 N.I. Katalevsky and V.Ya. Eremenko, *Gidrokhim. Mater.*, 61 (1974) 110.
413 R. Woodgriff and D. Siemer, *Appl. Spectrom*, 23 (1969) 38.
414 J. Dawson, D. Ellis, T. Hartley and M. Evans, *Analyst (London)*, 99 (1974) 602.

415 H. Heinrichs, *Fresenius' Z. Anal. Chem.*, 273 (1975) 197.
416 C. Fairless and A.J. Bord, *Anal. Lett.*, 5 (1972) 433.
417 V.Z. Krasil'shchik, A.F. Yakovleva and G.A. Shteinberg, *Zh. Anal. Khim.*, 26 (1971) 1897.
418 V.Z. Krasil'shchik, T.G. Manova and N.I. Kuznetsova, *Zh. Anal. Khim.*, 32 (1977) 837.
419 I.A. Berezin, *Zavod. Lab.*, 34 (1968) 1318.
420 S. Dogan and W. Haerdi, *Anal. Chim. Acta*, 76 (1975) 345.
421 J.J.A. Boslett, A. John, R.L.R. Towns, R.G. Megarle and K.H. Pearson, *Anal. Chem.*, 49 (1977) 1734.
422 E. Ando and M. Tanabe, *Shimadzu Rev.*, 30 (1973) 63.
423 A. Sobkowska and M. Balinska, *Microchim. Acta*, 2 (1975) 227.
424 E. Stitz and J. Simmer, *Anal. Chem.*, 26 (1954) 304.
425 F.L. Babina, A.G. Karabash, Sh.I. Peizulaev and E.F. Semenov, *Zh. Anal. Khim.*, 20 (1965) 501.
426 V.P. Gladyshev, D.P. Synkova, R.Sh. Enikeev, N.A. Kucherenko and V.V. Voiloko-va, *Tr. Kom. Anal. Khim. Akad. Nauk SSSR*, 15 (1965) 213.
427 H. Jaskolska, L. Rowińska and M. Radwan, *J. Radioanal. Chem.*, 20 (1974) 419.
428 A.L. Lipchinsky, *Zh. Anal. Khim.*, 13 (1958) 402.
429 Kh.Z. Brainina and T.A. Krapivkina, *Zh. Anal. Khim.*, 22 (1967) 1382.
430 Kh.Z. Brainina and E.Ya. Sapozhkova, *Zh. Anal. Khim.*, 21 (1966) 807.
431 E.Ya. Neiman, L.G. Petrova and V.I. Ignatov, *Zh. Anal. Khim.*, 33 (1978) 1496.
432 E.Ya. Neiman and Kh.Z. Brainina, *Zh. Anal. Khim.*, 30 (1975) 1073.
433 L.I. Lipchinskaya, M.S. Zakharov and V.V. Pnev, in *2nd All-Union Conference on Concentration Methods Used in Analytical Chemistry*, Nauka, Moscow, 1977, p. 139 (in Russian).
434 E.Ya. Naiman and G.M. Dolgopolova, *Zh. Anal. Khim.*, 35 (1980) 976.
435 Kh.Z. Brainina, L.S. Fokina and N.D. Fedorova, *Zh. Anal. Khim.*, 33 (1978) 225.
436 K.S. Parubochnaya, V.V. Pnev and M.S. Zakharov, in *2nd All-Union Conference on Concentration Methods Used in Analytical Chemistry*, Nauka, Moscow, 1977, p. 140 (in Russian).
437 A.A. Kaplin, T.I. Khakhanina, V.G. Voevodin, V.S. Morozov, A.I. Zinovyev and G.E. Tkachenko, in *Electrochemical Methods of Analysis, Abstracts of Reports Presented at 2nd All-Union Conference on Electrochemical Methods of Analysis, Tomsk, June 4–6, 1985*, Part I, Tomsk, 1985, p. 52 (in Russian).
438 V.A. Kosukhin and A.V. Bashkatov, ibid., p. 58.
439 H. Gunasingham, K.P. Ang, C.C. Ngo, P.C. Thiak and B. Fleet, *J. Electroanal. Chem.*, 186 (1985) 51.
440 A.W. Mann and M.J. Lintern, *Aust. Water Resour. Counc. Tech. Pap.*, No. 83, X (1984) 1.
441 J. Wang, *Am. Lab. (Fairfield, Conn.)*, 15 (1983) 14, 16 and 20.
442 T. Yamada, S. Okazaki and T. Fujinaga, *Bull. Inst. Chem. Res., Kyoto Univ.*, 56 (1978) 225.
443 R. Nakata, S. Okazaki and T. Fujinaga, *J. Chem. Soc. Jpn, Chem. Ind. Chem.*, No. 10 (1980) 1615.
444 V.V. Tenkovtsev, *Zavod. Lab.*, 21 (1955) 525.
445 S.A. Tarafdar and M. Rahman, *Fresenius' Z. Anal. Chem.*, 316 (1983) 715.
446 B. Fu, A.M. Ure and T.S. West, *Anal. Chim. Acta*, 152 (1983) 95.

193

447 S.L. Dobychin and V.B. Aleskovsky, *Tr. Leningr. Tekhnol. Inst. im. Lensoveta*, No. 48 (1958) 22, 34, 45, 49.
448 J. Czakov, in *Proc. Colloq. Spectrosc. Int. 19th*, (1967) 219.
449 M.T. Kozlovsky, *Tr. Kom. Anal. Khim. Akad. Nauk SSSR*, 15 (1965) 132.
450 V.A. Tsimergakl and R.S. Khaimovich, *Zavod. Lab.*, 14 (1948) 1289 and 1313.
451 M.V. Virtsavs and O.E. Veveris, *Activation Analysis*, Zinatne, Riga, 1976, p. 53 (in Russian).
452 V.P. Gladyshev, in *Uspekhi Polyarografii s Nakopleniem (Successes of polarography with Accumulation), Materials of the All-Union Conference Amalgam Polarography with Accumulation and its Application in Scientific Studies*, Tomsk University Press, Tomsk, 1973, p. 184 (in Russian).
453 V.F. Toropova, Yu.N. Polyakova and L.N. Soboleva, *Zh. Anal. Khim.*, 30 (1975) 1881.
454 A. Mizuike, K. Fukuda and T. Sakamoto, *Bull. Chem. Soc. Jpn.*, 46 (1973) 3596.
455 V.F. Toropova, Yu.N. Polyakov and L.N. Sobolev, *Zh. Anal. Khim.*, 27 (1972) 2395.
456 V.F. Toropova, Yu.N. Polyakov and E.A. Naumova, in *2nd All-Union Conference on Concentration Methods Used in Analytical Chemistry*, Nauka, Moscow, 1977, p. 146 (in Russian).
457 V.F. Toropova, Yu.N. Polyakov and G.N. Zhdanova, *Zh. Anal. Khim.*, 38 (1983) 238.
458 N.M. Kuz'min and M.S. Matyukhina, *USSR Pat.* 160900, class 42L, 353 MPK G01 n, published in 1964, *Bull. Izob.* No. 5 (1964).
459 W. Koch, *Metallkundliche Analyse*, Verlag Chemie, Düsseldorf, 1965.
460 *Modern Methods of State Analysis of Steels*, AGNE, Tokyo, 1979.
461 E. Jackwerth and A. Kulok, *Fresenius' Z. Anal. Chem.*, 257 (1971) 28.
462 E. Jackwerth, F. Doring, J. Lohmar and G. Schark, *Fresenius' Z. Anal. Chem.*, 260 (1972) 177.
463 R. Höhn and E. Jackwerth, *Spectrochim. Acta, Part B*, 29B (1974) 225.
464 E. Jackwerth and J. Messerschmidt, *Fresenius' Z. Anal. Chem.*, 274 (1975) 205.
465 E. Jackwerth and J. Messerschmidt, *Anal. Chim. Acta*, 87 (1976) 341.
466 R. Höhn and E. Jackwerth, *Fresenius' Z. Anal. Chem.*, 282 (1976) 21.
467 E. Jackwerth, J. Messerschmidt and R. Höhn, *Anal. Chim. Acta*, 94 (1977) 225.
468 R. Höhn and E. Jackwerth, *Fresenius' Z. Anal. Chem.*, 289 (1978) 47.
469 Sun-Tak Hwang and K. Kammermeyer, *Membranes in Separation*, Wiley, New York, 1975.
470 Yu.I. Dytnersky, *Membrane Processes of Separating Liquid Mixtures*, Khimiya, Moscow, 1975 (in Russian).
471 V.P. Dubyaga, L.P. Perepechkin and E.E. Katalevsky, *Polymer Membranes*, Khimiya, Moscow, 1981 (in Russian).
472 L.Kh. Khaeva, V.A. Zarinsky and V.A. Ryabukhin, *Zh. Anal. Khim.*, 29 (1974) 66.
473 L.Kh. Khaeva, V.A. Ryabukhin and V.A. Zarinsky, *Zh. Anal. Khim.*, 29 (1974) 2023.
474 L.N. Moskvin, G.G. Martysh and Yu.L. Chereshkovich, *Zh. Prikl. Khim.*, 50 (1977) 430.
475 J. Cox and R. Carlson, *Anal. Chim. Acta*, 130 (1981) 313.
476 L.N. Moskvin, N.N. Kalinin and L.A. Godon, *Atom. Energ.*, 36 (1974) 198.
477 L.N. Moskvin, N.N. Kalinin and L.A. Godon, *Atom. Energ.*, 39 (1975) 94.
478 L.N. Moskvin, N.N. Kalinin, L.A. Godon, S.B. Tomilov and G.I. Kizym, *Zh. Anal. Khim.*, 31 (1976) 2396.
479 L.N. Moskvin, N.N. Kalinin and L.A. Godon, *Zh. Anal. Khim.*, 32 (1977) 1899.
480 R. Bock, *Methoden der Analytischen Chemie, Eine Einführung, Band 1, Trennungsmethoden*, Verlag Chemie, Weinheim, 1974.

481 G.J. Moody and Y.D.R. Thomas, in J.G. Cook (Editor), *Practical Electrophoresis*, Merrow, Durnham, 1975.
482 A.V. Stepanov and E.K. Korchemnaya, *Electromigration Method in Inorganic Analysis*, Khimiya, Moscow, 1979 (in Russian).
483 L.P. Bochkova, A.V. Stepanov, E.K. Korchemnaya and A.N. Ermakov, *Zh. Anal. Khim.*, 39 (1984) 142.
484 L. Ribeiro, E. Mitidieri and O.R. Affonso, *Paper Electrophoresis, A review of methods and results*, Elsevier, Amsterdam, 1961.
485 E. Blasius and T. Ehrhardt, *Talanta*, 26 (1979) 713.
486 K. Aitzetmüller, K. Buchtela and F. Hecht, *Mikrochim. Acta*, 1101 (1966) 1164.
487 V.A. Mikhailov and M.V. Korneivich, *Zh. Anal. Khim.*, 27 (1972) 2184.
488 V.A. Mikhailov, M.V. Kornievich and D.D. Ochironapova, *Tr. Khim. Khim. Tekhnol. (Gorky)*, No. 4 (35) (1973) 86.
489 V.A. Mikhailov and D.D. Bogdanova, *Electron Transfer in Liquid Membranes. Theory and Application*, Nauka, Novosibirsk, 1978 (in Russian).
490 I. Yudelevich and I. Shelpakova, in *6th Int. Symp. High-Purity Mater. Sci. and Technol., Dresden, May 6–10, 1985, Proc. 2: Characterization*, Oberlungwitz, 1985, p. 268.
491 K. Bächmann, *Talanta*, 29 (1982) 1.
492 T. Nakahara, *Prog. Anal. At. Spectrosc.*, 6 (1983) 163.
493 V.Em. Sachini, M. Craiu and E. Ivana, *Rev. Roum. Chim.*, 19 (1974) 165.
494 F. Dolinšek and J. Stupar, *Analyst (London)*, 98 (1973) 841.
495 J.D. Sheaffer, G. Mulvey and R.K. Skogerboe, *Anal. Chem.*, 50 (1978) 1239.
496 C.J. Pickford and G. Rossi, *Analyst (London)*, 97 (1972) 647.
497 G. Torok, L.G. Nagy and J. Ruip, in *Abstr. Pap. Pittsburgh Conf. Anal. Chem. Appl. Spectrosc., Atlantic City, NJ, 1980*, Pittsburgh, PA, 1980, p. 703.
498 N.M. Kuz'min, *Zh. Anal. Khim.*, 22 (1967) 451.
499 N.A. Kershner, E.F. Joy and A.J. Barnard, *Appl. Spectrosc.*, 25 (1971) 542.
500 Z. Kuzma, M. Oldak, J. Rzeszotarska and B. Zawadzki, *Chem. Anal. (PRL)*, 18 (1973) 447.
501 N.V. Larin, V.N. Shishov, V.A. Krylov and E.N. Mishina, *Zh. Anal. Khim.*, 31 (1976) 2193.
502 G.G. Devyatykh, V.N. Shishov, V.G. Pimenov, A.N. Egorochkin and Yu.V. Revin, *Zh. Anal. Khim.*, 33 (1978) 464.
503 S.H. Harrison, P.D. La Fleure and W.H. Zoller, *U.S. Dep. Commer. Nat. Bur. Stand. Spec. Publ.*, No. 492 (1977) 148.
504 K.H. Lieser, W. Calmano, E. Heuss and V. Neitzert, *J. Radioanal. Chem.*, 37 (1977) 717.
505 M. Takahashi, Z. Yoshida, H. Aoyagi and K. Izawa, *Mikrochim. Acta*, 2 (1974) 329.
506 M.D. Erickson, M.T. Giguere and D.A. Whitaker, *Anal. Lett.*, 14A (1981) 841.
507 Sh.I. Peizulaev, *Spectral Methods for Analytical Control of High-Purity Substances with Preliminary Concentration, D.Sc. degree thesis*, Moscow State University, Moscow, 1969.
508 V.M. Barinov and T.K. Aidarov, *Methods of Analysis of High-Purity Substances*, Nauka, Moscow, 1965, p. 402 (in Russian).
509 Z.G. Fratkin, M.I. Volokhova and N.G. Polivanova, *Zavod. Lab.*, 27 (1961) 846.
510 J.H. Yoe and H.J. Koch (Editors), *Trace Analysis*, Wiley, New York, Chapman & Hall, London, 1957, p. 637.

511 S.K. Ng and P.T. Lai, *Appl. Spectrosc.*, 26 (1972) 369.
512 F.M. Farhan and H. Pazanden, *Analusis*, 3 (1975) 201.
513 G.I. Postogvard and V.D. Fedtsova, *Zavod. Lab.*, 50 (1984) 37.
514 L.B. Gorbunova, A.F. Kuteinikov, M.A. Avdeenko and V.N. Murashkina, *Zavod. Lab.*, 41 (1975) 178.
515 B.D. Joshi, T.R. Bangia and A.G.I. Dalvi, *Microchim. Acta*, 5 (1974) 829.
516 J.E. Patterson, *Anal. Chem.*, 51 (1979) 1087.
517 M.A. Volodina, T.A. Gorshkova, A.S. Arutyunova and T.M. Repenyuk, *Zh. Anal. Khim.*, 36 (1981) 144.
518 E.V. Williams, *Analyst (London)*, 107 (1982) 1006.
519 J.T. Brenna and G.H. Morrisson, *Anal. Chem.*, 56 (1984) 2791.
520 G. Knapp, *Trends Anal. Chem.*, 3 (1984) 182.
521 G. Kaiser, D. Götz, P. Schoch and G. Tölg, *Talanta*, 22 (1975) 889.
522 G. Kaiser, D. Götz, G. Tölg, G. Knapp, B. Maichim and H. Spitzy, *Fresenius' Z. Anal. Chem.*, 291 (1978) 278.
523 V.S. Sukhnevich, *Zh. Prikl. Spektrosk.*, 27 (1977) 780.
524 A. Hulanicki and R. Karwowska, *Chem. Anal. (PRL)*, 22 (1977) 637.
525 S. Bajo, U. Suter and B. Aeschliman, *Anal. Chim. Acta*, 149 (1983) 321.
526 E. Jackwerth and S. Gomišček, *Pure Appl. Chem.*, 56 (1984) 479.
527 R. Nakashima, E. Kameta and S. Shibata, *Bunseki Kagaku*, 33 (1984) E343.
528 C.J. Jackson, D.G. Porter, A.L. Dennis and P.B. Stockwell, *Analyst (London)*, 103 (1978) 317.
529 W. Lautenschläger, *Beckman Bull.*, No. 2 (1976) 3.
530 *Manufacture of Rare Metals and Semiconductor Materials*, Obzornaya informatsiya, issue 3; T.M. Malyutina, E.G. Namvrina and O.A. Shiryaeva, *Chemical Preparation of Samples in Analysis of Materials of Rare-Metal and Semiconductor Industry*, TsNII Ekonomiki i Informatsii Tsvetnoi Metallurgy, Moscow, 1985 (in Russian).
531 M.S. Chupakhin, A.I. Sukhanovskaya, V.Z. Krasil'shchik, S.U. Kreingold, L.A. Demina, V.I. Bogomolov, E.V. Dobizha, M. Przhibyl, Z. Slovak, I. Borak and M. Smrzh, *Methods of Analysis of Pure Reagents*, Khimiya, Moscow, 1984 (in Russian).
532 A.G. Karabash, Sh.I. Peizulaev and L.I. Slyusareva, in *Materials of Xth All-Union Conference on Spectroscopy, 1956*, L'vov State University Press, L'vov, 1958, p. 556 (in Russian).
533 A.G. Karabash, Sh.I. Peizulaev, N.P. Sotnikov and S.K. Sazonova, *Tr. Kom. Anal. Khim. Akad. Nauk SSSR*, 12 (1960) 108.
534 N.P. Sotnikov, L.S. Romanova, Sh.I. Peizulaev and A.G. Karabash, *Tr. Kom. Anal. Khim. Akad. Nauk SSSR*, 12 (1960) 151.
535 V.N. Muzgin, V.L. Zolotavin and F.F. Gavrilov, *Zh. Anal. Khim.*, 19 (1964) 111.
536 V.V. Malakhov, N.P. Protopopova, V.A. Trukhacheva and I.G. Yudelevich, *Tr. Kom. Anal. Khim. Akad. Nauk SSSR*, 16 (1968) 89.
537 V.V. Malakhov, N.P. Protopopova, V.A. Trukhacheva and I.G. Yudelevich, *Izv. Sib. Otd. Akad. Nauk SSSR, Ser. Khim. Nauk*, 9 (4) (1968) 78.
538 V.V. Malakhov, I.G. Yudelevich, N.P. Protopopova and V.A. Trukhacheva, *Zh. Anal. Khim.*, 24 (1969) 575.
539 V.Z. Krasil'shchik, *Zavod. Lab.*, 42 (1976) 153.
540 Kim Guk Hong, Li Yong Sik and Kim Gwang Sung, *Punsok Khvakhak*, 9 (1971) 111.
541 Z.G. Fratkin, *Zavod. Lab.*, 30 (1964) 170.
542 Kh.I. Zil'bershtein and O.N. Nikitina, *Methods of Analysis of High-Purity Substances*, Nauka, Moscow, 1965, p. 78 (in Russian).

543 L.S. Vasilevskaya, S.A. Sadofyeva, M.A. Notkina and A.I. Kondrashina, *Zavod. Lab.*, 28 (1962) 678.
544 L.L. Baranova, L.D. Berliner, B.Ya. Kaplan, T.M. Malyutina, M.G. Nazarova and L.S. Razumova, *Zavod. Lab.*, 50 (1984) 10.
545 A. Kato, Y. Osumi and Y. Miyake, *Bull. Govt. Ind. Res. Inst., Osaka*, 23 (1972) 167.
546 A.I. Maslova, N.V. Trofimov, L.V. Romanova and L.M. Novoselova, *Zavod. Lab.*, 48 (1982) 6.
547 B.D. Brodskaya, M.A. Notkina and N.P. Menshova, *Zh. Anal. Khim.*, 27 (1972) 151.
548 Z.G. Fratkin and N.G. Polivanova, *Methods of Analysis of High-Purity Substances*, Nauka, Moscow, 1965, p. 462 (in Russian).
549 E.B. Sandell, *Colorimetric Determination of Traces of Metals*, Interscience, New York, 1950, p. 25.
550 L.B. Kuznetsov, V.N. Belyaev and G.G. Kovalev, *Zh. Anal. Khim.*, 38 (1983) 1800.
551 L.B. Kuznetsov, V.N. Belyaev and V.P. Baluda, *Zh. Anal. Khim.*, 39 (1984) 215.
552 D.J. Nicolas, *Anal. Chim. Acta*, 55 (1971) 59.
553 J. Gulmont, A. Bouchard and M. Pichett, *Talanta*, 23 (1976) 62.
554 C.L. Luke, *Anal. Chem.*, 27 (1955) 1150.
555 J. Olafsson, *Anal. Chim. Acta*, 68 (1974) 207.
556 M. Nishimura, K. Matsunaga and S. Konishi, *Jpn. Anal.*, 24 (1975) 655.
557 A.N. Sidorenko and A.A. Tumanov, *Methods of Analysis of High-Purity Substances*, Nauka, Moscow, 1965, p. 52 (in Russian).
558 J. Ramírez-Muñoz, *J. Flame Notes, (Beckman)*, No. 1 (1957) 45.
559 V.I. Rigin and G.N. Verkhoturov, *Zh. Anal. Khim.*, 32 (1977) 1965.
560 V.I. Rigin, *Zh. Anal. Khim.*, 34 (1979) 1569.
561 R.G. Godden and D.R. Thomerson, *Analyst (London)*, 105 (1980) 1137.
562 J.E. Goutler, in *Abstr. Pap. Pittsburgh Conf. Anal. Chem. Appl. Spectrosc., Atlantic City, NJ, March 9–13, 1981*, Monroeville, PA, p. 129.
563 E. Pruszkowska, P. Barrett, R. Ediger and G. Wallace, *At. Spectrosc.*, 4 (1983) 94.
564 R.R. Liversage and J.C. van Loon, *Anal. Chim. Acta*, 161 (1984) 275.
565 B. Welz and E. Wiedeking, in *Int. Symp. Microchem. Tech., Davos, 1977*, Abstr. S. 1, p. 167.
566 P.J. Whiteside, T.C. Dymott, G. Küllmer and P.M. Green, *Chem. Labor. Betr.*, 36 (1985) 21.
567 T.W. May and J.L. Johnson, *At. Spectrosc.*, 6 (1985) 9.
568 P.N. Vijan and G.R. Wood, *At. Absorpt. Newslett.*, 13 (1974) 33.
569 A. Kuldvere, *At. Spectrosc.*, 1 (1980) 138.
570 M.H. Arbab-Zavar and A.G. Howard, *Analyst (London)*, 105 (1980) 744.
571 N.W. Barnett, L.S. Chen and C.T. Kirbright, *Spectrochim. Acta, Part B*, 34B (1984) 1141.
572 S. Tanaka, M. Kaneko, Y. Konno and Y. Hashimoto, *Bunseki Kagaku*, 32 (1983) 535.
573 J.G. Crock and F.E. Lichte, *Anal. Chim. Acta*, 144 (1982) 223.
574 S. Terashima, *Bunseki Kagaku*, 33 (1984) 561.
575 H. Agemian and E. Bedek, *Anal. Chim. Acta*, 119 (1980) 323.
576 K.S. Subramanian, *Fresenius' Z. Anal. Chem.*, 305 (1981) 382.
577 H. Narasaki and M. Ikeda, *Anal. Chem.*, 56 (1984) 2059.
578 C.C.Y. Chan and M.W. Baig, *Anal. Lett., Part A*, 17 (1984) 143.
579 I.A. Brovko, A. Tursunov, M.A. Rish and A.D. Davirov, *Zh. Anal. Khim.*, 39 (1984) 1768.

580 K.S. Subramanian and V.S. Sastri, *Talanta*, 27 (1980) 469.
581 T. Nakahara, *Appl. Spectrosc.*, 37 (1983) 539.
582 P.P. Chou, P.K. Jaynes and J.L. Balley, *J. Anal. Toxicol.*, 8 (1984) 158.
583 O. Aström, *Anal. Chem.*, 54 (1982) 190.
584 Yan Du, Yan Zhang, Cheng Guang-Shen and Li An-Mo, *Talanta*, 31 (1984) 133.
585 G.A. Hambrick, P.N. Froelich, M.O. Andreae and B.L. Lewis, *Anal. Chem.*, 56 (1984) 421.
586 K. Matsumoto, T. Ishiwatari and K. Fuwa, *Anal. Chem.*, 56 (1984) 1545.
587 F. Mohr, B. Luft and H. Bombach, *Jenaer Rundsch*, 28 (1983) 120.
588 W.G. Pfann, *Zone Melting*, Wiley, New York, 1966.
589 M. Zief and W.R. Wilcox (Editors), *Fractional Distillation*, Vol. 1, Marcel Dekker, New York, 1967.
590 V.N. Vigdorovich, A.E. Volpyan and G.M. Kurdyumov, *Directed Crystallization and Physico-Chemical Analysis*, Khimiya, Moscow, 1976 (in Russian).
591 A.N. Kirgintsev, L.I. Isaenko and V.A. Isaenko, *Impurities Distribution in Directed Crystallization*, Nauka, Moscow, 1977 (in Russian).
592 A.N. Kirgintsev, V.A. Isaenko, I.I. Kisil, A.B. Blank, L.A. Prokhorov and N.A. Pyl'nova, *Controlled Crystallization in Tubular Container*, Nauka, Novosibirsk, 1978 (in Russian).
593 A.B. Blank, *Analysis of pure Substances using Crystallization Concentration*, Khimiya, Moscow, 1986 (in Russian).
594 A.B. Blank, *Zh. Anal. Khim.*, 38 (1983) 1939.
595 A.N. Kirgintsev, *Fiz. Met. Metalloved.*, 26 (1968) 101.
596 A.B. Blank, *Monocrystals and Technics,* VNII Monokristallov Press, Khar'kov, No. 2, 1973. p. 189 (in Russian).
597 A.B. Blank, L.I. Afanasiadi, E.S. Zolotovitskaya and B.M. Fidel'man, *Monocrystals, Scintillators, and Organic Luminophores*, VNII Monokristallov Press, Khar'kov, No. 5, Part I, 1969, p. 78 (in Russian).
598 A.B. Blank and L.I. Afanasiadi, *Methods of Analysis of High-Purity Alkali and Alkaline-Earth Metal Halides*, VNII Monokristallov Press, Khar'kov, Part I, 1971, p. 30 (in Russian).
599 A.B. Blank, E.S. Zolotovitskaya, L.I. Afanasiadi and B.M. Fidel'man, *Ukr. Khim. Zh.*, 37 (1971) 70.
600 A.N. Kirgintsev, *Zh. Anal. Khim.*, 26 (1971) 1719.
601 A.B. Blank and L.I. Afanasiadi, *Zh. Anal. Khim.*, 25 (1970) 2085.
602 A.B. Blank, L.I. Afanasiadi, E.S. Zolotovitskaya and B.M. Fidel'man, *Zh. Anal. Khim.*, 25 (1970) 2291.
603 A.N. Kirgintsev and S.G. Gryaznova, *Zh. Anal. Khim.*, 26 (1971) 1725.
604 A.N. Kirgintsev, S.G. Gryaznova and Z.V. Zil'berfain, *Zh. Anal. Khim.*, 28 (1973) 1069.
605 A.N. Kirgintsev and I.I. Gorbacheva, *Zh. Anal. Khim.*, 23 (1968) 336.
606 E.S. Zolotovitskaya, A.B. Blank, L.I. Afanasiadi and L.V. Glushkova, *2nd All-Union Conference on Concentration Methods Used in Analytical Chemistry*, Nauka, Moscow, 1977, p. 155 (in Russian).
607 A.B. Blank, L.I. Afanasiadi and L.P. Esperiandova, *Zh. Anal. Khim.*, 27 (1972) 221.
608 A.B. Blank, L.P. Esperiandova, N.I. Komishan and V.G. Potapova, *2nd All-Union Conference on Concentration Methods Used in Analytical Chemistry*, Nauka, Moscow, 1977, p. 154 (in Russian).

609 A.B. Blank, L.P. Esperiandova, L.S. Manzhely and L.I. Gorodilova, *Zh. Anal. Khim.*, 36 (1981) 437.
610 V.G. Chepurnaya and A.B. Blank, *Zh. Anal. Khim.*, 28 (1973) 1016.
611 A.B. Blank and L.P. Esperiandova, *Zh. Anal. Khim.*, 37 (1982) 940.
612 A.B. Blank and V.G. Chepurnaya, *Zh. Anal. Khim.*, 27 (1972) 2035.
613 A.B. Blank and V.G. Chepurnaya, *Zh. Anal. Khim.*, 29 (1974) 1006.
614 A.B. Blank, V.G. Chepurnaya, L.P. Esperiandova and V.Ya. Vakulenko, *Zh. Anal. Khim.*, 29 (1974) 1705.
615 A.B. Blank, N.I. Shevtsov and E.S. Zolotovitskaya, *Zh. Anal. Khim.*, 30 (1975) 2036.
616 A.B. Blank, N.I. Shevtsov and I.G. Kogan, *Zh. Anal. Khim.*, 31 (1976) 1018.
617 A.B. Blank and N.I. Shevtsov, *2nd All-Union Conference on Concentration Methods Used in Analytical Chemistry*, Nauka, Moscow, 1977, p. 153 (in Russian).
618 E.E. Konovalov and Sh.I. Peizulaev, *Tr. Kom. Anal. Khim. Akad. Nauk SSSR*, 15 (1965) 375.
619 V.N. Vigdorovich, *Zavod. Lab.*, 31 (1965) 568.
620 H. Schildknecht, *Anal. Chim. Acta*, 38 (1967) 261.
621 J. Downarowicz, *Chem. Anal. (PRL)*, 4 (1959) 643.
622 E.E. Konovalov, Sh.I. Peizulaev, G.P. Pinchuk, I.E. Larionova and L.I. Kondratyeva, *Zh. Anal. Khim.*, 18 (1963) 624.
623 E.E. Konovalov and Sh.I. Peizulaev, *Zh. Anal. Khim.*, 22 (1967) 736.
624 E.E. Konovalov, Sh.I. Peizulaev and V.P. Emelyanov, *Zh. Anal. Khim.*, 18 (1963) 1500.
625 F.H. Horn, *U.S. Pat.*, 3432753 (1969).
626 B.S. Krasulina, Yu.M. Ivanov, V.G. Yakovleva and T.V. Rudneva, *2nd All-Union Conference on Concentration Methods used in Analytical Chemistry*, Nauka, Moscow, 1977, p. 157 (in Russian).
627 I.F. Baryshnikov (Editor), *Sampling and Analysis of Noble Metals, A Handbook*, Metallurgiya, Moscow, 1978 (in Russian).
628 J. Haffty, L.B. Riley and W.D. Goss, *Geol. Surv. Bull. (U.S.)*, No. 1145 (1977) V, p. 58.
629 E.P. Zdorova, N.N. Popova and M.A. Kondulinskaya, *Zavod. Lab.*, 43 (1977) 926.
630 A. Diamantatos, *Anal. Chim. Acta*, 91 (1977) 281.
631 L.P. Kolosova, N.V. Novatskaya and E.G. Vinnitskaya, *Zavod. Lab.*, 42 (1976) 508.
632 N.L. Fishkova, E.P. Zdorova and N.N. Popova, *Zh. Anal. Khim.*, 30 (1975) 806.
633 E.L. Grinzaid, L.P. Kolosova, L.S. Nadezhina and N.V. Novatskaya, *Zavod. Lab.*, 44 (1978) 682.
634 E.L. Grinzaid, L.S. Nadezhina, L.P. Kolosova, G.S. Loginova, N.V. Novatskaya and G.F. Nikolaeva, *Zavod. Lab.*, 42 (1976) 420.
635 F.E. Beamish, *The Analytical Chemistry of the Noble Metals*, Pergamon Press, Oxford, London, 1966.
636 S.M. Anisimov, *Methods of Analysis of Platinum Metals, Gold and Silver*, Metallurgizdat, Moscow, 1960 (in Russian).
637 G.H. Faye and P.E. Moloughney, *Talanta*, 19 (1972) 269.
638 V.V. Lichadeev, A.K. Dementyeva, M.P. Lukicheva and L.G. Krasnova, *Nauchn. Tr. Sib. Nauchno-Issled. Proektn. Inst. Tsvetn. Metall.*, No. 4 (1971) 225.
639 G.N. Verkhoturov, T.I. Lekhtarnikova and A.M. Zelentsova, *Nauchn. Tr. Sib. Nauchno-Issled. Proektn. Inst. Tsvetn. Metall.*, No. 5 (1972) 16.
640 E.L. Hoffman, A.J. Naldrett, J.C. Loon, R.G.Y. Hancock and A. Manson, *Anal. Chim. Acta*, 102 (1978) 157.

641 A.P. Kuznetsov, Yu.N. Kukushkin and D.F. Makarov, *Zh. Anal. Khim.*, 29 (1974) 2155.

642 A.P. Kuznetsov, D.F. Makarov and Yu.N. Kukushkin, *2nd All-Union Conference on Concentration Methods Used in Analytical Chemistry*, Nauka, Moscow, 1977, p. 164 (in Russian).

643 F.I. Danilova, V.A. Orobinskaya, V.S. Parfenova and R.M. Nazarenko, *Zh. Anal. Khim.*, 29 (1974) 2142.

644 N.A. Artem'yev, D.F. Makarov, V.A. Shestakov and Yu.N. Kukushkin, *Zh. Prikl. Khim.*, 47 (1974) 1010.

645 D.F. Makarov, Yu.N. Kukushkin and T.A. Eroshevich, *Zh. Prikl. Khim.*, 47 (1974) 1215.

646 V.P. Kirillov, S.K. Kalinin, E.N. Pokrovsky, A.I. Stratyev and V.A. Sidorov, *2nd All-Union Conference on Concentration Methods Used in Analytical Chemistry*, Nauka, Moscow, 1977, p. 163 (in Russian).

647 O.I. Artem'yev, V.M. Stepanov and G.E. Kovel'skaya, *Zh. Anal. Khim.*, 33 (1978) 493.

648 O.I. Artem'yev, B.G. Kiselev and V.M. Stepanov, *Zh. Anal. Khim.*, 33 (1978) 2163.

649 H. Djojosubroto, O.B. Liang, Y. Sumaryani and R.A. Hartono, in *Abstr. Pittsburgh Conf. Anal. Chem. Appl. Spectrosc., Cleveland, OH., 1977*, Pittsburgh, PA, 1977, p. 421.

650 N.Ya. Chalkov and A.M. Ustinov, *Zavod. Lab.*, 33 (1967) 447.

651 V.M. Kuligin, E.P. Zdorova, N.N. Popova and E.E. Rakovsky, *Zh. Anal. Khim.*, 31 (1976) 1702.

652 D.I. Leipunskaya, S.I. Savosin, V.I. Drynkin, A.I. Aliev, Ya.B. Fil'kenshtein, N.N. Popova and E.S. Zemchikhin, *Zavod. Lab.*, 37 (1971) 1471.

653 P.A. Gurin and L.I. Chechulin, *Nauchn. Tr. Irkutsk. Nauchno-Issled. Inst. Redk. Tsvetn. Met.*, No. 28 (1976) 54.

654 G.T. Georgiev, *Zh. Anal. Khim.*, 33 (1978) 740.

655 G.T. Georgiev, *Zh. Anal. Khim.*, 33 (1978) 745.

656 P.W. Alexander, R. Hohn and L.E. Smythe, *Talanta*, 24 (1977) 549.

657 V.N. Nikitin and V.P. Polun, *Sb. Nauchn. Tr. Krasnoyarsk. Politekh. Inst.*, No. 15 (1973) 32.

658 L.A. Obolentsev, V.A. Sidorov, A.I. Stratyev and N.V. Barokha, *Tr. Tsentr. Nauchno-Issled. Geologorazved. Inst. Tsvetn. Blagorodn. Met.*, No. 121 (1975) 93.

659 St. Aleksandrov, *Zh. Anal. Khim.*, 30 (1975) 322.

660 V.A. Shestakov, N.A. Arkhipov, D.F. Makarov and Yu.N. Kukushkin, *Zh. Anal. Khim.*, 29 (1974) 2176.

661 E.G. Koleva, S.Kh. Arpadzhyan and Z.T. Georgieva, *God. Sofii. Univ. Khim. Fak.*, 1969/70 (1972) 64 and 73.

662 A. Diamantatos, *Anal. Chim. Acta*, 90 (1977) 179.

663 O.P. Berezkin, M.P. Lukichev, T.I. Karashuba and A.K. Dementyeva, *Nauchn. Tr. Sib. Gos. Nauchno-Issled. Proektn. Inst. Tsvetn. Metall.*, No. 6 (1973) 21.

664 R.J. Coombes, A. Chow and R. Wageman, *Talanta*, 24 (1977) 421.

665 A. Diamantatos, *Anal. Chim. Acta*, 165 (1984) 263.

666 P.G. Sim, *Appl. Spectrosc.*, 28 (1974) 23.

667 A.M. Harris, G.J.B. Len and F. Farell, *Talanta*, 25 (1978) 257.

668 S. Lupan, M. Protopopescu and T. Ponta, *Rev. chim. (PSR)*, 25 (1974) 833.

669 N.F. Chaskova, *Analysis and Technology of Noble Metals*, Metallurgiya, Moscow, p. 232, 1971 (in Russian).

200

670 P.E. Moloughney, *Talanta*, 24 (1977) 135.
671 P.E. Moloughney and G.H. Fage, *Talanta*, 23 (1976) 377.
672 E.L. Hoffman, A.J. Naldrett and J.C. van Loon, *Anal. Chim. Acta*, 102 (1978) 157.
673 O.I. Artem'yev and V.M. Stepanov, *Zavod. Lab.*, 48 (1982) 1.
674 V.N. Mikshevich, K.E. Moiseev and V.A. Antonov, *Activation Analysis*, FAN, Tashkent, 1971, p. 127 (in Russian).
675 A.I. Samchuk and I.K. Latysh, *Ukr. Khim. Zh.*, 48 (1982) 638.
676 V.P. Khrapai and G.M. Gusev, *Spectral Analysis in Non-Ferrous Metallurgy*, Gos-metallurgizdat, Moscow, 1960, p. 308 (in Russian).
677 A.G. Karabash, *D.Sc. Thesis*, GEOKhI AN SSSR, Moscow, 1967.
678 N.Ya. Chalkov, *Can. Sc. Thesis*, Institute of Inorganic Chemistry of the Siberian Division of the USSR Academy of Sciences, Novosibirsk, 1973.
679 Yu.K. Delimarsky and B.F. Markov, *Electrochemistry of Molten Salts*, Metallurgiz-dat, Moscow, 1960, p. 186 and 190 (in Russian).
680 K. Fukuda and A. Mizuike, *Jpn. Anal.*, 17 (1968) 319.
681 A. Mizuike and M. Hiraide, *Pure Appl. Chem.*, 54 (1982) 1555.
682 K. Sekine and H. Onishi, *Chemistry (Kyoto)*, 29 (1974) 141.
683 P. Somasundaran, *Sep. Sci.*, 10 (1975) 93.
684 M. Hiraide and A. Mizuike, *Bunseki*, 38 (1978) 96.
685 S. Nakashima, *Analyst (London)*, 103 (1978) 1031.
686 Sik Kim Young and H. Zeitlin, *Anal. Chem.*, 43 (1971) 1390.
687 Sik Kim Young and H. Zeitlin, *Sep. Sci.*, 7 (1972) 1.
688 D. Voyce and H. Zeitlin, *Anal. Chim. Acta*, 69 (1974) 27.
689 Tseng Jau-Hwan and H. Zeitlin, *Anal. Chim. Acta*, 101 (1978) 71.
690 Liang Shu-Chuan, Zhong You-Lan and Wang Zhi, *Fresenius' Z. Anal. Chem.*, 318 (1984) 19.
691 S.M. Nemets, Yu.I. Turkin and V.L. Zyeva, *Zh. Anal. Khim.*, 38 (1983) 1782.
692 L.M. Cabezon, M. Caballero, R. Cela and J.A. Perez-Bustamante, *Talanta*, 31 (1984) 597.
693 S. Nakashima and M. Yagi, *Anal. Lett., Part A*, 17 (1984) 1693.
694 S. Nakashima and M. Yagi, *Bunseki Kagaku*, 33 (1984) T1.
695 E.N. De Carlo and H. Zeitlin, *Anal. Chem.*, 53 (1981) 1104.
696 Y. Urihira, *J. Chem. Soc. Jpn., Pure Chem. Sec.*, 84 (1963) 642.
697 K. Sekine, *Fresenius' Z. Anal. Chem.*, 273 (1975) 103.
698 K. Sekine, *Microchim. Acta*, 4 (1975) 313.
699 K. Sekine and H. Onishi, *Fresenius' Z. Anal. Chem.*, 288 (1977) 47.
700 I.Yu. Andreeva, L.I. Lebedeva and O.L. Drapchinskaya, *Zh. Anal. Khim.*, 40 (1985) 694.
701 T. Nozaki, K. Kato, K. Okamura, Y. Soma, N. Ikuta and N. Mise, *Bunseki Kagaku*, 32 (1983) 479.
702 Z. Skorko-Trubula and E. Kozińska, *Chem. Anal. (PRL)*, 26 (1981) 581.
703 M. Aoyama, T. Hobo and S. Suzuki, *Anal. Chim. Acta*, 129 (1981) 237.
704 T. Hobo, Y. Sudo, S. Suzuki and S. Araki, *Jpn. Anal.*, 27 (1978) 104.
705 K. Sekine and H. Onishi, *Anal. Chim. Acta*, 62 (1972) 468.
706 E.W. Berg and D.M. Downey, *Anal. Chim. Acta*, 120 (1980) 237.
707 Z. Marczenko, *Pure Appl. Chem.*, 57 (1985) 849.
708 Z. Marczenko, M. Balcerzak and H. Pasen, *Mikrochim. Acta*, II (5–6) (1982) 371.
709 K. Oikawa, Y. Hashimoto and S. Yanagisawa, *J. Spectrosc. Soc. Jpn.*, 23 (1974) 111.

710 I. Tesařová, *Chem. Listy*, 68 (1974) 12.
711 G. Krishnamurty, K.M. Saraswathy and O. Thomas, *Government of India Atomic Energy Commission (Rept.)*, No. 749 (1974).
712 M. Janssens and R. Dams, *Anal. Chim. Acta*, 70 (1974) 25.
713 A. Sigimae and R.K. Skogerboe, *Anal. Chim. Acta*, 97 (1978) 1.
714 J. Forest and L. Newman, *Anal. Chem.*, 49 (1977) 1579.
715 B. Magyar and H. Vonmont, *Fresenius' Z. Anal. Chem.*, 280 (1976) 115.
716 N.N. Aleksandrov, G.S. Guniya, A.I. Gunchenko and Yu.I. Turkin, *Tr. Gl. Geofiz. Obs.*, No. 314 (1974) 104.
717 B.N. Noller, H. Bloom and C.R. Parker, *Int. Conf. At. Spectrosc., Abstr. Pap. 5th, 1975*, A12 (1975).
718 S. Oda, H. Kubo and H. Kamada, *Jpn. Anal.*, 24 (1975) 66.
719 W. Wiedner, H. Nathansen, G. Thümmler and H. Litke, *Z. Gesamte Hyg.*, 20 (1974) 4.
720 P.N. Vijan and G.R. Wood, *At. Absorp. Newslett.*, 13 (1974) 33.
721 R.E. Lacey and S. Loeb (Editors), *Industrial Processing with Membranes*, New York, London, Sydney, Toronto, Wiley Interscience, 1972.
722 Yu.I. Dytnersky, *Reverse Osmosis and Ultrafiltration*, Khimiya, Moscow, 1978 (in Russian).
723 B.Ya. Spivakov, K. Geckeler and E. Bayer, *Nature (London)*, 315 (1985) 313.
724 K.S. Subramanian and V.S. Sastry, *Sep. Sci. Technol.*, 15 (1980) 145.
725 J.E. Di Nunzio, R.L. Wilson and F.P. Gatchell, *Talanta*, 30 (1983) 57.
726 J.A. Cox and J.E. Di Nunzio, *Anal. Chem.*, 49 (1977) 1272.
727 J.A. Cox and J. Carnahan, *Appl. Spectrosc.*, 35 (1981) 447.
728 G.L. Lundquist, G. Washinger and J.A. Cox, *Anal. Chem.*, 47 (1975) 319.
729 J.A. Cox and Z. Twardowski, *Anal. Chem.*, 52 (1980) 1503.
730 R.L. Wilson and J.E. Di Nunzio, *Anal. Chem.*, 53 (1981) 692.
731 J.A. Cox, E. Olbrych and K. Brajter, *Anal. Chem.*, 53 (1981) 1308.
732 J.E. Di Nunzio and M. Jubara, *Anal. Chem.*, 55 (1983) 1013.
733 E. Martins, M. Bengtsson and G. Johansson, *Anal. Chim. Acta*, 169 (1985) 31.
734 J.A. Cox, T. Gray, S. Yoon Kyung, Kin Yeon-Taik and Z. Twardowski, *Analyst (London)*, 109 (1984) 1603.
735 G.G. Devyatykh and Yu.E. Illiev, *An Introduction into the Theory of Deep Cleaning of Substances*, Nauka, Moscow, 1981 (in Russian).
736 H. Schäfer, *Chemische Transportreaktionen*, Verlag Chemie, Weinheim, 1962.
737 M.M. Seidler, *Reinststoffe in Wissenschaft und Technik*, Berlin, 1963, p. 125.
738 V.F. Dorfman, K.A. Bol'shakov and I.P. Kislyakov, *Materials of the All-Union Conference on Methods of Obtaining High-Purity Substances*, IREA Press, Moscow, 1967, p. 12.
739 Kuo Yue, *Chem. Eng. Prog.*, 80 (1984) 37.
740 P.A. Michalik and R. Stephens, *Talanta*, 28 (1981) 37.
741 P.A. Michalik and R. Stephens, *Talanta*, 28 (1981) 43.
742 T. Uchida, M. Nagase, I. Kojima and C. Iida, *Anal. Chim. Acta*, 94 (1977) 275.
743 T. Golembski, *Talanta*, 22 (1975) 547.
744 T. Rains and O. Menis, *J. Assoc. Off. Anal. Chem.*, 55 (1972) 1339.
745 A.V. Vanaeva, N.I. Kuznetsov and T.I. Moroshkina, *Vestn. Leningr. Univ.*, No. 16 (1971) 154.
746 H.S. Mahanti and R.M. Barnes, *Anal. Chem.*, 55 (1983) 403.

747 R. Neeb and F. Wahdat, *Fresenius' Z. Anal. Chem.*, 269 (1974) 275.
748 E. Jackwerth, J. Lohmer and G. Schwark, *Colloq. Spectrosc. Int., Plenary Lect. Rep. 16th, 1971*, (1971) 49.
749 I.G. Yudelevich, L.I. Buyanova, N.P. Protopopova and B.K. Dzhakueva, *Zavod. Lab.*, 35 (1969) 426.
750 V.V. Nedler, B.D. Brodskaya and M.A. Notkina, *Methods of Analysis of High-Purity Substances*, Nauka, Moscow, 1965, p. 207 (in Russian).
751 G.I. Spielholtz, G.C. Toralballa and J.J. Willsen, *Mikrochim. Acta*, 4 (1974) 649.
752 L.N. Moskvin, N.N. Kalinin, L.A. Godon, S.B. Tomilov and G.I. Kizym, *Zh. Anal. Khim.*, 31 (1976) 2396.
753 G.V. Myasoedova, G.I. Malofeeva, O.P. Shvoeva, E.V. Illarionova, S.B. Savvin and Yu.A. Zolotov, *Zh. Anal. Khim.*, 32 (1977) 645.
754 L.I. Pavlenko, G.I. Malofeeva, L.M. Simonova and O.Yu. Andryushenko, *Zh. Anal. Khim.*, 29 (1974) 1122.
755 S.B. Savvin, I.I. Antokolskaja, G.V. Myasoedova, L.I. Bolshakova and O.P. Shvoeva, *J. Chromatogr.*, 102 (1974) 287.
756 I. Yoshida, H. Kobayashi and K. Ueno, *Talanta*, 24 (1977) 61.
757 A. Diamantatos, *Anal. Chim. Acta*, 98 (1978) 315.
758 I.Yu. Davydov, A.P. Kuznetsov, I.I. Antokolskaya and N.N. Nikolskaya, *Zh. Anal. Khim.*, 34 (1979) 1145.
759 L.V. Aizenberg and O.I. Savel'zon, *Zh. Anal. Khim.*, 35 (1980) 374.
760 P.S. Fishbern, *Utility Theory for Making a Decision*, Nauka, Moscow, 1978 (in Russian).
761 Yu.A. Zolotov, *Pure Appl. Chem.*, 50 (1978) 129.
762 H. Monien, D. Bohn and P. Jacob, *Angew. Chem.*, 83 (1971) 921.
763 N.P. Kuz'mina, V.D. Krylova and L.V. Lukashevich, *International Conference on Physical Aspects of Atmospheric Pollution*, Vilnius, 1974, p. 26 (in Russian).
764 N. Dancheva, K. Tsvetanov and E. Ognyanova, *God. NII Tsvet. Metall.*, 7 (1969) 115.
765 D.E. Leyden, W. Wegscheider and W.B. Bodnar, *Int. J. Environ. Anal. Chem.*, 7 (1979) 85.
766 D.E. Leyden, W. Wegscheider, W.B. Bodnar, E.D. Sexton and W.K. Nonider, in J.B. Albarges (Editor), *Analytical Techniques in Environmental Chemistry*, Pergamon Press, Oxford, 1980, p. 469.
767 I.G. Yudelevich, *Izv. Sib. Otd. Akad. Nauk SSSR, Ser. Khim. Nauk*, 9 (4) (1978) 17.
768 I.G. Yudelevich, P.I. Artyukhin, L.S. Chuchalina, N.P. Protopopova, L.M. Skrebkova, E.N. Gilbert and V.A. Pronin, *Zh. Anal. Khim.*, 21 (1966) 1457.
769 V.V. Malakhov, N.P. Protopopova, V.A. Trukhachev and I.G. Yudelevich, *Izv. Sib. Otd. Akad. Nauk SSSR, Ser. Khim. Nauk*, 9 (4) (1968) 78.
770 I.G. Yudelevich, L.M. Buyanova, I.R. Shelpakova, G.V. Afanas'heva and O.I. Shcherbakova, *Tr. Khim. Khim. Tekhnol. (Gorky)*, No. 4 (1973) 76.
771 I.G. Yudelevich, A.N. Kirgintsev and A.S. Prokhorov, *Zh. Anal. Khim.*, 24 (1969) 1090.
772 A.B. Blank, *Zh. Anal. Khim.*, 30 (1975) 1423.
773 W.J. Williams, *Handbook of Anion Determination*, Butterworth, London, 1979.

Chapter 3

Combining preconcentration with the determination methods

3.1. Combined and hybrid methods

Let us consider possible combinations of preconcentration techniques with methods of determination. Of course, it is reasonable to combine multi-elemental determination methods with group concentration techniques. A collector (matrix for the concentrate) should meet as fully as possible the demands of the selected determination method: graphite powder for atomic emission analysis, organic extracts for flame atomic absorption spectrometry, etc. There is no point in combining a high-precision determination method with a concentration technique that does not ensure good precision and *vice versa*.

The combinations of separation (concentration) and determination methods can be divided into two groups. In the first group the concentrate can be analysed virtually by any suitable method; the separation technique here has not to be strictly tied to any particular method of determination. Such combinations are known as combined methods.

The combined methods include all possible arbitrary combinations of the methods of separation (concentration) with subsequent determination; often the separation product of the concentrate has to be additionally converted into a form suitable for determination. In other words, separation (concentration) and determination are merely two successively used stages of analysis. Here, for subsequent determination it makes no difference by which method the separation product or the concentrate has been obtained.

The combinations of the second group are called the hybrid methods [1–7] (see also refs. 8–11). At one stage it was difficult to decide with which group the methods like gas chromatography or inverse voltam-

TABLE 3.1

Classification and occurrence of hybrid methods of analysis

Method of separation/concentration	Method of determination	Occurrence*
Liquid–liquid extraction	Atomic absorption	+ + +
	Photometric	+ + +
	Atomic fluorescence	+ +
	Voltammetric	+ +
	Determinations with the use of flame ionization, electron-trapping and other detectors employed in GC	+ +
	Determinations with the use of conductometric and other detectors employed in ion chromatography	+
	Fluorimetric	+ +
	Activation	+ +
	Atomic emission	+ + +
	Kinetic	+
	Titrimetric	+
	Electron spin resonance	+
Extraction with melts	X-ray fluorescence	+ +
	Photometric	+ +
	Voltammetric	+ +
	Electron spin resonance	−
Sorption methods	Determinations with the use of flame ionization, electron-trapping and other detectors employed in GC	+ + +
	Determinations with the use of refractometric, spectrophotometric, and other detectors employed in HPLC	+ + +
	Photometric	+ +
	X-Ray fluorescence	+ + +
	Activation	+ +
	Determinations with the use of conductometric and other detectors employed in ion chromatography	+ +

	Method		
	Atomic absorption	+	+
	Mass spectrometry (MS)	+	+
	Atomic emission	+	+
	Atomic fluorescence	+	+
	Spark source MS	+	−
	Electron spin resonance	−	−
Precipitation and co-precipitation	Gravimetric	+	+
	Activation	+	+
	Atomic emission	+	+
	Atomic absorption	+	+
	Spark source MS	+	−
	Electron spin resonance	−	−
Volatilization and sublimation	Activation	+	+
	Atomic absorption	+	+
	Atomic emission	+	+
	Gravimetric	+	+
	Atomic fluorescence	+	+
	Spark source MS	+	−
	X-Ray fluorescence	−	
	Electron spin resonance	−	
Volatilization with preliminary chemical transformation	Atomic absorption	+	+
	Atomic emission	+	+
	Gravimetric	+	+
	Titrimetric	+	+
	Determinations with use of flame ionization and other detectors employed in pyrolytic GC	+	
	Activation	+	+
	Atomic fluorescence	+	
	Spark source MS	−	
	X-Ray fluorescence	−	
	Electron spin resonance	−	

(Continued on p. 208)

208

TABLE 3.1 (*continued*)

Method of separation/concentration	Method of determination	Occurrence*
Dry mineralization (ashing)	Gravimetric	+ + +
	Activation	+ +
	Atomic emission	+ +
	Atomic absorption	+
	Spark source MS	+
	Atomic fluorescence	–
	X-Ray fluorescence	–
	Electron spin resonance	–
Wet mineralization	Activation	+
	Atomic emission	+
	Atomic absorption	–
	Atomic fluorescence	–
	Spark source MS	–
High-, average- and low-temperature extraction	Determinations with the use of flame ionization detector and katharometer (GC)	+ +
	IR spectrometry	+ +
	MS	+ +
Electrodeposition, electrodissolution	Voltammetric	+ + +
	Atomic absorption	+ + +
	Gravimetric	+ + +
	Atomic emission	+ + +
	X-Ray fluorescence	+
	Activation	–
	Atomic fluorescence	–
	Spark source MS	–

Electrophoresis	Photometric	++
	Radiometric	++
	Fluorimetric	+
Crystallization, zone melting	Atomic emission	+
	Activation	−
	Spark source MS	−
	X-Ray fluorescence	−
Filtration	Activation	++
	X-Ray fluorescence	++
	Atomic emission	++
	Atomic absorption	−
	Gravimetric	−
	Atomic fluorescence	−
	Spark source MS	−

* +++ = Widely used combinations; ++ = moderately used; + = infrequently used; − = not practically used.

References pp. 276–288

metry should be classed. These are simultaneously the methods of separation and determination. In this group, separation and determination are combined together in one analytical cycle; often these are combined in one instrument. In the combinations of this group the separation (concentration) product is analysed by a definite method, sometimes without any additional treatment. Often, these offer significant advantages, being the result of such a determined combination. Thus, hybrid methods are based on close, harmonious combinations of the methods of separation (including concentration) and subsequent determination, and this leads to the formation of stable and indissoluble combinations.

With hybrid methods may be classed, for instance, extraction photometric determination of elements, *i.e.* the photometric determination of the coloured compound extracted from the aqueous phase or formed in the extract by adding any reagent after the extraction is over. The extraction–kinetic methods can be grouped with hybrid methods if catalytic determination is done, not in the aqueous solution, but in the extract. Other hybrid methods are gas chromatography (GC), high-performance liquid chromatography (HPLC), ion chromatography, gas chromatography–mass spectrometry (GC–MS) and stripping voltammetry.

An approach similar to the given classification is available in the article by Laitinen [12] dedicated to integration of analytical operations. In his opinion, the classical approach to chemical analysis envisages a number of separate operations including sampling, pretreatment of the sample, adjustment of conditions, separation, measurement and processing of data. Despite the expediency of these stages (from the didactic and practical viewpoints), modern analytical methods show a distinct tendency towards combination of these operations. In fact, in the methods of local analysis, the sampling is combined with measurement without any intermediate operations.

The combinations of preconcentration and separation techniques can be classified both by the preconcentration and determination methods. Separation precedes determination; therefore, these combinations can be classified by the separation method. Table 3.1 contains such a classification and shows the incidence of various combinations [6]. Therein are listed not only the known but also the possible combinations; this enables prospective combinations of separation and concentration (with subsequent determination) techniques to be revealed.

Let us now consider concrete combinations of concentration with the most important determination methods, the discussion being based on the determination methods.

3.2. Atomic emission spectrometry

3.2.1 GENERAL REMARKS

Preconcentration is of great importance for atomic emission spectrometry (AES). In this combination, the important advantage of AES, *i.e.* the possibility of determining simultaneously a large number of elements, is largely retained. In fact, the best procedures of preconcentration permit up to 20–25 elements to be determined in a sample. Concentration facilitates a reduction of the relative detection limit for elements by one to two orders of magnitude in solid samples and by three to four orders of magnitude in liquid and gaseous samples. The detection limits of trace elements can amount to 10^{-5}–$10^{-8}\%$ and in record breaking cases can be even lower. Concentration reduces the probability of introducing a systematic error during atomic emission determination, which results from sample inhomogeneity and diversity of the forms of determinable elements. It simplifies calibration because the microcomponents are converted into a new matrix of known composition. Calibration is known to be highly complicated because of the considerable effect of the matrix on the analytical signal.

Numerous laboratories have to deal with very varied substances; a given element that today is the matrix of an analysed sample can tomorrow become the determinable microcomponent. In atomic emission analysis one has to reckon with the "memory" of the instrument, laboratory conditions, etc. This difficulty can be overcome to a large extent by concentration which ensures isolation of trace elements and removal of the matrix.

The elements having a multilinear spectrum (iron, tungsten, molybdenum, uranium, plutonium, lanthanum and others) are very inconvenient for the atomic emission analysis of their impurities because of the frequently observed superposition of the matrix lines on the lines of the determined elements. Removal of the matrix by concentration eliminates this constraint. As a result, it becomes possible to dispense with the high resolution spectrometer.

By employing preconcentration the number of determined elements can be increased (due to the removal of the multiline matrix) or de-

TABLE 3.2

Some methods of atomic emission analysis

Material analysed	Method of preconcentration	Elements concentrated	Detection limit (%)	Ref.
High-purity silver	Extraction of matrix	Al, As, Ba, Be, Bi, Ca, Cd, Co, Cr, Cu, Fe, Ga, In, Mg, Mn, Ni, Pb, Sn, Te, Tl, V, Zn, Zr	$2 \cdot 10^{-6} - 1 \cdot 10^{-7}$	16
Alloy steels	Extraction of trace elements	Bi, Cd, Cu, Pb, Sb, Sn, Zn	$10^{-4} - 10^{-6}$	17
Sodium iodide	Extraction of trace elements	Ag, Al, Bi, Cd, Co, Cr, Cu, Fe, Mg, Mn, Mo, Ni, Pb, Sb, Sn, Ti, Zr	$10^{-5} - 10^{-7}$	18
Solutions after decomposition of ores and concentrates	Extraction of trace elements	Ag, Au, Ir, Pd, Pt, Rh, Ru	$10^{-4} - 10^{-6}$	19
Arsenic trichloride	Extraction of macro-components	Ag, Al, Au, Ba, Bi, Ca, Co, Cr, Cu, Fe, Ga, In, Mg, Mn, Ni, Pb, Sb, Sn, Te, Tl, Zn	$10^{-5} - 10^{-7}$	20
Solutions after decomposition of ores and concentrates	Sorption on CuS and complex-forming sorbent	Ag, Au, Ir, Pd, Pt, Rh, Ru	$10^{-5} - 10^{-7}$	21
Aluminium, gallium, gallium arsenide, indium phosphide	Co-precipitation on Bi_2S_3, $Bi(OH)_3$, In_2S_3	Ag, Al, Cd, Co, Cr, Cu, Mn, Ni, Pb, Sn, Zn	$10^{-5} - 10^{-7}$	22

creased (because all the desired elements are not separated during concentration). The precision and accuracy of determination can be better or worse in comparison to direct analysis because preconcentration, by eliminating the effect of "third" elements, on the one hand produces a positive effect on these characteristics and, on the other hand, exerts an adverse effect on them by complicating the analysis procedure.

If the emission spectra of the matrix elements are sufficiently simple and the physico-chemical properties of the matrix are favourable for the determination of trace elements, then there is no need to separate the matrix completely. The matrix residue may be used as a collector or internal standard. For example, in the analysis of high-purity cadmium [13], after the extraction of macro-component from iodide solutions with diethyl ether, the residue of its iodide served as collector of the concentrated elements. In such cases the matrix of reference samples should in its bulk and phase composition correspond to the matrix of the sample analysed (if it affects the analysis results, of course). In all cases the obtained concentrate should be in a form convenient for direct excitation of spectra of the determinable elements.

After preconcentration the trace elements can be collected in a small-volume collector or can be localized at the end surface of an electrode if analysis is to be carried out using arc and spark. The demands to be made on the substances – collectors – have been formulated [14, 15]: the substances should be stable on storage, non-hygroscopic, have a simple emission spectrum, good sorption and adhesion properties, low volatility, be chemically inert and readily obtainable. In addition they should be highly pure. Graphite powder satisfies these requirements. Most often sodium chloride is added to it for improving the excitation conditions. Many chemical atomic emission methods are based on the use of this collector. The methods worked out by us (Table 3.2) can serve as examples.

The concentrate of elements may be organic completely or partially (in co-precipitation of trace elements with organic collectors; on using extraction or certain sorption methods). The presence of organic substances in the concentrate affects to some extent the excitation and, hence, the value of the analytical signal; therefore it becomes necessary to mineralize the concentrate by a suitable method [15]. Preconcentration can be effected in such a manner that the concentrate will contain minimum amounts of organic substances which will not affect the excitation of atoms. Thus, Pavlenko et al. [18] used a mixture of cupferron and sodium diethyldithiocarbamate as extractant and chloroform as solvent. On extracting from solutions with pH > 7, both the reagents remain in the aqueous phase, for they are in the dissociated form under such conditions. For simultaneous removal of alkaline earth metals, trioctylphosphine oxide is added to the extractant. It is possible to extract at least twenty elements in one step: Ag, Al, As, etc. (Table 3.2). The elements, after evaporating the extract on a graphite collector and mixing with sodium chloride, are excited in a dc arc. Small amounts of trioctylphosphine oxide present in the extract do not hinder the subsequent atomic emission analysis.

Sometimes it is not necessary to mineralize the concentrate containing organic substances. Nebesar [23, 24] and Kuz'min and co-workers [25, 26] usually volatilized the organic solvent and adapted the reference samples, used for obtaining calibrated dependences, to the composition of the analysed sample by introducing spectroscopic defects and employing other techniques.

Several books and reviews (see, for example, refs. 27–29) are dedicated to the use of concentration in AES. Concentration has long been used in AES [30, 31]. The preconcentration technique suggested back in the 1950s by Gorbach and Pohl [32–36] has found application in analysis of acids [32, 33], ashes of plants [37], natural waters [38], rocks and soils [39], and others. After dissolving the sample, iron interfering with the analysis was precipitated by the ammonium benzoate method, and Ag, Al, Au, Bi, Cd, Co, Cu, Ga, Hg, In, Mo, Mn, Ni, Pb, Pd, Pt, Sb, Sc, Sn, Ta, Th, Ti, V, Y, Zn, Zr and rare earth elements were extracted at different pH values (3, 5, 7 and 9) with chloroform as chelates, with 8-hydroxyquinoline, dithizone and diethyldithiocarbamate.

Which concentration techniques have found extensive use in combination with AES? These are volatilization, solvent extraction, sorption methods, co-precipitation and electrochemical methods.

3.2.2 FRACTIONAL VOLATILIZATION

There is a concentration technique specific for AES – the method of fractional volatilization, the prototype of which can be found in ref. 40. This technique has long been developed in two variants [41, 42] for analysing pure nuclear materials: volatilization in air [41, 43] and under vacuum [44]. Fractional volatilization is often used in the analysis of high-purity substances and was developed to fit mainly the AES method. Preconcentration is based on the difference in the evaporation or sublimation temperatures of matrix elements and of the elements being determined. A refractory powder sample is placed in a small graphite vessel heated by a high-intensity current in a graphite furnace. The vapours of relatively readily volatile compounds of trace elements condense on the cooled graphite or copper capsule which subsequently serves as an electrode of the arc in atomic emission analysis. The use of two-stage evaporation ensures a more efficient removal of the matrix.

An example of this is the analysis of manganese [45]. Here, manganese was first converted into dioxide, mixed with graphite powder, and placed in a graphite crucible; Bi, Cd, Cu, Pb and Sn were volatilized at 1400–1500°C and condensed at the end surface of the graphite electrode (receiver). In another work [46] a method has been suggested for analysing molybdenum trioxide for Al, As, B, Ba and nineteen other elements; it involves fractional volatilization of impurities at 2000–2300°C. The sample is first mixed with graphite powder; relatively involatile molybdenum carbide and metallic molybdenum are formed on heating; the impurities are concentrated on a cooled receiver (carbon electrode). This is how one manages to attenuate the effect of the multiline spectrum of molybdenum.

Fractional volatilization is used also when the matrix is more volatile than the trace elements to be determined. Into the category of such matrixes fall iodine, cadmium, zinc, tellurium, antimony, selenium dioxide, ammonium chloride, mercury chloride, calomel, chlorides of arsenic, antimony, zinc or cadmium. The matrix element is volatilized or sublimated depending on its chemical form.

For analysis of comparatively large weighed amounts (5–10 g), it is expedient to use fractional volatilization of traces in a current of inert carrier gas [47]. In this case the sample carried in a boat of inert material is placed in a quartz tube pulled (from the side of the gas exit) into a capillary which is cooled. The impurities concentrated on the walls of the capillary are dissolved in a suitable acid and analysed. The detached capillary together with the condensate may be ground and analysed in the form of a powder [48]. Use is made also of carrier gases which by reacting with trace impurities convert them into highly volatile compounds. Thus, the trace impurities Al, Be, Co, Ga, Fe, In, Mn, Ni, Sn and Ti were volatilized in a current of hydrogen chloride in analysing synthetic silicon dioxide and natural quartz [49]. In the analysis of refractory metals [50] the trace elements were chlorinated with silver chloride which liberates atomic chlorine by thermal dissociation.

Fractional volatilization is also applicable when use is made of inductively coupled plasma (ICP) [51]. The determination of B, Be, Cd, Co, Cr, Cu, Fe, K, Li, Mn, Na and Pb in U_3O_8 powder is based on evaporating the traces from the solid sample placed in the recess of a graphite electrode and closed from above with a graphite lid with an opening. The upper end of the electrode was placed in the plane of the top coil, with the plasma ignited. In two seconds the electrode got heated to about 1800°C, after which emission was measured. In these conditions the matrix hardly evaporated at all and caused no hindrances.

3.2.3 OTHER VOLATILIZATION AND RELATED METHODS

Simple volatilization, sublimation, and driving-off after chemical transformations are widespread methods of preconcentration in AES; generally speaking, fractional volatilization also enters this group.

Concentration of trace elements can be achieved by simple volatilization of a highly volatile matrix – water, inorganic and organic acids, halides of As, Bi, Ge, Pb, Si, Sn and W, petroleum products and inorganic solvents. During fast volatilization of the matrix the determined elements may disappear as highly volatile forms and with the aerosol. However, a decrease in the volatilization rate is attended by an increase in the analysis time and may cause the correction and the fluctuation of the blank experiment to grow.

A proposal was made [52] to carry out fractional volatilization of the analyzed solution on a graphite collector whose temperature was 60–70°C higher than the boiling point of the sample. The solution was evaporated drop by drop in the device shown in Fig. 3.1. An amount of 50 g of graphite powder was taken in a crucible (diameter 8 mm, depth 5 mm) and the analyzed liquid feed rate was adjusted in such a way so as to give time for a drop to evaporate completely before a fresh drop got on the collector. The technique has been tested in the analysis of water, ethanol and amyl acetate. By ensuring a sufficiently high rate of volatilization, this method made it possible to eliminate the losses of Cu, In, Mn, Ni and V whose contents were $1 \cdot 10^{-6}\%$ of the mass.

To decrease background contaminations which determine the detection limits of trace impurities in analysis of high-purity silicon, the samples were decomposed with vapours of hydrofluoric acid and the matrix (silicon) was volatilized directly in graphite electrodes at atmospheric pressure [53, 54]. The method has been used for autoclave

216

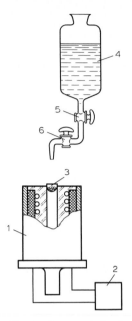

Fig. 3.1. Apparatus for fractional evaporation of liquid. 1 = Electric furnace for heating the graphite collector; 2 = collector of crucible heat temperature; 3 = crucible; 4 = dropping bottle; 5 and 6 = shut-off and adjusting stopcocks.

vapour phase decomposition of high-purity quartz glass (weighed amounts of 50–500 mg) in hydrofluoric acid vapours directly in the cavities of electrodes [55, 56]. The autoclave is schematically shown in Fig. 3.2. The matrix was volatilized completely in the autoclave and the level of background impurities of prevalent elements did not exceed $1 \cdot 10^{-10} - 1 \cdot 10^{-12}$ g. The reactor (11) made of polytetrafluoroethylene resin was placed in the steel body (1) which was made air tight by tightening the nuts (2). Between the steel body and the protective casing (7) are installed the heater (4), the cooler (5) and the thermal insulation (10). The sample was heated to about 220°C and the auto-clave was cooled down to room temperature in 30 min. As cooling agents first air was used and then water. The temperature was meas-ured with a Chromel–Copel thermocouple placed in the opening (6). The electric supply to the heater was controlled through a poten-tiometer. The shape and dimensions of the reactor made it possible to place it in the cup (3) containing 60 cm³ hydrofluoric acid and eighteen graphite electrodes (8 and 9). Complete decomposition of samples (and

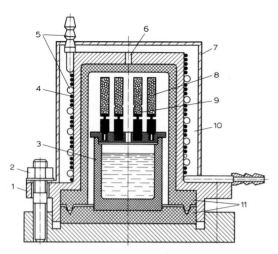

Fig. 3.2. Scheme of an autoclave for vapour phase sample decomposition. 1 = Steel body; 2 = fastening nut; 3 = Teflon cup with hydrofluoric acid; 4 = heating element; 5 = cooler; 6 = hole for thermocouple; 7 = protective housing; 8 and 9 = graphite electrodes; 10 = heat insulation; 11 = Teflon reactor.

volatilization of the matrix) in acid vapours at 220°C was reached in 3 h. The impurities concentrated in the cavities of graphite electrodes were determined by using dc arc discharge or a hollow cathode.

The same technique has been used to determine sixteen impurities in high-purity germanium [57]. Vapour phase autoclave decomposition of samples directly in graphite electodes eliminates the use of a collector and the contact of samples and concentrates with the vessel material and the laboratory atmosphere in the course of concentration.

Concentration after the formation of hydrides is a variant of volatilization used after chemical transformations. The system of continuous generation of hydrides goes well with ICP and other plasma sources – microwave plasma, plasma, and glow discharge of direct current, and others. Nakahara [58] has systematized numerous works dedicated to developing a method of analysis (including those which involve the use of automated systems) based on obtaining As, Bi, Ge, P, Pb, Sb, Se, Sn and Te hydrides in 10–30 s by means of the $NaBH_4$–acid reaction and by introducing a gaseous phase (or, rarely, solutions after the absorption of hydrides) into ICP. In particular, an automated system containing a reactor and a scanning spectrometer with ICP enables the hydride-forming elements and mercury to be determined at a

218

level of $n \cdot 10^{-7}\%$ [59]. In determining As, Bi and Sb in soils and precipitation, the combination of hydride generation and AES–ICP enables 100 samples to be analysed in one day [60]. The same combination ensured group determination of As, Sb and Se in sea samples (biological tissues and precipitation) [61].

Volatilization of boron fluoride (for determining 25 elements in boron and its compounds) [62] and wet mineralization of vegetation in a Teflon bomb (for determining 26 trace elements including sulphur in the needles of conifers) [63] have been suggested to fit ICP. In analysing biological materials their samples (0.3–1.0 g) were first mineralized (successively by concentrated nitric acid and 70% hydrochloric acid at atmospheric pressure or only with nitric acid under pressure) [64]. The solutions obtained were introduced into ICP and Ba, Ca, Cu, Fe, K, Mg, Mn, Na, P, Sr and Zn were determined in succession.

3.2.4 EXTRACTION

This method of concentrating trace elements is very widely used in AES. Extraction enables isolation of the determinable trace components and the matrix elements (see Chapter 2). Extraction of traces is the most extensively employed technique in the atomic emission analysis of multi-elemental and complex samples or of samples from which it is hard to extract the main components. In extracting trace elements attempts are made to retain one of the essential merits of AES: the possibility of multi-elemental determination. Of the variants of extraction that are suitable for the purpose, most important is the extraction of chelates, for example of dithiocarbamates, dithizonates, 8-hydroxyquinolinates or of complexes with 1-phenyl-3-methyl-4-benzoyl-pyrazolone-5.

The method of introducing the concentrate of trace elements into the excitation source is very important. Organic or aqueous solution containing the determinable elements may be directly introduced into the source or it may be first evaporated to dryness.

When solutions are introduced, better accuracy characteristics are obtained and the analysis time is reduced. Besides, it is easier to automate the analytical technique. However, the detection limits of elements become worse. In evaluating the prospects of this method, account must be taken of the ever increasing applications in which samples are introduced as solution. For direct introduction of solutions use is made of a rotating disc, vacuum cup, or porous electrode. Quite often solutions are introduced into the source as aerosol by pneumatic spraying. The methodical and technical aspects of this problem are discussed by Török et al. [65].

The technical difficulties faced in direct introduction of extracts into ICP have now been overcome. Thus, a method has been suggested for determining phosphorus in sea and river waters after it has been extracted as molybdoantimony–phosphoric acid with diisobutyl ketone [66]. In the analysis of impurities and precipitates, electrolytically separated out from steel and containing carbides, nitrites, and inter-metallic compounds, these were fused with $Na_2S_2O_7$, dissolved in tartaric acid, Cr, Fe, Mn, Mo and Nb were extracted as different complexes, and the extracts were introduced into a plasma [67]. Simultaneous determination of four to five elements took one hour, whereas spectrophotometric determinations lasted 18 h. A combined solvent (a mixture of o-xyloline and acetic anhydride) has been proposed [68] and used for direct introduction into ICP of the extract containing As, Bi, Cd, Co, Cr, Cu, Fe, In, Ni, Pb, Re, Se, Sn, V and Zn. This technique, besides eliminating the stages of solvent volatilization, mineralization of dry residue, and subsequent dissolution (or back extraction), simplifies the preparation of multi-elemental reference samples. Tao et al. [69], in the analysis of natural waters, extracted Cd, Co, Cr, Cu, Fe, Mn, Mo, Ni, Pb, V and Zn with o-xyloline as tetramethyldithiocarbamate and hexamethylenedithiocarbamate chelates and then introduced the extract into a plasma. The detection limits attained after 100-fold concentration ensured high-precision determination of trace contents of the enumerated elements.

Minimum detection limits are achieved by introducing the dry residue, obtained by evaporating the concentrate on a suitable collector (often on graphite powder), directly on the surface of an electrode or on a graphite base. In the given case the absolute detection limits of elements attain the values 1 ng to 1 pg and, sometimes, even less.

3.2.5 SORPTION METHODS

These methods are sufficiently widely used in AES. This is evidenced by numerous examples listed in Chapters 2 and 4 and also in refs. 29 and 70. Either the concentrate itself or the solution obtained after the desorption of determined elements is subjected to analysis. In the case of concentrate the sorbent should be so selected that it is no hindrance to determination. For example, such a convenient sorbent for AES is activated charcoal. In a number of cases use can also be made of synthetic organic sorbents, particularly if they have been obtained in a very pure form or they are purified prior to their application.

Shiryaeva *et al.* [71] concentrated Ir, Pd, Pt, Rh and Ru by sorption on polymeric thioether and, after elution, determined these elements by the AES-ICP method. The detection limit of $5 \cdot 10^{-5}\%$ was attained in analysis of solid products and a limit of $0.01\,\mu g/ml$ in analysis of ferrous ore processing liquid products with a relative standard deviation of no more than 0.06. It has been shown that it is possible to determine Au, Ir, Pd, Pt, Rh and Ru [72], concentrated on a Monivex ion exchanger, by introducing aqueous suspension of the sorbent (particle size $\leqslant 75\,\mu m$) directly into ICP. Pierce and Broun [73] have developed a method for isolating and determining barium and strontium in surface waters. Automatic isolation of these elements is effected in two Technicon devices adapted to carry out two cycles of operation in succession. The isolation cycle includes sampling of aliquot and transferring it into a cation-packed chromatographic column, addition of buffer solution, and subsequent elution of barium and strontium. Sodium chloride solution is added to the eluate containing barium and strontium to improve excitation of spectrum in the $C_2H_2-N_2O$ flame. By this method, 40 samples can be analysed in 90 min.

For determining platinum and palladium in geological samples [74] a sample (0.4–0.5 g) was treated with a (3:1) mixture of concentrated nitric and hydrochloric acids. And the solution, after necessary treatment, was passed through a column washed with HCl and packed with the ion exchanger Bio-rad AG-50Wx8 in the H^+ form. Then the column was washed with a ten-fold (with respect to the sample) volume of HCl. The isolated elements were determined by the AES–ICP method. According to the authors, the method is applicable also for the determination of gold.

3.2.6 ELECTRODEPOSITION

This method can be successfully combined with AES [75, 76]. Especially effective is the electrodeposition of trace elements on a graphite rod which is then used as electrode of a dc or ac arc. Ensuring significant lowering of detection limits of trace elements, electrolytic concentration enables satisfactory precision of analysis results to be attained: relative standard deviation in the working range of concentrations varies from 0.08 to 0.25; here, the main contribution to the value of this error is apparently made by the excitation source.

In the majority of cases the spectrum is excited, as said earlier, in an arc source. Examples of the use of hollow cathode discharge tube, spark discharge, plasma torch, and flame are also known. Since trace elements precipitate as a thin layer on the edge of the adjoining portion of the electrode, evaporation is very fast (10–15 s). As a result, it becomes possible to overlap spectra of several electrodes at one point of the photographic plate [77]. In all cases, carriers (preferably sodium chloride) are used in analysis. If trace elements deposit on the cylindrical surface of the graphite electrode, use is made of a support of specific design [78] which ensures rotational–reciprocating motion of the horizontally positioned electrode, the anode, relative to the fixed upper electrode, the cathode. Sometimes [79, 80] the upper layer of the graphite electrode containing trace elements is removed mechanically and analysed by the method of three standards. Reference samples are prepared using graphite powder as the base. The method of analysing concentrate using a rotating copper or graphite disc, which serves at the same time as electrode in the electrolytical cell, has also proved successful.

3.2.7 OTHER CONCENTRATION METHODS

These are rarely used in combination with AES. This refers in particular to co-precipitation of trace elements with inorganic and organic collectors, and especially to precipitation of matrix or trace elements. The main disadvantage of these methods is that concentration takes a long time.

In determining As, Cd, Cu, Fe, Mn, Ni, Pb, Sb and Zn in water samples [81] the trace elements were co-precipitated (pH 10–12) with lanthanum hydroxide. The precipitate that settled on a membrane filter was dissolved (1:1) in HCl and the solution obtained was introduced into an ICP.

Sugimae and Mizoguchi [82] have described a method of determining iron in air-dust particles, which involves collection of dust on polystyrene filters and preparation of suspensions suitable for introduction into an ICP.

3.3. Atomic absorption spectrometry

Initially, when AAS had just appeared, many analysts were of the opinion that AAS did not require preconcentration and separation of

macro- and trace elements owing to the absence of selective disturbances. However, it was soon ascertained that the application of this
method also often required separation of the matrix. Different concentration methods are used in combination with AAS, but most often
preference is given to extraction, sorption, distillation and related
methods.

3.3.1 EXTRACTION

Extensive use is made of the combination of extraction with flame
AAS. In addition to the advantages associated with preconcentration
proper (elimination of the matrix effect, simplification of calibration,
reduction in relative detection limits, etc.), this combination has one
more advantage, *i.e.* the absolute detection limit can be lowered by
substituting aqueous solution with an organic solvent. Probably, this
is mainly due to the formation of more finely dispersed aerosol when
extracts are atomized [83–85].

Residues, or an appreciable layer of emulsion, may sometimes form
on the walls of the syphon and the drainage system of the instrument
when organic solvents are sprayed. Besides, certain organic solvents,
for example aromatic and halogen substituted hydrocarbons, do not
burn completely and produce yellow smoking flames; low-boiling solvents (methanol, ethanol, diethyl ether, acetone) exert an adverse
effect on the burning stability due to intense evaporation in the spray-
er. High-viscosity organic solvents and benzol (due to its toxicity) are
also less suitable for the purpose. Methyl isobutyl ketone and butyl-
acetate are suitable solvents.

Chloroform is considered not to be a very good solvent. It burns
badly forming phosgene and hydrogen chloride; an increase in the
volume of gases in the flame and in detection limits is also observed.
Nonetheless, chloroform has valuable properties (formation of small
aerosols, small reciprocal solubility in water). Tsalev *et al.* [86] have
developed a system for drying aerosol with hot air flow. The system
succeeds in lowering the detection limit and accelerating the analysis
of chloroform extracts containing compounds of Co, Cu, Fe, Ni, Pb and
Zn with hexamethylene ammonium hexamethylenedithiocarbamate.
This technique also makes it possible to analyse other extracts that are
inconvenient for introduction into a flame (diethyl ether-, benzene-
based), as well as solvents which absorb strongly.

Studies aimed at improving the procedures for introducing aerosol into a flame have been made. Berndt and Jackwerth [87] have worked out an injection method which involves introduction of small volumes (*ca.* 200 μl) of solution into a flame. Conditions have been selected which enable $2 \cdot 10^{-9}$ g Ag, Bi, Cd, Co, Cu, Fe, In, Ni and Pb to be determined from 1 ml of sample. In comparison to the method of continuous introduction of sample into the flame, this method gives about a seven-fold gain in the absolute limit of detection. To increase the amount of substance reaching the flame in a readily atomized state, Bailey and Lo-Fa-Chun [88] suggest direct evaporation of organic chelates, in particular of acetylacetone and its fluorine substituted complexes. For this use is made of a specially heated chamber, from which the vapours are carried away into the flame by nitrogen flow; the detection limit of iron lowers in this case from 0.2 (usual pneumatic introduction) to 0.05 μg/ml. Kowamura *et al.* [89] have described a similar device for introducing β-diketonates of Cr, Cu and Fe in chloroform into a flame.

In the case of electrothermal AAS (ETAAS) the effect of the extract, *i.e.* the nature of solvent and extractant, the type of compound to be extracted, and the method and technique of introducing the extract, is interpreted with much difficulty and in a less unique manner [90].

Aggett and West [91] suggest that analysis of extract with the use of ETAAS would have definite advantages over flame atomization; removal of atomizer will make it possible to work with viscous extracts; complete removal of organic solvents from ETA upon drying will eliminate non-selective absorption of light at the atomization state and the effect of solvent on the value of analytical signal; the range of extracts suitable for atomic absorption determination will widen owing to the use of non-combustible chlorine-containing solvents. However, the greater sensitivity of extracts, as compared to aqueous solutions, to geometric parameters and the material of atomizer and to the composition of the shielding gas and the method of registering the atomic absorption signal [92] causes a significant effect of the organic matrix on the value and fluctuation of the analytical signal and, hence, on the metrological characteristics of AAS.

As it follows from the overview [90], in ETAAS sometimes the absolute detection limits decrease on going from aqueous solutions to extracts. The introduction of extract significantly changes the process of atomizing the determinable element compared to aqueous solutions: the conditions change under which drying, evaporation and transformation of the extracted (often altered) compound into a form that

TABLE 3.3

Values of atomic absorption signals of indium and gallium concentrations (0.1 and 0.08 g/ml), respectively, in aqueous solutions and extracts when atomized from the walls of an ETA (A_w) and from a graphite disc (A_d) in the presence ($A_{asc.}$) or absence of ascorbic acid

Composition of solution	Indium			Gallium		
	A_w	A_d	$A_{d, asc.}$	A_w	A_d	$A_{d, asc.}$
0.1 M HNO$_3$	0.101	0.177	0.127	0.154	0.209	0.276
0.1 M HCl	0.026	0.072	0.212	0.064	0.105	0.264
0.1 M HBr	0.022	0.085	0.205	0.057	0.094	0.213
Chloride complex in MIBK	0.032	0.130	0.221	0.061	0.138	0.271
Chloride complex in 2,2'-di-chlorodiethyl ether	–	–	–	0.012	0.109	0.224
Bromide complex in MIBK	0.025	0.072	0.126	0.045	0.130	–
8-Hydroxyquinolinate in MIBK	0.075	0.094	0.107	0.070	0.132	0.247
8-Hydroxyquinolinate in CHCl$_3$	0.008	0.021	0.310	0.006	0.710	0.228

yields analytical signal are effected. Difficulties are faced in metering because of spreading of the extract on the surface of the atomizer; also, partial absorption of sample into graphite takes place and uncontrolled evaporation of the solvent and the extracted compounds is noticed. All of these have a negative effect on the detection limit, precision and the accuracy of determination.

The nature of the solvent and the form of compound in the extract affect the absorption signal. The sensitivity of determination abruptly decreases when the elements are determined in the presence of halogen-containing solvents (chloroform, carbon tetrachloride, 2,2'-dichlorodiethyl ether). This is evidenced by the results given in Table 3.3 [92]. The main cause is the formation of readily volatile chlorides which are partially carried away from the atomizer in the undissociated state. It is true that halogen–graphite compounds of the type $C_n^+ X^- \cdot 3X_2$, which are stable up to ca. 2000°C, can also exist. The introduction of ascorbic acid eliminates the depressing effect of chlorine-containing solvents, while in a number of cases the analytical signal generated by the chlorine-containing organic matrix increases compared to aqueous solutions [90]. There also exists a problem of non-selective absorption of light, the absorption being caused by the residues of organic solvent.

Sometimes a less volatile compound, compared to an inorganic compound, is formed during extraction. This favourably affects the metrological characteristics. Thus, the volatility of mercury decreases

TABLE 3.4

Examples of atomic absorption determination of trace elements after extraction concentration

Trace elements determined	Material analysed	Main extraction reagent	Detection limit (%)	Ref.
Ag	Rocks, ores, concentrates	Diphenylthiourea	$1\cdot10^{-6}$	98
Ag, Cu, Tl	Metal halides	Diphenylthiourea	$10^{-5}-10^{-7}$	99
Ag	Rocks, ores, minerals	Triphenylphosphine	$1\cdot10^{-6}$	100
Au	Minerals, cyanide solutions	Triphenylphosphine	$1\cdot10^{-4}$ ($2\cdot10^{-3}\,\mu g/ml$)	101
Cu, Fe, Mo, V, Zn	Titanium tetrachloride	Trioctylamine	$2\cdot10^{-5}-2\cdot10^{-6}$	102
Bi, Cd, Cu, Pb	Alloy steels	Trioctylamine	$10^{-4}-10^{-5}$	103
Fe	Sodium and potassium halides	Dithizone, sodium tetraphenylborate	–	104
Ag	Rocks, minerals	O-Isopropyl-N-methyl-thiocarbamate	$1\cdot10^{-5}$	105
As, Bi, Cd, Co, Cr, Cu, Fe, Ir, Mn, Mo, Ni, Pb, Pd, Rh, Sb, Se, Te, Zn	High-purity silver	O-Isopropyl-N-methyl-thiocarbamate	$10^{-6}-10^{-9}$	106
Ag, Au	Geochemical samples	O-Isopropyl-N-methyl-thiocarbamate and mixture of petroleum sulphides	$5\cdot10^{-4}$ (Ag) and $7\cdot10^{-3}$ (Au) $\mu g/ml$	107
In	Aqueous solutions	Methyl isobutyl ketone and other solvents	–	108
Ga	Aqueous solutions	Methyl isobutyl ketone and other solvents	–	109
As	Copper	Dialkyltin dinitrates	$10^{-5}-10^{-6}$	110
Ir, Pd, Pt, Rh, Ru	Non-ferrous metals, ore processing products	Di-2(ethylhexyl)dithiophosphoric acid	$10^{-3}-10^{-5}$ ($0.05\mu g/ml$)	111

appreciably when the extract containing its (mercury) dithizonate is subjected to analysis [93].

On the whole, from the considered literature data it follows [90] that on going from aqueous solutions to extracts the detection limits determined by the ETAAS methods usually deteriorate. However, extraction ensures the removal of matrix and the interfering trace components; it is therefore an important stage of preparation of samples for analysis.

Now we shall talk about extraction systems used for preconcentration of traces in AAS. In general use is the extraction of chelates: 8-hydroxyquinolinates, β-diketonates and, especially, dithiocarbamates [9, 15, 83, 84, 94]. Very often use is made of ammonium pyrrolidinethiocarbamate (APDTC) [95]. It forms chelates with at least thirty elements; methyl isobutyl ketone (MIBK) is the best solvent for extraction of its chelates. Detailed information on the use of the APDTC–MIBK system in extraction–atomic absorption analysis is available in ref. 96. Many techniques involving the use of this system have been worked out for analysis of waters, biological and other samples. Instead of APDTC, hexamethylene ammonium hexamethylenedithiocarbamate has been successfully used [94, 97]. Yet AAS requires selective methods of preconcentration.

With the participation of the authors of this book a number of AA procedures have been worked out for determining small amounts of silver in natural and industrial materials (Table 3.4) [98]; the trace element, after decomposing the sample with a mixture of HCl and HNO_3, was selectively extracted from chloride solutions using chloroform solution of diphenylthiourea in the presence of large amounts of copper and other metals. Atomic absorption determination of silver was carried out by spraying the extract into a flame or by evaporating it from a graphite boat (in the "furnace-flame" type atomizer). For better precision of height and shape of absorption peaks, the extracts in both cases were preliminarily mixed with an equal volume of n-butanol. The determination sensitivity for 1% absorption was no less than 0.1 μg/ml in the spray variant; in the "furnace-flame" variant 0.0005 μg silver was determined.

A method has been proposed for determining gold and silver from a weighed amount. It involves separation of these elements by a mixture of petroleum sulphides and O-isopropyl-N-methylthiocarbamate in toluene [107]. Combining fifty-fold concentration with the use of a small volume of extract and its expeditious introduction into the atomizer ensures detection limits of $7 \cdot 10^{-3} \mu g/ml$ and $5 \cdot 10^{-4} \mu g/ml$ for gold and silver, respectively. Dinitrites of dialkyl tin (selective extractants of phosphate and arsenate ions) have been used for concentration of arsenic from aqueous solutions containing appreciable amounts of cations of various metals [110]. The extract was fed into a flame or, after volatilization of the solvent, into a furnace-flame atomizer. The latter has been used for determining $10^{-6}-10^{-5}\%$ arsenic in copper. Of the seventeen studied metals (Ag, Al, Bi, Ca, Cd, Co, Cr, Cu(II), Hg, Mg, Mn, Ni, Pb, Pd, Sb, Sn(IV) and Zn) only Cu(II) and partially Ag, Hg and Pd (in the presence of picrate ions) are extracted by the solutions of tetraazomacrocyclic extractants, XXXII and XXXIII [112]. Thousand-fold amounts of Co(II), Mn(II) and Zn and practically any amount of Ni are no hindrance to quantitative separation and to the determination of Cu(II). A technique that uses these reagents has been developed for extraction atomic absorption determination of Cu in metallic nickel.

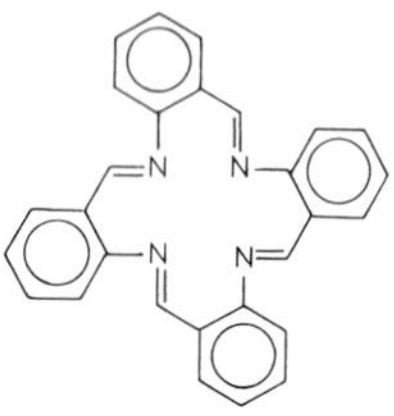

XXXII 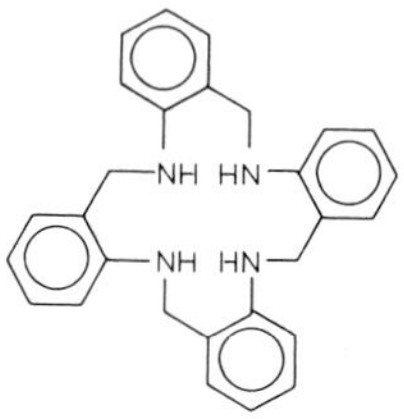 XXXIII

The combination of extraction and AAS can be automated. For instance, Pierce et al. [113] have proposed a highly efficient method for automatic extraction–atomic absorption determination of Cd, Co, Fe and Pb in waters, which involves separation of trace elements as pyrrolidinethiocarbamates with the use of methyl isobutyl ketone. The Technicon instrument used allows work with up to sixty samples per hour. The results are in satisfactory agreement with those obtained in unautomated conditions.

3.3.2 SORPTION

A concentrate (complex-forming sorbent or synthetic ion exchanger in the form of suspension) can be directly introduced into an electro-chemical atomizer [114–123]. This simplifies sample preparation and speeds up the analysis, although complications may arise in working with organic matrices (as in extraction AAS).

Isozaki et al. [116] have described the determination of copper,

228

present in the solutions to be analysed (volume 250 ml) at a level of $10^{-6}-10^{-7}\%$, after concentrating it on a chelate exchanger Chelex-100 (mass 0.1 g) under static conditions. The solid phase is mixed with 5 ml water and an aliquot of the suspension obtained (10 μl) is introduced into a graphite furnace. The residue is dried, reduced to ashes and atomized. Ion-exchange resins can also be introduced into a graphite atomizer for determining copper [115] as well as Co, Hg, Mo and Ni [120] sorbed on them. Direct atomization of the concentrate-sorbent (after thermal destruction of the organic matrix) was used to determine copper in solutions [117] and sea water samples [118], and platinum metals in complex natural samples [123].

Ref. 70 contains examples of using complex-forming sorbents for concentration of traces in analysis of natural waters, industrial materials, ores and rocks. The determination is often concluded with AAS.

3.3.3 VOLATILIZATION AND RELATED METHODS

Examples of successful application of fractional volatilization are known. Belyaev et al. [124] filtered the vapours of determinable elements through the bottom of a heated graphite crucible with a view to eliminating non-resonance absorption of matrix elements vapours. Then the vapours were condensed on the end of the auxiliary graphite electrode. The concentrate obtained was subjected to AA or AF analysis using an electrothermal atomizer. Independent selection of crucible temperature (for selective diffusion of the determinable element through the bottom), and also of the temperature build-up rate of the electrode with condensate, enabled Ag, Cd and Tl to be determined in carbonate sedimentary rocks. The sample was mixed with graphite powder and 15 ml of the mixture was taken in a graphite crucible clamped between the brushes of the atomizer. Bächmann et al. [125] separated Cd and Zn from aluminium and Os, Re and Zr from gold by the fractional volatilization method in the atmosphere of an inert gas, hydrogen and oxygen. Cd and Zn separate well from aluminium in an atmosphere of nitrogen and argon containing 10% hydrogen; Os, Re and Zr were separated from the matrix in a current of oxygen. Atomic absorption analysis was carried out after condensing trace elements in a graphite cuvette. The procedure for determining Bi, Cd and Tl in rocks [126] involves concentration of trace elements by volatilizing them at 1200°C in a flow of hydrogen and nitrogen and by atomization in a graphite cuvette.

For determining Cd, Cu and Pb in different samples, the samples were evaporated directly in a cup-type graphite atomizer allowing work with volumes up to 30 ml [127]. On this principle is based the technique for AA determination of silver in rain water and snow [128].

In common use is the combination of the method of hydride generation with AAS [58], which enables the hydride-forming elements with detection limits up to $1 \cdot 10^{-7}\%$ to be determined. Argon–hydrogen and nitrogen–hydrogen flames are widely used for this purpose. At present, they are being replaced by flame-tube atomizers, by atomizers in the form of a quartz tube which is heated from outside by a flame, and by electrically heated quartz tubes. Often, graphite furnace atomizers are employed for burning hydrides.

For determining As, Bi, Pb, Sb, Se, Sn and Te in coals and fly ashes of coals [129] the sample (200 mg) was decomposed in a Teflon bomb with a mixture of $HCl–HNO_3–HF$ at 100°C for 2 h. Then boric acid was added to the solution obtained. An aliquot of the solution was placed in a generator of hydrides; after adding an $NaBH_4$ pellet, the hydrides were volatilized at 3–5 M acidity and were carried over by nitrogen current into a quartz absorbing cell heated with a flame. A similar technique has been used to determine As, Bi, Sb, Se, Sn and Te in steels [130] and for semi-automatic atomic absorption determination of arsenic and selenium in geological materials [131].

The combination of generation of hydrides and their subsequent atomic absorption determination can be successfully automated. Some examples are the automatic methods of determining arsenic and antimony in geological materials [132], an automated system of generating Plasma-Therm hydrides with continuous supply of solutions of samples and its use for determination of arsenic and selenium by the flame AAS method [133], a method for automatic determination of arsenic [134] present in the form of water-soluble compounds, and others.

In concentrating mercury use is made of the fact that its elemental form is highly volatile. Thus, in analysing waters Welz and Melcher [135] passed through the sample (250 ml), after adding a reducing agent ($SnCl_2$) and HCl, a current of helium which, upon passing through a glass filter, entered into the tube containing porous alloy of gold and platinum. After absorption of mercury, the tube was heated and the metal vapours were carried over to the atomizer by a flow of helium.

In the atomic absorption determination of trace elements in organic substances and materials, petroleum products, food products and biological materials wide use is made of wet and dry mineralization. Hassan [136] generalizes data on the determination of Ag, Al, As, B, Ca,

230

Cd, Cl, Co, Cr, Fe, Hg, K, Li, Mg, Mn, Mo, Na, Ni, P, Pb, Rh, Ru, Se, Sn, Sr, V and Zn.

Analysing differing petroleum products (crude oil, mazut, gasoline, etc.) for Cu, Mn, Ni, Pb and V, Kägler [137] compared their detection limits determined in an HGA-72 atomizer and acetylene–nitrous oxide flames. Flameless atomization ensured a gain in the detection limit by one to two orders. A combination of preliminary ashing with flameless atomization assured determination of the indicated trace elements at their concentrations $\leqslant 1 \cdot 10^{-5}\%$. The determination of mercury in various organic and inorganic substances [138] involves direct burning of sample in an oxygen flow and precipitation of the trace element on a gold filter. After the burning is over, the filter is heated, the cuvette is filled with vaporous mercury, and absorption is measured. The detection limit of mercury in organic substances is $5 \cdot 10^{-8}\%$ and in inorganic substances, $ca.$ $1 \cdot 10^{-8}\%$.

3.3.4 OTHER METHODS

In electrodeposition the electode can be directly used for obtaining an electrothermal AA signal. If the electrode has the form of a thin filament or a coil, it is first heated by electric current to the required temperature and then the atomic absorption is measured [139–141], or the electrode is directly introduced into a tubular atomizer [142], particularly into tubular graphite atomizers [141, 143, 144]. Thus, mercury was volatilized after electrochemical concentration and the vapours formed were introduced by nitrogen flow into the atomizer [145]. Electrodeposition of trace elements in combination with AAS has been considered by Krasil'shchik et $al.$ [75] and by Chupakhin et $al.$ [76]. Therefore, we shall restrict ourselves to only two examples.

In determining gold in samples with a complex matrix [146] the trace element, after decomposing the sample, was electrolytically precipitated from the solution on a tungsten filament. The filament was then placed in a spherical absorbing cell and atomization was carried out in an inert atmosphere. A highly sensitive method of determining selenium (up to 0.1 μg/l) involves electrolytic concentration of selenium on a platinum wire which is then heated in an argon–hydrogen flame [147].

Co-precipitation is very rarely used in combination with AAS. In the analysis of mineral waters for Cd, Co, Cr, Cu, Ni and Pb the trace elements were co-precipitated with iron hydroxide or iron tetramethylenedithiocarbamate [148]. Aliquots of the solution obtained, after dissolving the residue in 0.1 M H_2SO_4, were introduced into the HGA-70 atomizer and the enumerated trace elements were determined successively. Of interest is the co-precipitation of heavy metals with ammonium pyrrolidinedithiocarbamate in a porous graphite crucible, and the use of the latter as ETA [149].

There have been attempts at substituting co-precipitation of trace elements with flotation, as this speeds up concentration. This question has been dealt with in a large number of studies conducted by Japanese analysts, especially Mizuike [150] and Hiraide and Mizuike [151] who have summed up the possibilities of this method. For instance, nanogram amounts of cadmium in waters were determined as follows [152]: from the sample (volume one litre, pH 9.1) cadmium was co-precipitated with zirconium hydroxide; a solution of sodium oleate was added and air was passed to float the precipitate; the concentrate was separated, dissolved in 6 M HCl and analysed by the ETAAS method.

In analysis of complex matrices two methods of concentration are applied in succession. Thus, for the decomposition of ores and rocks containing platinum metals use has been made of xenon tetrafluoride and chlorine trifluoride [153] which, upon heating, decompose with the formation of atomic fluorine. Free fluorine changes platinum metals into higher fluorides. Decomposition of ores and rocks by fluorination causes the determinable elements to partially concentrate at the cost of transformation of silicate, sulphide, sulphate, oxide and carbon-containing components into gaseous compounds. An important point is

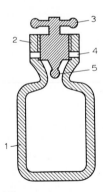

Fig. 3.3. Autoclave for sample decomposition. 1 = Body; 2 = neck with cutting; 3 = packing nut; 4 = gas outlet; 5 = packing.

232

that the excess reagent decomposes to gaseous compounds and therefore does not increase the mass of concentrate.

The autoclave used to decompose ores and rocks is shown in Fig. 3.3. Through a hole in the lateral wall of the threaded portion of the neck the gases escape on opening the autoclave. On the autoclave surface, which is made of pure nickel, a thick film of nickel fluoride is formed, which can withstand the action of fluorine up to 900 K. A fluorinating agent (in six to eight fold amounts compared to the sample mass) was placed in the autoclave. The autoclave was cooled with liquid nitrogen, ground sample was added, and the autoclave was air sealed. A shaft furnace was heated to 700 K at the rate of 10°/min and was kept at this temperature for 40–50 min. The autoclave was cooled to room temperature and then down to 230–240 K with a solution of solid CO_2 in acetone. The air tight cap was unscrewed until the discharge hole opened, and the autoclave was heated to room temperature. After the gases were released, the autoclave was opened and the dry residue was dissolved in 3 M HCl containing 0.01 M H_3BO_3 and 0.01 M $BeCl_2$. From the solution obtained, platinum metals (excepting osmium) were extracted with 1 M solution of n-octylamine in diisobutylketone. After back-extraction in 7 M HCl, platinum metals were determined by the AAS method using a flame or electrothermal atomiser. Detection limits of 10^{-5}–10^{-8}% were attained with a weighed amount of 50 g.

3.4. Atomic fluorescence spectrometry

This technique is developing rapidly. In a number of cases, atomic fluorescence spectrometry (AFS) has advantages over AAS [154–158]. Concentration makes it possible to eliminate the negative influence of quenching collisions in the atomizer and the matrix effect. Information on the use of electrodeposition and of volatilization after chemical transformation can be found in ref. 158. The generation of halides, as applied to AFS, has been considered by Nakahara [58].

The hydride generation method has been applied for determination of arsenic and selenium in coal (after decomposition), soil extracts, river and sea water samples [159]. The solutions of the sample and $NaBH_4$ were supplied into a 28-coil spiral tube with peristaltic pumps, and the isolated arsine and hydrogen selenide, which are separated in the separator, were carried over by a current of argon into a flame torch where it was mixed with hydrogen. Detection limits of 0.34 and 0.13 ng/

ml were attained respectively for arsenic and selenium. A device has been worked out for determining As, Bi, Sb, Se and Te (after converting them into hydrides) by a dispersionless method of AFS; it ensures detection limits of 0.1–0.2 ng [160]. Hydrochloric acid was poured into the solution and the mixture was placed in the reactor through which argon was blown; $NaBH_4$ was used as a reducing agent. The hydride formed was transferred to the cell heated to 1000°C. After atomization, the useful signal was measured. As it is inferred by Nakahara [58], hydride generation was used in the AF determination of hydride-forming elements in aluminium, lead, phosphoric acid, siliceous epitaxial structures, flying coal dust, sulphide ores, natural and waste waters, plants and food products.

Rigin [161] employed high-temperature volatilization of As, Hg and Se in an oxygen current from a complex matrix. A schematic diagram of the device used to decompose samples and to volatilize the deter-mined trace elements is shown in Fig. 3.4. The sample placed in a porcelain boat was burnt in a quartz furnace heated to 1500 K. The volatile products together with residual oxygen were transferred to an

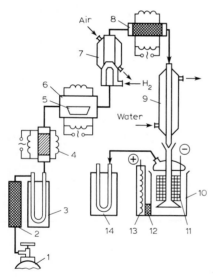

Fig. 3.4. Schematic diagram of an apparatus for test sample decomposition. 1 = Oxygen bottle; 2 = columns packed with zeolites and activated charcoal; 3 = cooler-trap with aerosol filters; 4 = heater for oxygen; 5 = boat with test sample; 6 = furnace for sample combustion in oxygen; 7 = hydrogen burner for reburning with air cooling; 8 = precipi-tation column; 9 = water cooler; 10 = electrolyzer trap; 11 = golden net cathode; 12 = porous glass plate; 13 = platinum anode; 14 = cooler-trap for arsine.

234

after-burning hydrogen torch. Thereafter, the gas phase was passed through a column packed with quartz beads and heated to 780 ± 10 K. In so doing the vapours of metallic mercury and the oxides of arsenic and selenium passed through the column while the oxides of other elements were held back. The determinable trace components entered into the electrolyzer and were absorbed by the electrolyte containing (by mass) 10% H_2SO_4, 5% Li_2SO_4, and 0.1% $K_2Cr_2O_7$. Mercury in the form of amalgam was electrodeposited on a gold cathode and was determined by dispersionless AAF. Arsenic was reduced to arsine which was concentrated in a liquid nitrogen trap. Thereafter, determination was carried out. Selenium was determined in the electrolyte. The method is capable of determining As, Hg and Se in different matrices like food products, soils and rocks.

An example of the combination of electrodeposition and AFS is the work by Rigin and Rigina [162] dedicated to the determination of lead in natural waters, soils, plants and blood plasma. From the solution to be analysed, lead was electrodeposited on a siliconized graphite rod. After electrolysis, the rod was washed, dried and placed in the current-conducting terminals of the atomizer. Atomization was performed in a flow of helium. A method has been proposed for dispersionless AF determination of mercury in natural waters and food products [163]; it involves electrodeposition of mercury from the solution to be analysed onto a gold cathode, transfer of mercury from the formed amalgam into the gas phase by heating and electrothermal atomization of the trace element in the gas phase in helium atmosphere.

Jones et al. [164] have developed an automated laboratory-scale system which, in combination with a six-channel atomic fluorescence spectrophotometer, was used for determining Co, Cr, Cu, Fe, Mn and Zr traces in sea water. The trace elements were separated as diethyldithio-carbamates using a mixture of n-butyl alcohol and MIBK, and the extract was sprayed into a flame. The throughput was 25 samples per hour.

3.5. Spectrophotometry and fluorimetry

Preconcentration is traditionally used in spectrophotometry and fluorimetry; it significantly widens the scope of these methods. Besides lowering the relative detection limits, preconcentration ensures separation of foreign substances which get absorbed or are luminescent in the required range of the spectrum and makes it possible to determine the elements which, under normal conditions, are determined only with difficulty.

The use of preconcentration in spectrophotometry and fluorimetry is so common that here it will not be considered in detail. Special sections or chapters of numerous monographs, reviews, etc. are devoted to this problem. Nonetheless, we would like to draw the reader's attention to the works which characterise the main features of such combinations.

3.5.1 EXTRACTION

Extraction has found extensive application. Normally, the intention is to achieve selective separation of the determinable element, although sometimes the matrix elements are extracted as well [15]. Most often, use is made of the same reagent for concentration which gives a coloured complex with the determinable element. However, two reagents may also be applied: first, the most selective reagent is used for separating the trace element, and then a reagent which may not be selective but is most suitable for photometry is employed. Thus, phosphorus(V) and arsenic(V) can be selectively extracted with dinitrites of dialkyl tin, and then, after adding molybdate, these are determined by the colour of hetero-polycompounds [165]. A highly selective method of determining mercury(II) involves its extraction separation from other elements in the form of HgI_2, and subsequent contact of the extract with aqueous solution of a basic dye [166].

To lower the detection limits of antimony determined by the extraction–fluorimetric method, Blyum et al. [167] suggest substitution of the crystalline violet cation in the ion associate to be separated by ethylrhodamine B or butylrhodamine B. In the analysis of geological samples the sample, after treatment first with hydrofluoric acid and then with sulphuric acid, was evaporated until the vapours of sulphuric anhydride appeared; the residue was then dissolved in $9\,M$ HCl and the ion associate, $SbCl_6^-$, with crystalline blue was extracted with benzene. Thereafter, the extract was brought into contact with a solution of ethylrhodamine B in $6.5\,M$ H_2SO_4, the organic phase obtained was removed and its fluorescence was measured. By this method, up to $0.3\,g$/ton antimony was determined in natural substances.

Reduction in the absolute limit of detection can be achieved by carrying out amplification reactions [168, 169]. An example of this is the conversion of As, Ge, P and Si into corresponding hetero-12-molybdenum complexes and subsequent determination of molybdenum in them by a sensitive reaction, in particular with phenyl fluorine.

Sometimes use is made of melt extraction for preconcentration prior to photometric determination. For example, an extraction–photometric method has been developed for determining nickel and lead [170]. The method involves extraction of nickel and lead chelates with dimethyl glyoxime, α-benzyldioxime, α-furyldioxime and 1,2-cyclohexanedioxime from water–alcohol solutions by using molten naphthalene. After the solution had cooled down, the solidified naphthalene was separated, dissolved in dimethylformamide or chloroform, and the optical density of the solution obtained was measured. This method is convenient, particularly, for quantitative and rapid extraction of elements that form kinetically inert complexes. For instance, in analysing synthetic solutions, Wasey *et al.* [171] extracted a palladium–phenanthraquinone monoxime complex with molten naphthalene. After cooling, naphthalene was separated, dissolved in chloroform, the solution was dried over anhydrous sodium sulphate, and the optical density was measured.

Macrocyclic compounds are promising for extraction–photometric determinations, the reason being their selective interaction with metal ions. Alkali metals are determined from absorption of a coloured counter-ion, *e.g.* of picrate, which goes into the extract in stoichiometric amounts with respect to the cationic complex of the metal with the macrocycle. Thus, a spectrophotometric method has been suggested for determination of potassium in blood by extracting it with benzene solution of 18-crown-6 (XXXIV) using Bromocresol Green as counter-ion [172]. By this method, 5 μg/ml potassium can be determined in the presence of 500 μg/ml sodium. In the analysis of tap water for lead (in the concentration range of $3 \cdot 10^{-3}$ to 0.5 μg/ml) the trace element was extracted with dichloromethane as a cationic complex with 18-crown-6 in the presence of eosin anion [173]. Then the fluorescence intensity of the extract was measured. The extraction–spectrophotometric method of determining traces of sodium in blood serum [174] involves its selective extraction with cryptand 2.1.1 (XXXV) solution in toluene in the presence of picrate ions. Determination of sodium is not hampered by 350-fold amounts of potassium and any quantities of other alkali metals. Polyvalent ions are masked with EDTA.

XXXIV

XXXV

An attractive method is the extraction–spectrophotometric determination of metals using macrocyclic compounds containing a chromophore group in the side chain or directly in the ring. An extraction–spectrophotometric method has been worked out for determination of sodium in blood serum using [2-hydroxy-3,5-dinitrophenyl]oxymethyl-15-crown-5 [175].

Nitrogen-containing macrocyclic compounds show promise for selective extraction of ions of transition and post-transition metals. Zolotov et al. [176] have studied four macrocyclic Schiff bases (XXXVI–XXXIX) having different cavity sizes:

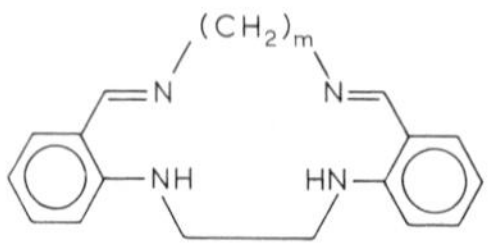

XXXVI   m = 2
XXXVII m = 4
XXXVIII m = 5
XXXIX   m = 6

Tetraazamacrocyclic compounds extracted only mercury from thirteen trace elements in the presence of picric acid. An extraction–photometric technique has been developed for determining copper in the extract; it involves selective extraction of copper in the presence of picrate ion using chloroform solution of the reagent XXXVII. Copper can be determined in the background of thousand fold and larger amounts of Ag, Ca, Co(II), Fe(III), Mn(II), Cd, Cr, Na, Pb, Tl, Zn and of hundred fold amounts of Ni and Sn(IV).

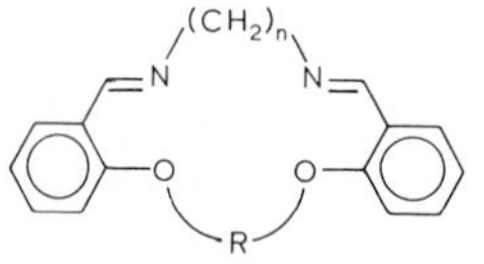

XL   n= 3; R = $-CH_2-CH_2-$
XLI n= 4; R= $-CH_2-CH_2-CH_2-CH_2-CH_2-$
XLII n= 4; R= $-CH_2-CH_2-O-CH_2-CH_2-$

In another work [177] the use of nitrogen-containing macrocycles of common formula as extracting agents has been studied. Counter-ions of picrylaminate or Bromothymol Blue extract only silver, while fourteen other elements are not extracted. A selective extraction–spectrophotometric method has been suggested for determining silver with the use of XLII and from the dipicrylaminate absorption. Determination is not hindered by seven hundred fold amounts of Hg(II) and hundred fold amounts of Cu(II); Cd, Co, Fe(III), Mn, Ni, Pb and many other elements are also of no hindrance.

238

Practically all concentration methods, for instance sorption techniques, can be combined with fluorimetry and photometry [178]. In analysing sea water for copper [179] the sample was acidified with HCl, filtered, and ammonium thiocyanate was added to the filtrate. The solution obtained was passed through an ion-exchange column packed with Amberlite CG400 in the thiocyanate form. Copper was eluted with $2\,M$ $HClO_4$ and, after adding 4-(2-pyridylazo)-resorcin and the masking agent, the eluate was measured photometrically. Nevoral [180] has developed a spectrophotometric method for determining rare-earth elements in mineral waters. The method involves three-stage isolation and separation of elements in ion-exchange columns packed successively with resins of the type Dowex 50Wx12, Dowex 1x8 and Dowex 50Wx8. Yoshimura et al. [181] have indirectly measured the optical density of the ion-exchange resin that had sorbed the chromium complex with diphenylcarbazide. In a similar manner they determined Co(II), Cu(II) and Fe(II) using respectively thiocyanate, zincon and 1,10-phenanthroline as reagents; as ion-exchange resins use was made of Dowex 50Wx12 (Fe) and Dowex 1x12(Cu and Co). The optical density was measured at 514 nm (Fe) and 630 nm (Cu and Co). The detection limit was less by one order than in the case of normal photometry. This method was recommended for analysis of natural waters.

Ruishi et al. [182] have described a spectrophotometric method for determining palladium in natural and sewage waters; it is based on forming a coloured complex of palladium with 5-chloro-2-pyridylazodiaminophenyl, sorbing the complex with a strong cation-exchange resin in static conditions, and measuring the optical density at 580 nm directly in the solid phase. The graduated curve was linear right up to the palladium concentration of $4.8 \cdot 10^{-2}\,\mu g/ml$. Direct spectrophotometric determination of concentrated trace elements in a solid phase sorbent is becoming a common technique [183].

Other methods of concentration are also used. A method has been worked out for isolation and concentration of phosphorus contained in metals and their salts [184]. The method involves reduction of P(V) to phosphine in the solid phase with the use of magnesium. The sample is mixed with magnesium powder in a boat, placed in a tube-like furnace (argon atmosphere), and heated. The mixture is then calcinated for 1 h at 560°C, cooled and transferred to the volatilization device filled with carbon dioxide, to which water is added in drops. The phosphine formed is collected in a receiver containing 25 ml of 10% solution of sodium hydroxide and the same amount of bromine so that the solution retained its yellow colour after absorption of phosphine. The distillate containing phosphoric acid is acidified with concentrated HCl, and phosphate ions are determined by the photometric method using yellow or blue molybdophosphoric heteropoly acid.

Conditions have been studied for group concentration of heavy metals by directed crystallization of water–salt solutions of eutectic composition with subsequent extraction of the carbamates of the determinable elements (Cd, Cu, Fe, Hg, In, Mn, Ni and others) by replacing them in these complexes with mercury(II) ions, and by measuring the optical density of diethyldithiocarbamate of mercury [185]. An example of this is the use of precipitation. A simple express method has been suggested [186] for concentrating phosphorus as the blue form of phosphorus–molybdenum heteroacid, which settles on the filtering membrane in the presence of a cationic surfactant (*i.e.*, dodecyltrimethylammonium bromide). Then the membrane was dissolved in a small volume of dimethylsulphoxide and the optical density of the solution was measured at 710 nm. The method is applicable for determination of phosphorus in water in concentrations 2.6–73 μg/l. Determination is not hindered by SiO_3^{2-}, anionic surfactant [di-(2-ethylhexyl)sulphosuccinate of sodium], non-ionogeneous surfactant (Triton X-100), and high concentrations ($\leqslant 0.5\,M$) of sodium chloride.

Flotation spectrophotometric determination of nitrite ion in water (concentrations being 3–40 g/l [187]) is based on its (nitrite ion) interaction with *n*-aminobenzenesulphonate of sodium and N-(1-naphthyl)-ethylenediamine. The formed azo dye was floated in the presence of sodium lauryl sulphonate, the foam was collected in a collector containing *n*-propanol, diluted with HCl and determined photometrically. Numerous examples of the application of flotation in spectrophotometry are available in the monograph by Mizuike [150].

3.6. X-ray fluorescence spectrometry

X-ray fluorescence spectrometry (XRFS) combines well with the concentration methods that can yield solid homogeneous samples. Among these methods are precipitation and co-precipitation, electrodeposition, sorption techniques and extraction with melts [188–194]. Direct analysis of solutions presents certain difficulties caused by corrosion of spectrometer units in analysis of corrosive liquids, precipitation of trace elements from the sample by radiation, abrupt increase in the absorption characteristics of the sample with increase in the concentration of the determinable element, and, as a result, an increase in detection limits.

3.6.1 PRECIPITATION AND CO-PRECIPITATION

Relatively low detection limits compared to those obtained by using inorganic compounds are attained if precipitation or co-precipitation is accomplished with organic reagents [188]. A rapid method for group determination of eleven elements in natural waters [195] involves precipitation of the elements in the form of pyrrolidinethiocarbamates at pH 3. The residue (on the filter) was analysed after drying in air. Iessuè and Tarali [196] have worked out an XRFS method for determining As, Bi, Cd, Co, Cu, Hg, Ni, Pb, Sb, Se, Sn and Zn in various matrices after their dissolution. The method involves precipitation of the elements with thioacetamide as sulphides in the presence of cellulose. As, Bi, Cd, Cu, Hg, Pb, Sb, Se and Sn were precipitated at pH 2.2 and at a temperature of 70–80°C; other elements were precipitated at pH 7–8. The precipitates after washing and drying were assembled, homogenized by grinding, converted into a tablet, and analysed. For determining microgram amounts of As, Cd, Co, Cu, Fe, Hg, Mn, Ni, Pb, Sb, Se and Zn in pharmaceuticals [197] the sample was dissolved in sulphuric acid in the presence of hydrogen peroxide and the trace elements were precipitated as dibenzyldithiocarbamates (excess reagent serves as collector) at pH 2–5. Very low solubility in aqueous solutions is the advantage of these chelates. In another work [198], trace amounts of Cu, Mn, Pb and Zn were co-precipitated with iron hydroxide at pH 8.5. Nanogram amounts of Co, Cr, Cu, Fe and Ni were co-precipitated as hydroxides and diethyldithiocarbamates at pH 3.8–4.5 using titanium as collector [199]. The precipitate was separated by filtering the solution under pressure through a Millipore cellulose filter. The precipitate that was collected on the filter was dried and analysed.

From natural waters, vanadium was co-precipitated with chromium-(IV) diethyldithiocarbamate and the precipitate that settled on the membrane filter was analysed [200]. For determining Cd, Co, Cu, Fe, Pb and Zn in water, Pik et al. [201] co-precipitated the trace elements with molybdenum pyrrolidinethiocarbamates. The residue was collected on a thin-film polycarbonate filter and the trace elements were determined. Analysing drinking, river and waste water samples for Cd, Cu, Fe, Hg, Pb, Se, Sn, Te and Zn, Panayappan et al. [202] co-precipitated the trace elements with thionalide in the presence of polyvinylpyrrolidone at pH 4. The residue was collected on a filter, dried, placed between polymer discs, inserted in the specimen holder of a spectrometer, and analysed.

Electrodeposition is also suitable for preconcentration of trace elements prior to X-ray fluorescence analysis. Marshall and Page [203] determined heavy metals in aqueous solutions after their deposition on an electrode made of graphite cloth. Carbon is convenient for subsequent analysis owing to its small atomic number and weak absorption of X-radiation. In the XRF analysis of fresh water for Co, Cr, Cu, Hg, Ni and Zn, Vassos et al. [204] deposited the trace elements as a thin film on pyrolytic graphite electrodes.

3.6.2 SORPTION METHODS

Sorption methods like precipitation and co-precipitation make it possible to obtain a concentrate in a form which may be used without complex preparation for analysis, i.e. in the form of a thin layer of powder-like substance.

The general requirement for sorbents is that they in their composition should not have elements with large atomic numbers. From this viewpoint, organic sorbents are generally better than inorganic ones. Among organic sorbents, wide use is made of ion exchangers and synthetic or cellulose-based complex-forming sorbents. Of the inorganic sorbents, silicas and glasses with attached complex-forming groups are quite suitable for X-ray fluorescence spectrometry. In most cases it is desirable that the sorbent is easily compressed; this requirement is not always obligatory.

For determining Ag, Al, Bi, Cd and many other elements in natural water samples, Vanderborght et al. [205] first converted the trace elements into 8-hydroxyquinolinates and then adsorbed the formed chelates on activated charcoal at 60°C. The latter was filtered out, forming a thin homogenized layer on a porous membrane filter, and analysed by the energy dispersion XRF method. The concentration coefficient was $1 \cdot 10^4$. Practically the same technique was employed in analysing chlorine electrolyte liquors for Mo, V and W [206]. The concentrate dried on activated charcoal was placed together with the membrane on a Teflon disc and then analysed. For determining $\geqslant 1 \cdot 10^{-7}\%$ silver and $2.5 \cdot 10^{-6}\%$ copper in pure lead, Lorder and Müller [207] passed the sample, after dissolving it in nitric acid in the presence of hydrogen peroxide, through a column packed with Chromosorb W-HP preliminarily dipped into o-dichlorobenzene solution of dithizone. The determinable elements were then concentrated as AgI_4^{3-} and CuI_4^{2-} on discs of the ion-exchange paper Serva SB-2, and analysed.

In another work [208] the metal ions were concentrated from aqueous solutions at pH 2–2.5 by using the cation-exchange paper Reeve-Angel SA-2; the suspended metal compounds were then separated by filtering through Millipore filters. This technique has been used for analysis of water samples of the Mediterranean sea.

Extensive use is made of complex-forming sorbents. Of these, Chelex-100, which contains chelate-forming groups of iminodiacetic acid, is convenient for concentrating heavy metals from different media [209–213], mostly from natural waters. On the basis of this sorbent, membrane filters have been developed which, after sorption, washing and drying, can be used for X-ray fluorescence spectrometry. Sorption may be employed in the static method also. The technique combines sampling with concentration of trace elements, enables these operations to be carried out under field conditions, and simplifies conservation, storage and delivery of samples to the laboratory. For group concentration of Cd, Cu, Fe, Mn, Ni, Pb and Zn, Brykina et al. [214] used the complex-forming sorbent Spheronoxine which contains 8-hydroxyquinoline groups. Sorption was carried out under static conditions at pH 5–6. The sorbent was analysed after drying.

A series of interesting works dedicated to combination of sorption of trace elements (with the use of cellulose exchange filters) and XRFS have been performed by Lizer and others (see Chapter 2, section 2.3).

Polymeric thioether has been used [215] to sorb Cd, Co(II), Cu(II), Fe(III), Hg, Ni, Pb and Zn (pH 4–7), In (pH 3–7), Ag, As(III), Au, Bi(III), Hg(II), Sb(III) and Sn(IV) (Ph 1), As(III) (1–3 M HCl), Au, Se and Te (1–10 M HCl) under static conditions. The sorbent makes it possible to isolate a large number of elements with different groups by varying the composition of the analysed solution. Subsequent determination is done directly in the obtained concentrate after compressing it. The emitters are strong and durable; the concentration techniques are simple and ensure good metrological characteristics. The method has been recommended for analysis of technological solutions of metallurgical production and environmental samples.

For sorption concentration of platinum metals from HCl solutions obtained upon decomposition of different composition samples, a polymeric tertiary amine of general formula $(-CH_2)_3-N]_n$, which is very selective towards platinum metals, has been used [216]. A sorbent (1.2 g) was added to 50–500 ml of 1–3 M HCl solution and the mixture was shaken for 1 h. The sorbent was then isolated and washed with 1 M HCl, acetone and diethyl ether. The air-dried concentrate was compressed into a tablet and analysed. The method ensures a relative standard deviation of 0.02–0.04.

Disam et al. [217] propose to concentrate Ag, As, Bi, Cd, Co, Cu, Hg, Ni, Pb, Se, Sn, Te and Zn by passing the solution to be analysed (different waters) through a thin layer of Mn, Zn and other sulphides; as a result of exchange reactions, the less soluble sulphides are held back in the precipitate which is then analysed.

Taylor and Zeitlin [218] have described an automated system for analytical control of composition of wash and waste waters which comprises an energy dispersion spectrometer with a semi-conductor detector, a minicomputer, and an automatic sample preparation block. As sorbent, use is made of an ion-exchange tape through which samples of volume 200–250 ml are discretely filtered at regular intervals. The subsequent operations – washing, drying and sample-emitter feeding into spectrometer – are performed automatically. Besides controlling the system operation, the computer processes the analysis results.

3.6.3 EXTRACTION

The use of extraction with melts [192, 193] is most interesting. When the concentrate is solidified, a solid solution is formed whose homogeneity is not inferior to that of glass. The low-melting-point organic substances that are used have in most cases similar and small values of X-radiation absorption coefficients. This simplifies accounting for the matrix effect and calibration and reduces the detection limits. In addition, the solidified compact phase concentrate is easily compressed and melted. That is why an emitter sample is readily obtained from it.

In extraction with melts two groups of organic compounds are used: substances acting simultaneously as reagent and solvent, e g. molten 8-hydroxyquinoline, and substances which act only as low-melting-point inert solvents, in particular, naphthalene and diphenyl. Extraction with low-melting-point solvents has certain advantages over conventional extraction. The degree of formation of low-stability compounds and the extraction rate (especially for kinetically inert complexes) increase and often the hydrolyzable elements are easily extracted quantitatively.

A method has been proposed for determining Cd, Co, Cu, Ni, Pb and

244

Zn [219] which involves extraction of these elements in the form of pyrrolidinedithiocarbamates with molten steric alcohol from hot solutions (0.5 M for sodium acetate or 0.1 M for ammonium citrate). The extract, after cooling, is converted into a disc and analysed.

Conventional liquid extraction is less convenient when used in combination with XRFS because the concentrate in this case is in the liquid phase. However, an extract can be prepared by XRFS by evaporating it to dryness on a suitable solid matrix: paper, cellulose, graphite, quartz, etc. [188]. Thus, the chloroform extract containing mercury(II) dithizonate was applied drop by drop onto a filter paper continuously blown with hot air [220]. Determination of traces of Co, Cu, Fe, Mn, Ni, Pb and Zn in solutions [221] is based on extraction of these components as diethyldithiocarbamates with methyl isobutyl ketone. An aliquot of the extract (100 μl) was applied drop by drop at the centre of a round target made of filter paper, and, after drying, it was analysed using a tungsten tube, a crystal of lithium fluoride and a scintillation counter.

Much less frequently the extract is directly analysed. An example is rapid determination of uranium after its extraction with 3-n-octylphosphine from industrial solutions [222].

3.6.4 OTHER METHODS

Volatilization and evaporation after chemical transformations were also employed in combination with XRFS; the concentrate is often obtained in a form convenient for analysis. Numerous examples illustrating the use of mineralization and volatilization in analysis of vegetable substances, animal tissues and human secretions, and natural waters can be found in refs. 188, 190, 194. Kato and Murano [223] have worked out a procedure for determining selenium in steels. The procedure involves reduction-oxidation volatilization of the trace component as H_2Se and its fixation on paper impregnated with silver nitrate. The paper is used as a radiator.

Thanks to the large number of nuclear particle accelerators X-ray spectral analysis with ion excitation (often of H^+, $^2H^+$, He^+ and He^{2+}) has found extensive application in the last few years. Unfortunately, liquids or biological samples cannot be placed in vacuum chambers without preliminary treatment. Bearse et al. [224] mixed the sample (0.1–5.0 ml) with potassium oxalate for determining the trace element composition in human and mouse blood. The mixture was chilled with liquid nitrogen and evacuated for several hours; as a result, the sample mass decreased by about five times. The sample was then mineralized by calcination in a stream of oxygen, and palladium chloride was added to the residue obtained as internal standard. Irradiating the sample with protons with an energy of 2 MeV, Ca, Cu, Fe, K, Pd, Rb, Se, Ti and Zn were determined by the X-ray spectral method.

Filtration is used in the analysis of atmospheric and industrial aerosols. In common use are cellulose and glass-fibre filters of which the latter are preferred for better mechanical properties. Dzubay and Stevens [225] have described a system for selecting aerosol samples and for separating dust particles into two granulometric fractions of sizes 2–10 μm and less than 2 μm. The automated device of Tosiuki and Kontiro [226] can analyse 1000 aerosol samples per 36 h.

3.7. Spark-source mass spectrometry

For mass spectrometric (MS) determination of trace metals a spark source of ions is widely used. Very low detection limits have been attained by using spark-source mass spectrometers with double focusing: absolute limits at the level of 10^{-12}–10^{-11} g and relative limits up to $n \cdot 10^{-7}\%$. With spark-source mass spectrometry (SSMS) it is possible to determine simultaneously more than sixty elements using very small weighed amounts (down to 10–20 mg). The precision is, however, not very high, but the relative error in a number of cases can be lowered to 10–20%. It is mainly due to the errors introduced by the photometric method of registration and the sample inhomogeneity. SSMS is used mainly for analysis of compact samples and other materials which can be converted into a tablet and have electroconductivity. Non-conducting substances including frozen liquids and dielectric materials can also be analysed by using specific techniques [227, 228]. The analysis in this case involves significant technical difficulties.

It is not always possible to make use of the advantages of SSMS because, owing to the small amounts of the specimen analysed, the sample may prove to be non-representative. Besides, the matrix affects the analysis results. The relative sensitivity coefficient (RSC) becomes unstable. Therefore, the precision and the accuracy of determination decrease, and the detection limits of elements may increase due to overlapping of the lines of the complex and polyatomic matrix ions on the analytical lines of trace elements. The use of reference samples eliminates these limitations, but the number of these samples is limited.

Finally, it is not always easy to utilize SSMS for analysis of liquid samples.

Suitable preparation of the sample, which involves preconcentration, makes it possible in a number of cases to overcome these difficulties. The trace elements may be separated on a suitable collector, precipitated and sprayed; the concentrate may be evaporated directly on the edge of the electrode (in analogy with atomic emission analysis). The calibration process may be appreciably simplified and an internal standard may be introduced.

Not many publications are devoted to the combination of preconcentration with spark-source MS, but they suggest that it is expedient to use SSMS with concentration methods.

Gladskoi et al. [229] analysed the concentrates deposited on a graphite electrode. The solutions of chlorides or nitrates of Ag, Al, Bi, Co, Cr, Cu, Fe, Ga, Mg, Ni, Pb, Sb, Se, Sn, Te, Ti, Tl and Zn and a solution of the internal standard (cadmium) were evaporated on the edges of spectrally pure electrodes (after impregnating the working surface with polystyrol solution in benzene and treating with low power spark discharge). A satisfactory agreement was observed between the measured and the initial concentrations of a majority of elements. This enabled the RSC of these elements to be computed in reference to cadmium. The values for Al, Fe and Mg showed a significant scatter; this was apparently owing to the uncontrolled effect of the laboratory air and to the contaminations introduced by carbon, water and the salts of determinable elements. The absolute limit of detection was 10^{-10}–10^{-11} g. The described procedure may be used for analysis of concentrates obtained by different methods.

A number of works are devoted to extraction preconcentration. Murozumi et al. [230] used the MS variant of the isotope dilution method for determining Cd, Cu and Pb in sea water after extracting the trace element dithizonates with chloroform and back-extracting them with a mixture of nitric and hydrochloric acids. Aliquots of the obtained solution (on an instrument with surface ionization) were subsequently analysed for each metal.

In analysing steels, Zolotov et al. [231] extracted the trace amounts of As, Bi, Cd, Cu, Pb, Sb, Sn and Zn with MIBK from iodide–sulphate solutions obtained by dissolving the sample. Dilute sulphuric acid was added to the extract and the obtained back-extract was evaporated with the collector (high-purity grade ammonium oxide). The concentrate obtained was calcinated at 800–900°C and analysed. Non-conducting aluminium oxide (5–10 mg) was pressed into a metallic aluminium tablet for the purpose (ref. 228, p. 142). Owing to the removal of matrix the mass spectrum of the concentrate was found to be much simpler than that of the steel sample, for the lines of polyatomic or complex ions of the collector did not overlap the analytical lines of the trace components. As a result, it became possible to increase the number of determinable elements and to lower the detection limits. Statistical processing of the analysis results, analysis being carried out by direct method, and of the data obtained on combining extraction with mass spectrometry revealed that the relative standard deviation in the results of Pb, Sb and Sn determination varied between 0.188 and 0.677 in the former case, and from 0.059 to 0.164 in the latter case. The detection limits of As, Bi, Cd, Cu, Pb, Sb, Sn and Zn in the analysis of steels (and also of sulphuric acid) were 10^{-5}–10^{-7}%.

By direct SSMS it is difficult to determine noble metals in rocks, ores and their treatment products, the reasons being non-uniform distribution of these metals in the sample, very low concentration of the elements to be determined, and overlapping of intense lines of complex ions on the platinum-group metal lines. This is the reason why Petrukhin et al. [232] used solvent extraction as a method of concentration of platinum metals in combination with their subsequent determination. N,N-Hexamethylene-N'-phenylthiourea (HMPTU) was used as group extractant. The compounds of platinum metals were preliminarily labilized with tin(II) chloride. 1% solution of HMPTU in acetone was added to the solution, the mixture was heated for 40 min, cooled, and then the formed complex was extracted with chloroform for 15–20 min. Simultaneously with this, copper was also extracted until the stoichiometric HMPTU to Cu ratio of 1:1 was not attained in the extract. The obtained extract was washed, ashed, and calcinated for 1 h at 600°C; 5–10 mg of the residue, principally composed of copper oxide, was compressed and analysed. For a concentration coefficient of 50 the lower limit of determinable contents was $1.6 \cdot 10^{-6}$%.

The examples given below will enable the reader to judge about the significance of other methods of concentration used in SSMS.

For determining about forty elements in natural water samples [233], the concentration level being 0.1 μg/l, a solution of 8-hydroxyquinoline in acetone was added to 1 l of the sample (pH of ca. 8) and the trace components were sorbed with activated charcoal under static conditions. The concentrate was mineralized in a current of oxygen by applying a magnetic field. Under these conditions the activated mole-

248

cules, radicals, ions and oxygen atoms completely oxidised the weighed amount of activated charcoal (containing the determinable elements) in 1 h at 40–50°C. The residue was dissolved in a minimum amount of concentrated nitric acid, an internal standard (indium) and high-purity graphite powder were added, and the solution was evaporated to dryness. The dry residue was compressed into an electrode and analysed.

Ion-exchange chromatography was used in combination with SSMS for determining rare-earth elements in rocks [234]. The sample after suitable treatment (0.5 M solution of HCl) was passed through a column packed with the cation exchanger AG50Wx8 in the H^+ form. The elements hindering the determination were eluted with 1.75 M solution of HCl, and the rare-earth elements together with Ba, Hf, Sc, Y and Zr, with 4 M HCl. The eluate, after necessary treatment, was evaporated on a graphite collector, compressed into an electrode and analysed. In the analysis of sea water samples [235, 236], trace elements were also concentrated with ion exchangers.

Volatilization has been used for analysing highly pure water, nitric, hydrochloric and acetic acids for 40–60 elements with detection limits of 10^{-9}–10^{-12}% [237]. The sample was carefully evaporated to about one drop and final evaporation was done on a high-purity silicon substrate. In analysing highly pure HNO_3, HCl, HF and acetic acid [238] the matrix was volatilized by slight heating (collector, graphite powder). Volatilization of matrix was also employed for determining trace elements in highly pure germanium tetrafluoride [239].

Sometimes two different methods of concentration are combined. For example, the isotope dilution method was used for simultaneous determination of Ag, Cu, Ni, Pb and Pd in platinum [240]. The sample (250 mg) was dissolved in *aqua regia*, known amounts of suitable isotopes of the determinable elements were added, platinum was removed with a cation exchanger, and the trace impurities were then electrodeposited on a gold electrode. The electrode was then analysed. For determining trace amounts of uranium and thorium in silicate rocks, meteorites and other natural materials [241], lead was isolated by electrolysis from the sample after dissolution and the solution was then passed through a chromatographic column packed with the inert carrier Kel-F on which tributylphosphate was deposited. Uranium and thorium were eluted from the column and, after necessary treatment, were determined by the isotope dilution method. Thus, it became possible to determine up to $1 \cdot 10^{-9}$ g uranium and thorium. For determining rare-earth elements in rocks, Puymbroeck and Gijbels [242] melted the samples with sodium peroxide, decomposed the alloy with hydrochloric acid, separated and cleaned the fractions of rare-earth elements with subsequent precipitation of hydroxides and fluorides. The elements were then converted into the nitrate form and the solution was evaporated on a graphite collector or was separated on ion exchangers. Best precision was attained on using anion-exchange chromatography.

3.8. Electron spin resonance

3.8.1 DETERMINATION OF PARAMAGNETIC IONS

The electron spin resonance (ESR) method is suitable for determination of paramagnetic substances: ions of a number of transition metals, free radicals, and compounds with an even number of electrons in the triplet state. Analytical applications of ESR are reviewed in refs. 243–248. Earlier, direct application of this method for qualitative detection and quantitative determination of elements was generally based on paramagnetism of transition metal ions. This limited the possibilities of the method because the number of metal ions giving an ESR signal is relatively small. Among them are copper(II), manganese(II), cobalt(II), molybdenum(V), vanadium(IV), gadolinium(III) and some other metals. A series of methods have been worked out for determining these elements after their concentration by various methods, $e.g.$ extraction.

Yamamoto et $al.$ [249] have worked out a method for determining copper(II) as diethyldithiocarbamate. Benzene extract gives a characteristic ESR spectrum whose intensity enables up to $5 \cdot 10^{-3} \mu g/ml$ Cu to be determined, and which varies linearly with the concentration of the determinable element in the range $0.005–1.0 \mu g/ml$. Mg, Fe(II), Ba, Mn(II), Co(II), Zn, Pb(II), Ni(II), Br^-, I^- and NO_3^- do not affect the copper determination; Hg(II) and CN^- interfere with the measurement.

Noble metals in the usual oxidation state [Pt(II, IV), Pd(II) and Ag(I)] are diamagnetic in contrast to Pt(III), Pd(III) and Ag(II) which have paramagnetic properties. Solozhenkin et $al.$ [250] have proposed to determine these elements by the ESR method after converting them into the paramagnetic state. Diamagnetic platinum(II) and palladium(II) bisdithio-α-diketonates present in the extract were oxidised to the paramagnetic state by adding n-phenylenediamine in pyridine. Quantitative determination amounts to finding the intensities of resonance lines and comparing them with the calibration plot. The method makes it possible to determine up to $20 \mu g/ml$ Pt and $2 \mu g/ml$ Pd using $0.25 ml$ extract. In a similar manner silver(I) present as diethyldithiocarbamate in a benzene or chloroform extract was reduced with tetra-ethylthiuram disulphide to Ag(II) which has paramagnetic properties. The method proposed [251] for determination of titanium in the chlorination products of titanium-containing minerals involves reduction of Ti(IV) to Ti(III) using amalgamated zinc in a $6\,M$ HCl medium and measurement of the ESR signal; Ti(III) has paramagnetic properties. The inhibiting effect of Nb(V), which changes into paramagnetic Nb(IV) under these conditions, is eliminated by keeping the solution to be analysed for $3\,h$ prior to reduction, $i.e.$ by forming hydrolysed forms of Nb(V). The ESR spectrum is recorded at liquid nitrogen temperature. The lower limit of determinable contents of titanium equals 0.1 $\mu g/ml$.

Solozhenkin et al. [252] have developed a method for determining silver in ores and their process products containing 10–5000 g/ton of metal. First the sample is quickly dissolved in an autoclave and then, after a suitable treatment at pH 4, silver is extracted with benzene solution of copper diethyldithiocarbamate. The silver in the extract is then determined from the variation in the concentration of copper.

3.8.2 USE OF SPIN-LABELLED REAGENTS

The analytical possibilities of the ESR method have significantly widened after the synthesis of organic compounds that contain stable radical fragments in combination with complex-forming groups capable of bonding metal ions without the loss of the properties of a free radical by the formed molecule [246]. It became possible to determine diamagnetic ions of metals. Besides, this technique ensured significant lowering of detection limits, as free radicals are determined with high sensitivity by the ESR method. Systematized information on stable free radicals of different classes can be found in refs 253–256. The most promising are nitroxyl radicals containing an unpaired electron in the $>N \doteq O$ fragment [257–259]; these exhibit high stability with practically complete localization of the unpaired electron at the N–O bond.

The ESR spectrum of the compounds containing one nitroxyl group consists of three narrow equidistant lines of similar intensity (Fig. 3.5). The spectrum parameters are less dependent on the chelate-forming groups. The line amplitude, which is proportional to the concentration of radicals present in the solution, may be used as an analytical signal. Compounds with two nitroxyl groups yield a similar spectrum, provided two paramagnetic centres do not interact with one another. If they interact, either along the chain that connects them or indirectly

[260], the spectrum will contain five or more lines. A more typical spectrum is observed when the interactions of paramagnetic centres are significant. Then, the ESR spectrum of the complex differs from that of the reagent.

Many complex-forming reagents containing nitroxyl radical – more than a hundred compounds – have been synthesized [261–269]. Among them are β-diketones (XLIII), xanthate (XLIV), oxyazo compound (XLV) and enaminoketone (XLVI).

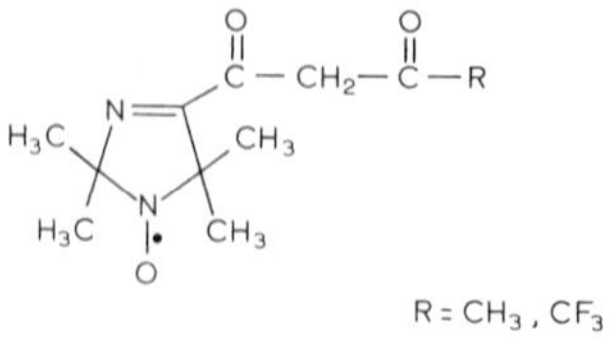

XLIII 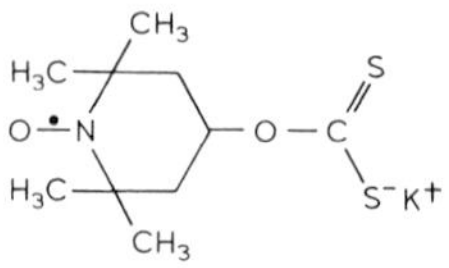   XLIV

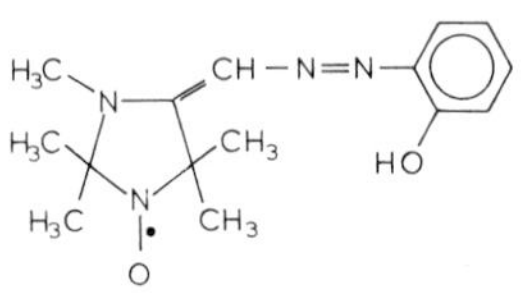

XLV 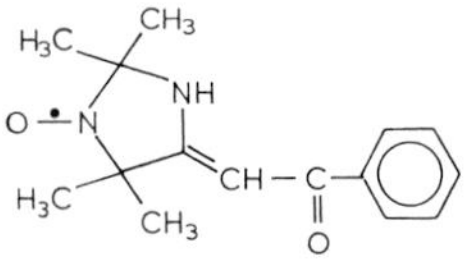   XLVI

It cannot be said that the application of spin-labelled reagents is devoid of difficulties. The main difficulty consists of eliminating the effect of ESR spectrum of the excess reagent upon the spectrum of the complex, if the reagent and the complex have similar spectra. Therefore, the excess reagent should be removed, say, by an extraction or sorption method before measuring the ESR signal. The fact that the reagent is mainly in the aqueous and the complex is in the organic phase facilitates the separation of ESR signals of the reagent and complex of course. The amount of excess reagent needed for complete removal of metal depends on the stability of the formed complex. The chemical stability of the reagents–radicals proper attains special significance because the spectroscopically active impurities present in the starting reagent, or formed during extraction, create a background and will hinder the determination of metals. These requirements, formulated by Zolotov et al. [270], should be taken into consideration in the synthesis and study of analytical radical-containing reagents.

Zolotov et al. [270] studied extraction of zinc with the use of 2,2,6,6-tetramethylpiperidine-1-oxyl-4-xanthate (TOX) (XLIV). This reagent

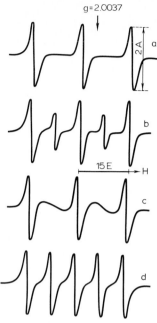

g = 2.0037

2A

a

b

15 E

H

c

d

Fig. 3.5. Typical spectra of nitroxyl-containing reagents and their complexes. a = Reagent with one nitroxyl group; b = reagent with two nitroxyl groups; c = spectrum of zinc 2,2,6,6-tetramethylpyridine-1-oxyl-4-xanthate; d = spectrum of palladium(II) 4-hydroxy-iminomethyl-2,2,5,5-tetramethyl-3-imidazoline-1-oximate.

readily dissolves in water and therefore enables metal complexes to be extracted in such a manner that excess reagent remains in the aqueous phase. Complexes with Co, Cu, Fe, Ni and Zn are extracted with chloroform. A detailed study has been made of extraction of zinc; at pH 7 and the reagent to metal ratio 1:10, 96% zinc is recovered, probably as ZnA_2. The impurities present in the starting reagent and formed during extraction have a detrimental effect on the determination results, therefore they are preliminarily separated chromatographically in a thin layer or in a column. The detection limit of zinc equals 0.01 $\mu g/ml$. However, this method cannot be employed in practice because of the decomposition of the reagent and the complex.

Solozhenkin et al. [271] have obtained complexes with paramagnetic properties by reacting Ag(I), Hg(II) and Pd(II) with the same reagent. Maximum extraction of Pd(II) is observed at pH 6 (acetate buffer solution) and of Ag and Hg at pH 10 (ammonium buffer solution) using chloroform as organic solvent. The extracted complexes had the following composition: Ag to A, 1:1, Hg to A 1:2, Pd to A, 1:2. For determining silver in zinc concentrate, aqueous solution of iminooxylxanthate was added to the solution obtained after decomposition of the sample (pH 10), the formed complex was extracted with chloroform, and then the ESR signal intensity was measured. The silver content was determined using the calibration curve. In a similar manner (but for different sample conditions) they determined mercury and palladium. A method has been worked out for determining copper in the products of non-ferrous metallurgy with this reagent [272, 273].

The extraction properties of 4-acetoacetyl-2,2,5,5-tetramethyl-Δ^3-imidazolin-1-oxyl (XLIII) have been studied [274]. Chloroform was used as solvent. The properties of the reagent were studied by extraction paper chromatography with respect to ions of 26 metals. Separation was maximum for Ag, Cd, Cu, Hg and Pb; Fe(III) was extracted only in the presence of caproic acid. Cu and Fe(II) were also extracted by a normal method. The ESR spectrum has a triplet shape. This reagent, after the removal of excess, was used for radiospectroscopic determination of copper, cobalt and nickel in extracts.

Zolotov et al. [275] have proposed a spin-labelled hydroxyazo compound, i.e. 4-(o-hydroxyphenylazomethyledene)-2,2,3,5,5-pentamethyl-imidazolin-1-oxyl (XLV). Conditions have been found for quantitative extraction of Co(II), Cu(II), Hg(II) and Pd(II) using this reagent and chloroform. Complexes of these metals have been isolated and the parameters of the ESR spectrum of palladium have been determined. An extraction radiospectroscopic method has been proposed for determination of palladium with a detection limit of $1 \cdot 10^{-7} M$. A complex of palladium and the reagent were quantitatively separated on a column packed with silica gel using a mixture of chloroform and acetone as mobile phase. Another spin-labelled reagent, enaminoketone (XLVI), has been proposed for radiospectrometric determination of Cu(II) [276]. Complexes of this reagent with Ag, Co(II), Ga, Hg(II), Mo(VI), V(V) and Zn are also extracted with high coefficients of distribution. This is the reason why these compounds hold promise for radiospectrometric determination.

3.9. Activation analysis

3.9.1 GENERAL REMARKS

This is one of the highly sensitive methods of modern analytical chemistry. At a neutron flux of about 10^{13}–10^{14} neutrons cm^{-2}s^{-1} more than seventy elements in the concentration range of 10^{-4}–$10^{-9}\%$ (sometimes to $10^{-12}\%$) can be determined by neutron activation analysis (NAA). The use of semiconductor Ge(Li) detectors, multichannel analyzers, two-crystal coincidence and anticoincidence spectrometers, computers to control the process of identifying radionuclides by their spectra, mathematical processing, and data storage and comparison has widened the possibilities of the instrumental version of NAA. The nature of radiochemical procedures has also undergone marked changes, especially from the viewpoint of their simplification.

One would think that intensive development of the instrumental variant of NAA should reduce to naught the use of concentration. However, at present it is hard to imagine an activation analysis of multicomponents and complex objects, say of rocks, ores and minerals without concentration and separation. The methods used to concentrate and separate macro- and trace elements have great importance in the NAA of samples with a strongly activating matrix: As, Au, Ga, Mn, Na, Sb, Sc, Ta and others. There is also a trend towards partial or complete use of automated schemes of analysis which appreciably reduce the radiation dose received by the analyst. At the same time, efforts are aimed at finding selective, radiation stable extractants, sorbents and co-precipitants that would make it possible to simplify the analysis and to develop unified and readily adaptable schemes of separation and concentration techniques.

NAA places a specific requirement on concentration techniques, *i.e.* the mixture of radioisotopes should be separated into the desired (minimum) number of fractions, commensurate with identification and quantitative determination of radioisotopes.

Any additional stage of analysis, including preconcentration, can lead to uncontrollable losses of trace elements or, alternatively, impurities may be added by the reagents used, the vessel, and the air of the laboratory. This holds true for all methods excepting the activation method. In the considered method there is no need to apply a blank correction. When the sample is irradiated before separation and concentration, only radioactive contaminations are undesirable, and their appearance is either less probable or can easily be detected. Quantitative separation of the determinable element is not obligatory in activation analysis because its losses can be easily accounted for. These are the main advantages of activation analysis. However, the application of concentration in activation analysis is made more difficult by the radiation hazard and the difficulties faced in the analysis of short-lived isotopes.

Of the concentration methods used in activation analysis, the most important are extraction and sorption methods; less important are precipitation, co-precipitation and electrodeposition. Reagentless evaporation methods are effective and simple; these are indispensable in the activation analysis of waters, mineral acids and organic solvents.

There are two methods of combining preconcentration and activation analysis: in the former (rarely used) variant the components to be determined are separated from the matrix before irradiation, and in the latter, after irradiation.

3.9.2 PRECONCENTRATION BEFORE IRRADIATION

Concentration before irradiation deprives NAA of its important advantage: the absence of blank correction. However, concentration prior to irradiation makes it possible to analyse substances that contain matrix elements which are strongly activated or intensively absorb neutrons. It also reduces the radiation hazard. Recently, interest in this has increased; at least four reviews have appeared, dedicated specifically to the use of concentration prior to irradiation in analysis of ores and minerals, environmental samples, inorganic compounds and biological materials [277–280].

One of the first works was related to the use of extraction concentration. In the analysis of sodium iodide for calcium and magnesium [281] the matrix (sodium) becomes highly active, and this requires the work to be carried out in special laboratories for radioactivity; in a number of cases this hampers the measurement of radioactivity and the interpretation of data. That is why preference is given to preliminary extraction concentration of determinable elements as chelates with 1-phenyl-3-methyl-4-benzoylpyrazolone-5 and to subsequent irradiation of the evaporated extract in a reactor.

Concentration prior to irradiation is not often used in analysis of environmental samples. In field conditions, large volumes of samples (natural and sewage waters, air) are taken, very dilute liquid and gaseous media are concentrated, and highly sensitive determination of trace elements is carried out under proper conditions, *i.e.* in well-equipped laboratories.

In analysing natural waters containing 21 cations, gold was concentrated with 0.5 M benzol solution of petroleum sulphides, coated on porous Teflon tablets [282].

Besides extraction, other methods are used. For determining small amounts of Ag, Au, Co, Mo, Sb, Sc and W in sewage waters the samples (solutions with different pH) were passed through columns packed with AV-17-8 and KU-2-8 exchangers, and then the sorbents were directly irradiated [283, 284]. This ensured the removal of sodium, bromine and other elements that hinder the determination. Akaiwa *et al.* [285] concentrated As, Cd, Co, Cu, Hg, Mn, Sb and Zn from natural waters using an ion exchanger treated with excess 8-hydroxyquinoline-5-sulphonic acid. The ion exchanger concentrate was washed with water, dried, and irradiated with a neutron flux. Complex-forming sorbents Rexyn 201 were used to concentrate As, Cd and Zn from river water (at pH 10) prior to irradiation [286]. Under similar conditions, Br, Cu, Eu, Hg, K, La and Na, which are quantitatively sorbed in a column, can be concentrated.

Of course, in concentrating trace elements prior to irradiation preference must be given to reagentless techniques in order to reduce the correction and the blank experiment fluctuation. For instance, using radioactive indicators Harrison *et al.* [287] studied the behaviour of 21 elements (present in natural waters) in concentrating them by lyophilic drying under vacuum in the frozen state. All the studied elements, excepting mercury and iodine, quantitatively remain in the dry residue. This technique was applied [288] for determining Au, Ba, Br, Ca, Ce, Co, Cr, Eu, Fe, K, La, Mo, Na, Sb, Sc, Se, U and Zn in natural waters.

3.9.3 PRECONCENTRATION AFTER IRRADIATION (RADIOCHEMICAL SEPARATION)

Previously, in the absence of high-resolution gamma spectrometers and, to some extent, of selective techniques of concentration and separation, the radiochemical procedures were cumbersome and complex. Now the situation has changed; the radiochemical part of activation analysis has become more efficient.

The matrix may get strongly activated and the formed ratioisotopes of macro- and micro-elements may have close half-life periods. Other cases are also possible where the matrix does not significantly affect the determination results of trace elements, but creates difficulties in separating and concentrating them. Therefore, it is necessary to remove the matrix elements to a content of about 10^{-3}–10^{-5} g. Sometimes, different trace elements affect the determination results of the impurity or alloying elements. For instance, arsenic and antimony hinder the determination of alloying elements in semiconductor materials. In all these cases, radiochemical separation is desirable or necessary.

As a result, diverse schematics have been developed for separation and concentration of elements (see, in particular, examples given in ref. 289). These schematics yield optimal results when effective techniques of concentration and separation of trace components are skilfully combined with multielemental instrumental determination.

3.9.3.1 Extraction methods

These methods have great significance for activation analysis. Several radiation stable extractants, which can be used to remove the matrix elements hindering determination, are available to separate the determinable trace elements in the organic phase and, if necessary, to divide them into groups suitable for γ-spectrometric determination. Extraction concentration can be effected automatically and the process can be remote controlled.

The selectivity of radiochemical separation can be appreciably increased by employing substoichiometric exchange extraction [290]. Stepanets et al. [291] have studied the subsequent displacement of determinable elements present in the organic phase as complex metal chloride anions which enter into the composition of anion associates with cation extractants. Reagents with the doubly charged anion $MeCl_4^{2-}$ include trioctylamine, trilaurylamine and trialkylbenzyl ammonium chloride (TABAC), with single-charged anions ($MeCl_4^-$), tetraphenylarsonium chloride (TPAC). The possibility of substoichiometric separation of elements forming sufficiently stable anion complexes has been ascertained theoretically and by experiment. It has been shown that stoichiometric displacement is possible in pairs Cd–Zn, Hg(II)–Cd (TABAC in nitrobenzene) and Ga–Fe(III), Au(III)–Ga (TPAC in chloroform). A procedure has been developed for determining impurities of Au, Fe and Ga in metallic aluminium using the HCl–TPAC–chloroform system. Kolotov et al. [292] used substoichiometric extraction of arsenic with dialkyl tin dinitrates (DATDN) for separating this element from solutions obtained after dissolving volcanic gas condensates or metallic zinc and the conversion of sample into a form convenient for analysis. The samples were irradiated with neutrons in a nuclear reactor and, after a suitable treatment, arsenic(V) was separated from aqueous solution (pH 2–3) with a solution of DATDN in chloroform containing 10% n-decyl alcohol. The extract was taken out and the radioactivity of ^{76}As was measured.

258

Of interest are the publications of Gilbert and co-workers [293–296] who have employed selective extraction of matrix in the neutron activation analysis of high-purity silver [293, 294], gold [294, 295], palladium [295] and mercury [296]. In analysing silver, [109]Pd, which was formed and hindered the determination, was extracted simultaneously. For determination of gold and palladium in rocks, ores and technological solutions [297] the elements were selectively extracted with a solution of dioctylsulphide or petroleum sulphides in benzene. Unlike other NAA techniques, the procedure requires less time and labour and ensures separation of both elements in one operation with high radiochemical purity and reliable measurements. Selective extraction of the matrix element (in the form of SnI_2 and daughter isotope, [125]Sb) with toluene from $6\,M\,H_2SO_4$ solution was applied in the NAA of high-purity tin [298]. The impurities (21 elements) were determined in the concentrate without further separation.

There is a large number of works dedicated to the application of chromatography for extraction and, mainly, separation of trace elements in NAA.

Extraction chromatography is very convenient for NAA. Besides its traditional advantage of adaptability to the concentration and separation methods, it ensures safe and remote manipulation with highly active solutions. It is suitable for laboratories where the objects of analysis change constantly: in a unified column packed with a suitable carrier, one can readily change the fixed phase, select an eluent and separate necessary elements. Several examples can be found in refs. 299 and 289. We shall restrict ourselves only to some works.

Using Diflon (a variety of Teflon) as carrier, nineteen elements were divided into six groups: (1) Fe, Hg and Sn, tributylphosphate–$2\,M$ HCl system; (2) Ag, Bi, Cd and Zn, trioctylamine in CCl_4–$2\,M$ HCl; (3) Sb, tributylphosphate–$6\,M$ HCl; (4) Al and Cr, acetylacetone–chloroform; (5) Co, Cu, Mg, Mn, Ni and Pb, 2-thenoyltrifluoroacetone–methyl isobutyl ketone at pH 5; (6) Ba, Ga and Sr, the same system but at pH 7 [300].

Radioisotopes with similar energies of γ-quanta, say P_m–S_m–Eu–Gd [301], Sc–Zn, Hg–Se, As–Br–Sb [302], were separated by extraction chromatography. The method is suitable for selective separation of noble metals [303] and actinides [304, 305] because in the presence of impurities the detection limits of determinable elements increase appreciably, even if radiation is registered by liquid scintillators and alpha-detectors [304].

Extraction chromatography can be applied even when the matrix has to be separated from the trace elements. In this case, care must be taken that the column is not overloaded: the stationary phase has limited capacity; also, it may "break away" from the carrier. However, this technique is applicable within reasonable limits. For instance, in analysing high-purity antimony, gallium and gallium antimonide [306] the highly activating matrix, after decomposition and suitable sample preparation, was absorbed in a column packed with graphite, silica gel or Teflon powder and containing 2,2′-dichloroethyl ether as fixed phase.

3.9.3.2 Sorption methods

In combination with NAA, extensive use is made of sorption methods, in particular, of chromatographic techniques. These methods ensure selective separation and concentration of elements, and ease of process control; also, they make it possible to automate the analysis and to carry it out from a distance. The importance of sorption methods used for NAA can be judged from numerous examples systematized in ref. 289.

The main trends of analytical application of ion exchange chromatography are associated with the use of complex-formation processes in solutions mixed with water–organic eluents, modified sorbents and inorganic ion exchangers. Formation of complexes improves separation selectivity owing to the difference created in the stability constants of complexes, and this makes it possible to separate individually even the rare-earth and transplutonium elements [307, 308]. Many schematics for the separation and concentration of elements involve the use of simplest complex-forming agents: anions of mineral and organic acids. The formation of chloride and fluoride complexes is fundamental for the schematics of separation of elements or the desired fractions in neutron activation analysis of high-purity substances [309–312].

Thus, in analysis of high-purity silicon dioxide, Ivanova et al. [310] combined volatilization of the matrix element as tetrafluoride with subsequent separation of trace components on synthetic ion exchangers. The sample after neutron irradiation was dissolved in a mixture of hydrofluoric and perchloric acids on heating slightly, the matrix was volatilized and the residue, concentrated by evaporation to moist salts, was dissolved in four to five drops of 0.01 M hydrofluoric acid. The determinable elements were divided into fractions in two stages: in column 1 packed with the cation exchanger KU-2-8 in the H^+ form, and in columns 2 and 3 packed with the anion exchanger AV-17-8 in the Cl^-

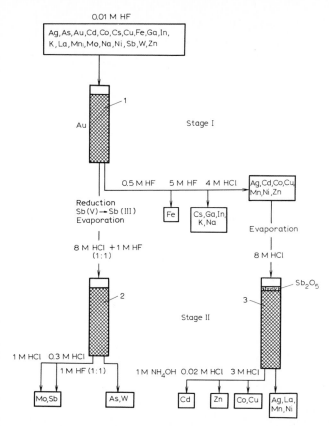

Fig. 3.6. Group chromatographic separation of trace elements in analysis of silicon dioxide. 1 = Column packed with the cation exchanger KU-2-8 in H⁺ form; 2 and 3 = columns packed with the anion exchanger AV-17-8 in the Cl⁻ form.

form. Fig. 3.6 shows distribution of elements by fractions and their subsequent yield from the columns. When the sodium content in the samples exceeded $5 \cdot 10^{-4}\%$, hydrated antimony pentaoxide was placed over the anion exchanger layer in one column, which acted as selective sorbent of this element. The procedure has been used for model mixtures containing 21 elements.

The addition of an organic solution into an eluent affects the stability of formed complexes and the order of their elution. Eluting systems consisting of mixtures of inorganic and organic acids with organic solvents are quite useful for NAA [313, 314].

Sometimes, trace elements are converted into chelates (dithiocarbamates, dithizonates, 8-hydroxyquinolinates, cupferonates and others).

Thereafter, they are concentrated and separated on activated charcoal. In particular, this procedure was used for separating trace amounts of As, Co, Cr, Fe, La, Mn, Ni, Pb, Sb, Se, Sn, V, U and Zn from the interfering activity caused by ^{82}Br, ^{38}Cl, ^{42}K, ^{24}Na and ^{32}P, and for analysing biological samples [315].

3.9.4 AUTOMATION

In the neutron activation analysis of highly radioactive samples automated systems are normally used, *i.e.* separation and concentration are effected without the participation of the analyst. Thus, for series neutron activation analysis of biological samples for about forty trace elements an activated system (with necessary instrumentation) has been suggested [316]; it ensures breakdown of quartz ampoules, dissolution of samples with mixtures of H_2SO_4, H_2O_2 and HBr, volatilization of As, Cl, Ge, Hg, Ru, Sb, Se and Sn, subsequent separation of the remaining trace elements in three columns packed respectively with ion exchanger Bio-rad AG-100, kieselguhr, and complex-forming sorbent Chelex-100.

Török *et al.* [317] have described a schematic for automated neutron activation determination of about twenty trace elements in soils, natural and waste waters, and biological samples. The schematic involves drying of sample, its irradiation in a nuclear reactor, radiochemical separation of the matrix, division of trace elements into groups, gamma-spectrometric measurement of induced activity, and mathematical processing of γ-spectra in computers. This is how Co, Cu, Mn and Zn in amounts 10^{-6}–10^{-8} g/ton are determined. Another work [318] informs us about a microprocessor-controlled system for the separation and concentration of elements. A polyethylene cell (volume 30 ml) is equipped with valves for introducing solutions of five different reagents, a small sensor for measuring pH, and a replaceable filter. The programme is chosen by the researcher, depending on the type of problem in hand.

3.10. Electrochemical methods

There are two main ways of using preconcentration in electrochemical methods of analysis: (1) stripping voltammetry (SV) and (2) direct polarography of extracts containing trace elements after mixing them with a current-conducting background.

Trace elements are first separated on a solid or mercury electrode, and then their concentration is determined by recording the current that appears during electrodissolution of the precipitate or the amalgam as a function of potential. In this method, preconcentration and determination are combined in one cycle. Only an insignificant, 100th, 1000th or even a smaller, part of the fraction of trace element is electrodeposited on an indicator microelectrode from very dilute solutions. Concentration coefficients of 10^2–10^3 and more are attained in this case. This method does not demand large amounts of reagents (electric current is the main "reagent"). The method has limitations, of course: formation of intermetallic compounds and mutual interference of elements on determination results.

Solid phase and amalgam stripping voltammetry is widely used for determination of trace elements in diverse substances: natural and waste water samples, high-purity substances, metals and alloys, biological materials, etc. [319–322]. The method enables trace elements to be determined at a level of 10^{-7}–$10^{-5}\%$. High accuracy, selectivity, rapidity, and comparatively simple instrumentation and interpretation of recorded curves are important advantages of this method.

Amalgam stripping voltammetry (ASV) makes it possible to determine Bi, Cd, Cu, In, Pb, Sb, Sn, Tl and Zn (or, rarely, Ga and Ge). Solid-phase stripping voltammetry (SPSV), which started developing after ASV, has gained a wider scope as it enables the anions, the elements that do not form amalgam and do not reduce to the metallic state, and the more noble metals than mercury to be determined. With this method it is possible to determine more than forty trace elements, wide use being made of stripping voltammetry of metals (SVM) [323].

The following are some examples illustrating the possibilities of stripping voltammetry. This method ensures determination of only individual elements or of a small group of them. The point is that the conditions of determining individual elements differ as to the concentration method and the potential, the type of indicator reaction, and the composition of the background electrolyte. Kaplin and Pichugina [324] propose to widen the possibilities of stripping voltammetry by replacing the electrodes and background electrolytes. They have succeeded in determining successively 10^{-7}–10^{-8} g of Au, Bi, Cd, Cu, Mn, Pb, Sb, Sn, Te and Zn from one weighed amount. The procedure has been used for analysis of semiconductor materials.

Brainina *et al.* [325] used SPSV for determining manganese as periodate [by the reaction of Mn(IV) with KIO_4] after electrochemical preconcentration on a disc-like graphite electrode. Then a cathode polarization curve was drawn for linearly varying potential. Gillain *et al.* [326] have made use of anodic stripping differential pulse voltammetry on a stationary dropping mercury electrode for simultaneous determination of Bi, Cd, Cu, Pb, Sb and Zn in sea water.

Sometimes, stripping voltammetry is preceded by additional concentration of trace elements. An example of this is chromatopolarography (concentration on synthetic ion exchangers and polarographic analysis of the eluate) [327]. To get rid of the complications associated with indirect determination of 10^{-8}–$10^{-9}\,M$ Cd in waters by anodic stripping differential pulse voltammetry, the trace elements were concentrated in a chromatographic column packed with the cation exchanger Dowex 1X8; cadmium was eluted with the eluent, $2\,M$ in NH_4NO_3 and $0.2\,M$ in HNO_3, and analysed. Of interest is the co-precipitation of Bi, Pb, Sb and Te with iron hydroxide, employed in analysis of high-purity copper [328]. Anodic stripping differential pulse voltammetry has been employed for determining 0.2–0.5 $\mu g/g$ Cd and lead in hepatic tissues [329]. First, the sample was ashed in the presence of H_2SO_4, the dry residue was dissolved in $1\,M$ HCl, and then the trace elements were separated on a film mercury electrode (mercury electrodeposited on a glassy carbon substrate).

3.10.2 INDIRECT POLAROGRAPHIC ANALYSIS OF EXTRACTS

Solvent extraction is often used in voltammetry [15, 320, 330–332]. In so doing, undesirable effects on the analytical signal may appear, say, owing to adsorption of organic solvents and reagents by the electrode surface, the appearance of surfactants, or electrochemical side-reactions.

The usual routine of analysis is well known: extraction (back-extraction) or mineralization–determination. The procedure involving the use of multiplying chemical reactions to lower the detection limits by dc polarography [333] is of interest, in particular. For determining phosphorus, it was converted into molybdophosphoric heteropoly acid (multiplication factor 12) and extracted. Then, molybdenum was back-extracted in an alkaline solution and determined from the catalytic wave in the presence of nitrate ion (multiplication factor 20–50).

The direct polarographic analysis of extracts is more important.

This technique has found application because it cuts short the analysis time compared to the classic variant involving mineralization of the extract or back-extraction. It becomes possible to determine the trace elements that hydrolyze in aqueous media, and to analyse the compounds which do not dissolve in aqueous solutions. As the coefficients of diffusion in water–organic and aqueous solutions are different, as relative concentration is attained during extraction and the mineralization stage can be eliminated, it is possible to improve the metrological characteristics of the method. Extraction makes it possible to remove those elements which interfere with determination. Besides, not all the elements that go into the extract are electrochemically active, and this also increases the selectivity of analysis.

However, a difficulty arises in polarographic analysis of extracts; in most cases the extracts are non-conducting, therefore suitable electrolytes have to be added. Most often methanol or ethanol solutions of alkali metal and ammonium salts are introduced into the extract, which ensure quite a high conductivity. It has been ascertained that the conductivities of mixtures of chloroform–methanol, acetylacetone–methanol, diethyl ether–methanol, hexane–methanol and benzene–methanol, with ratio of components 1:1 and $0.1\,M$ concentration of ammonium nitrate, are close to the conductivity of $0.1\,M$ aqueous solution of potassium chloride [334]. The detection limit can be lowered by 5–7 times by raising the temperature in stripping voltammetry [335–337]; it can be further lowered by more than an order if use is simultaneously made of the effect of ammonium amalgam [334].

"Double" concentration, as used by Karbainov [338], is interesting. It combines extraction concentration with subsequent electrochemical concentration on a "hanging drop" mercury electrode. Such a method is called extraction stripping voltammetry (ESV). With this method it is possible to determine trace elements at a level of 10^{-10}–$10^{-11}\,M$.

The possibilities of ESV were widened by using solid electrodes instead of mercury electrodes. A large number of complex compounds upon oxidation on solid electrodes produce current which is proportional to the concentration of the depolarizer in the solution. The analysis time is cut short because there is no need to deaerate the solutions.

Extraction of chelates is most commonly employed for isolating trace elements and their subsequent determination in water–organic media [339]; use is made of other systems also. Table 3.5 contains examples of indirect polarographic analysis of extracts.

TABLE 3.5

Examples of indirect polarographic determination of extracts

Element	Extractant	Method of determination	Substance analysed	Ref.
Cadmium and copper	Sodium diethyldithiocarbamate, benzene	SV	–	340
Molybdenum	8-Hydroxyquinoline, dichloromethane	Differential pulse polarography	Steels, biological materials	341
Molybdenum	4-Methyl-8-mercaptoquinoline, molten naphthalene	SV	Titanium alloys	342
Silver	Dithizone, chloroform	SV	Solutions	343
Nickel	Dimethylglyoxime, $(C_4H_9)_4NOH$ in dichloromethane	Differential pulse polarography	Steels and alloys, river deposits	344
Mercury	Zinc dibutyldithiophosphate, chloroform	SV	Fertilizers	345
Bismuth and antimony	Diethyldithiophosphoric acid, ethyl acetate	SV	Pure tin	346
Lead and copper	O,O-Dialkyl-S-thiocarbamoyldithiophosphates, chloroform and acetone (2:3)	SV	Soils and alloys	347
Bismuth and antimony	Thiourea, ethyl acetate	SV	Pure tin	348
Uranium	Petroleum sulphoxides	SV	Solutions	349
Silver and gold	Petroleum sulphides, chloroform	SV	Ores	350
Cadmium	$(C_4H_9)_4NI$, acetonitrile	Differential pulse voltammetry	Lead, zinc, indium and their nitrates	351

Preconcentration is employed to improve analytical characteristics in other electrochemical methods. For increasing the determination selectivity of gold, Maistrenko *et al.* [352] extracted gold with toluene solution of dihexylsulphide from HCl solution obtained upon decomposing copper–zinc ores. The concentrated element was then determined coulometrically in the extract itself. Prior to determination, dimethylsulphoxide and potassium iodide were added to the extract. In analysis of sulphide ores, improvement in selectivity and decrease in the detection limit were attained by extracting silver with chloroform solution of dihexylsulphide [353]. The extract was mixed with ethanol, concentrated sulphuric acid was added and silver was titrated with ethanol solution of ferrocene at a potential of 0.8 V. Analysis was completed with amperometric titration.

A combination of directed crystallization and ionometric determination is also known [354]. For determining fluorides in caesium iodide and chlorides and bromides in potassium nitrate, crystallization was carried out in aqueous solutions of salts having eutectic composition. After recrystallization, the top portion of the ingot was dissolved in a minimum volume of hot water. The activity of determinable ions was measured with ion-selective electrodes having polycrystalline membranes ($AgCl–Ag_2S$ and $AgBr–Ag_2S$) and solid internal contact, and also with lanthafluoride electrodes EF-VI.

Electrolysis with chemically modified electrodes is being developed [355]. Electrodes are modified in different ways. Most often, an electrode is coated with polymeric film containing active functional groups. Immobilization of the working surface of the electrode or the application of a current-conducting paste containing an ion exchanger or a complex-forming sorbent also holds promise. Concentration is effected by a sorption method, although sorption can be combined also with electrochemical concentration. Analysis is completed by voltammetry, ionometry, coulometry and amperometry. High selectivity is the main advantage of electrolysis carried out with chemically modified electrodes. Thus, as per ion-exchange mechanism, chromium(VI) was preconcentrated on a platinum electrode modified with poly(4-vinylpyridine) [356]. Chromium was determined by the cyclic voltammetry method. Very selective voltammetric determination of traces of uranium was accomplished after preliminary concentration of uranium on a glassy carbon electrode coated with trioctylphosphine oxide (TOPO)

[357]. The TOPO layer masks the ions of metals, which are electrochemically active under these conditions.

3.11. Catalymetry

Catalymetry is a highly sensitive method of determining trace elements. The lower limit of determinable contents of most elements is 10^{-8}–$10^{-10}\,M$ or 10^{-3}–$10^{-4}\,\mu g/ml$, which is one to two orders lower than the values attained by spectrophotometry and voltammetry methods and by the flame variant of atomic absorption spectrometry [358, 359]. However, catalytic reactions do not feature high selectivity. The method's selectivity is improved by applying concentration techniques that have a positive effect on the accuracy and precision of analysis results and also on the detection limit. Monographs [358, 359] contain examples of determining Co, Cr, Cu, Fe, Mn, Mo, Ni, V and other elements; in these cases, the elements were preconcentrated by different methods.

Of the concentration methods, extraction is most generally employed. It aids in improving the selectivity of catalymetry and lowering the relative detection limit. In the majority of cases, extraction complicates the process as the organic solvent has to be volatilized. If the extracted complex is catalytically active in the extract the indicator reaction components can then be introduced into the organic phase and the catalytic reaction can be performed in it. Apparently, such an extraction–catalytic method was first proposed in ref. 360: copper(I) in the form of a complex with neocuproine was selectively extracted with chloroform; solutions of n-phenetidine and cumene hydroperoxide were introduced into the extract and the reaction rate was measured. Knowing the reaction rate the concentration of copper was determined.

Molybdenum, by the catalytic oxidation reaction of 1-naphthalene amine with bromate, was extracted in chloroform extract [361] after separating the element as 8-hydroxyquinolinate. The results obtained were used in determining molybdenum in sea water samples.

In pure aqueous solutions, 0.5–1 $\mu g/ml$ silver can be determined by carrying out catalytic oxidation of bromopyrogallol red with potassium persulphate. 1,10-Phenanthroline used as activator lowers the detection limit down to 1–13 ng/ml. Extraction of silver with nitrobenzene [362] enables determination to be carried out even in the presence of 200 μg Fe(III), Co(II) and Pd(II). A detection limit of 0.2–20 ng/ml was attained in the automatic variant of the method. The determination selectivity of iron, based on the catalytic oxidation of n-phenetidine with hydrogen peroxide, was abruptly increased by preliminary extraction of the trace component from $7\,M$ solutions of lithium chloride [363], the extract being mixed with water and ethyl alcohol for carrying out the indicator reaction. The method makes it possible to determine iron in the concentration range of $1 \cdot 10^{-3}$–$5 \cdot 10^{-6}\%$ in Co, Cu, Li, Mg and Ni salts.

Otto et al. [364] extracted copper with chloroform as a mixed pyridine–salicylate complex, $Cu(HSal)_2(Py)_2$, and determined it directly in the organic phase. Oxidation of sulphonic acid with hydrogen peroxide (activator, pyridine) was used as indicator reaction. The catalytic reaction was carried out in a mixture of chloroform and ethanol. As sulphonic acid does not dissolve in alcohols, it was first dissolved in pyridine, which mixes with alcohols. The effect of thirteen metal ions on catalytic determination of copper after extraction was studied. Cobalt(II), which has higher catalytic activity in aqueous media than copper, did not affect its determination even when it was present in 1000-times excess amounts. Unfortunately, extraction does not reduce the interfering effect of Ni, Zn and Hg because these elements are extracted under the same conditions as copper, but they can be masked with tartrate (Ni), thiocyanate (Zn) and bromide (Hg). This method enables 3–60 ng/ml copper to be determined.

In the presence of activators (8-hydroxyquinoline, pyrocatechin, Tiron and others) vanadium is catalytically active in the oxidation reaction of aromatic amines, in particular of o-phenylenediamine, with bromate ion [365]. However, the selectivity of this reaction is not sufficiently high: determination is hindered by hundred fold amounts of chromium and molybdenum and thousand fold amounts of Ag, Cu, Fe, Ti, W and Zr. The selectivity of this reaction was improved by extracting vanadium(V) as a complex with 8-hydroxyquinoline or as an ion associate with pyrocatechol and tertiary amine. As organic solvent, use was made of n-butanol which dissolved the desired (for carrying out the reaction) amounts of potassium bromate. Determination proper was done in the water–organic medium n-butanol–ethanol–water. The attained detection limit was almost similar to that in an aqueous medium. The method has been used for determining vanadium in high-purity organic salts.

For determining copper in the salts of lead, cadmium, nickel and other metals, the analysed weighed amount was dissolved in a mixture containing acetate-based buffer solution, hydroxylamine and sodium perchlorate [366]. Copper was extracted with chloroform solution of

2,2'-diquinoline. To the extract was added ethanol, ethanol solution of n-phenetidine and a solution of hydrogen peroxide. Then the kinetic curve was registered with the aid of a photometer and a recorder.

3.12. Chromatographic methods

Wide use is made of these methods (as determination methods) in analysis of organic compounds including complex composite mixtures of natural and industrial origin. In inorganic analysis these methods were employed later, and it seems that they will not become as important as in organic analysis. To some extent, this is because element analysis is the main aim of chromatographic methods used in the analytical chemistry of inorganic compounds; for this, many competitive techniques are available. Nonetheless, the sphere of application of chromatographic methods in inorganic analysis is widening.

Gas chromatography (GC), high-performance liquid chromatography (HPLC) (including ion chromatography) and thin-layer chromatography (TLC) are the methods that mainly effect separation, but do not concentrate trace elements. Here, we shall consider only those examples where use has been made of preconcentration. For more detailed information the reader is referred to the books and reviews [367–377].

3.12.1 GAS CHROMATOGRAPHY

Different analytical forms including gaseous elements and vapour phase of highly volatile elements (for example, sulphur and zinc), some oxides, halides, hydrides, metal chelates, elemental organic compounds, various volatile forms obtained upon interaction of determinable elements and organic compounds (piazoselenoles, triethylfluorosilane or trialkyl ethers of organic acids) are suitable for gas chromatography. Here, it must be pointed out that thermochromatography has also been developed for analysis of high-boiling halides, oxides and hydroxides (at temperatures up to 1000–1200°C and even at 1500°C) [373]. With GC are combined extraction and volatilization and related methods of concentration.

GC of chelates is noted for high selectivity and very low detection limits; chelates are not always stable and therefore they may get lost never to return in the separation process. Primarily chelates are obtained by extraction methods and, rarely, by precipitation. As mentioned in the review [378], the development of gas chromatography of chelates is restrained by appreciable interface absorption that causes asymmetric peaks to appear. Specific requirements are placed on the reagents and their chelates: they should be volatile, stable to heat, should readily form compounds and hardly give rise to coordinated unsaturated compounds. The latter, due to increased polarity, presence of water, and catalyzing decomposition of chelates, are a handicap to a successful GC determination [378].

Different reagents are used in the GC of chelates (see, for example, refs. 367, 368, 372 and 378). Most often use is made of β-diketonates, particularly their fluoro-substituted compounds which are highly volatile and stable: acetylacetone, trifluoro- and hexafluoroacetylacetone, benzoylacetone, thenoyltrifluoroacetone, isobutyrylacetone, pivaloylacetone, dibenzoylmethane, etc. As β-diketonates can be determined Al, Ba, Be, Ca, Cd, Co, Cr, Cu, Fe, Ga, Hf, In, Mg, Mn, Mo, Ni, Rh, Ru, Sc, rare-earth elements, Ti, U, V, Zn, Zr and others. The complexes with fluorinated β-diketonates are more volatile and stable to heat than the corresponding chelates with acetylacetone, its alkyl- and aryl-substitutes, but they are less stable in aqueous solutions as they hydrolyse. In the gas chromatography of chelates, use is made also of β-diketonates, β-ketoaminates, alkyldithiocarbamates and dithiophosphinates of metals. Volatile chelates of almost all metals have now been synthesized [368, 372]. Addition of reagent vapours into the carrier gas improves the chromatographic behaviour of chelates.

For rapid and exact GC separation and determination of chelates at the extraction stage, it is better to use low-volatile solvents so that the first peak of the solvent does not overlap the peaks of determinable trace elements. Besides, the solvent should not contain elements to which the detector is sensitive.

Along with flame ionization, electron-capture and catharometer detectors, atomic absorption spectrometers and atomic emission spectrographs are also now finding application [379, 380]. Such a combination ensures good metrological characteristics in determining metals in environmental samples and in the products of the petroleum refining industry. Atomic emission spectrometers with ICP [381–383] are very convenient for this purpose.

GC is employed also to determine anions [367, 371, 384] which, as a rule, are first converted into organic compounds suitable for subsequent GC separation and detection. For example, 4-fluoro-o-phenylenediamine and 4-trifluoromethyl-o-phenylenediamine were used for determining traces of selenium (in the form of corresponding piazo-selenoles) [385] which are thermally stable and volatile. A method has been developed for determining selenium with 4-trifluoromethyl-*o*-phenylenediamine in various biological samples. The sample was first mineralized by treating it with a mixture of concentrated nitric and hydrochloric acids, the excess HNO_3 was removed by treating with concentrated HCl on heating and with simultaneous reduction of Se(VI) to Se(IV). After cooling, the solution was found to have an acidity of about $0.01\,M$ in HCl; from its aliquot (after adding the reagent) the formed piazoselenole was extracted with toluene. The extract, after washing, was introduced into a column packed with SE-30 and Chromosorb W and subjected to chromatographic analysis.

In recent years, inorganic gas chromatography (IGC) has been developed extensively as a method of analysis of volatile compounds [373] which are obtained by chemical transformations or simple distillation of volatile compounds. Most often, chlorides are used; bromides and iodides often decompose at high temperatures; fluorides are more volatile and very corrosive. As detectors, it is recommended to use atomic absorption, atomic emission, atomic fluorescence methods and alpha-spectrometers. Electrochemical detectors and X-ray fluorescence spectrometers are used as off-line detectors in the same manner as atomic absorption spectrometers with electrothermal atomization. For off-line detectors it is necessary that the analyzed fractions should first be separated on a plate or disc or in a capillary tube.

For determining platinum metals in rocks, ores and minerals, Rigin [386] combined sample decomposition with preconcentration. To this end, he used xenon tetrafluoride as a fluorinating agent. On mixing the finely ground samples with xenon tetrafluoride in an autoclave at 700 K elementary fluorine was formed. The main components of rocks, ores and minerals (silicates, oxides, sulphides and sulphates) go completely or partially into the gaseous phase and are volatilized. Simple higher fluorides of platinum metals have comparatively low boiling points; therefore, they are separated from many accompanying metals that form non-volatile fluorides and are determined by the IGC method. For this they are separated on a column packed with cerium trifluoride, using an atomic fluorescence spectrometer as detector. Fig. 3.7 shows that separation of fluorides of all metals, excepting rhodium and palladium, is satisfactory. Maximum analytical signals of the determinable elements were observed at 1600–1700 K.

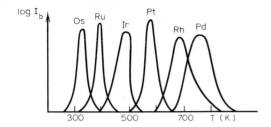

Fig. 3.7. Chromatogram of platinum metal fluorides mixture.

3.12.2 HIGH-PERFORMANCE LIQUID CHROMATOGRAPHY

High-performance liquid chromatography (HPLC) is suitable for analysis of a variety of samples [374–376]. Extensive use is made of HPLC of chelates of metals on synthetic ion exchangers and complex-forming sorbents. In the case of chelates, complexes are obtained by extraction. At this stage, trace elements can be concentrated, if necessary. Then, if need be, the chelates are separated; this is preferably done by employing the adsorption mechanism of separation [377]. Wide use is made of metal complexes with dithiocarbamate acids, dithizone, β-diketones, β-ketoamines, thiosemicarbazones, thiobenzhydrazones and hydrazones [374]. Chelates are conveniently separated with sorbents, say silica gel, aluminium oxide and cellulose.

For the determination of chromium(VI) and chromium(III) in water samples [387], they were extracted as pyrrolidinedithiocarbamates, the solvent was volatilized from the extract and the dry residue, after dissolving in acetonitrile, was subjected to chromatographic analysis in a column packed with Lichrosorb RP-18. A mixture of acetonitrile and water was used as mobile phase and a spectrometer was used as detector. Shibata *et al.* [388] have proposed an interesting method for determination of selenium ($\geqslant 3$ pg) in biological samples. Selenium was extracted by cyclohexane in the form of a complex with 2,3-diamino-naphthalene. An aliquot of the extract was introduced into a column packed with μBondapak C_{18} or UNISIL $5C_{18}$. The complex was eluted and the concentration of selenium was determined fluorimetrically.

A scheme has been proposed for concentration of trace amounts of metals ions and their subsequent separation and determination in waters [389]. The sample [volume $\leqslant 200$ ml in 0.001 M citrate solution (pH 4.8) or 0.01 M solution of nitric acid] was passed through the shell concentrator with cation Aminex A-5, the concentrate was washed with 0.06 or 0.08 M sodium citrate (pH 4.8 or 4.6) directly in the column packed with Aminex A-5. Thereafter, the trace elements were separated using the same solution, $2 \cdot 10^{-4} M$ solution of sodium salt of 4-(2-pyridylazo)-resorcin in 2 M solution of NH_4OH (1 M in ammonium acetate) was introduced into the eluate and determined photometrically. For example, cobalt can be determined in the reactor cooling water and Cu, Fe, Mn, Pb and Zn in distilled water.

Reversed-phase high-performance liquid chromatography (RP-HPLC) has been used for determination of nitrate and nitrite ions in water [390]. The method involves nitration of phenol added to 10–100 ml of water sample and determination of the nitrophenol obtained. Nitrite ions were determined after oxidising them to nitrate ions with hydrogen peroxide. The nitrophenol formed was extracted with dichloromethane. The extract was mixed with caustic soda solution, evaporated to dryness, the residue was dissolved in 0.5–1.0 ml of the mobile phase and an aliquot was analysed chromatographically in a column packed with RP-C_{18}, eluent being a mixture of phosphate-based buffer solution and methanol, using an amperometric detector.

3.13. Other methods

Photoelectron spectroscopy was employed for determination of traces of metals absorbed as per the ion-exchange mechanism on the surface of a graphite polypropylene disc [391]. The disc was placed in the solution to be analysed and the metals were determined. In determining silver the detection limit was found to be equal to 10 ng. The spectral lines for Ag, Ca, Cd, Cu(II), Fe(II) and Hg(II) have been found to have good resolution. The technique may be used in analysing sewage waters. By this method, Brinen and McClure [392] determined zinc with preliminary electrochemical concentration. Lead was electrolytically deposited on a polished surface of a disc-like electrode made of glassy carbon. The concentration of lead, which can be quantitatively determined, should be less than or equal to $1 \cdot 10^{-3} \mu g/ml$.

Some works are dedicated to the combination of methods of concentrating trace elements with molecular emission in a cavity (MECA spectroscopy). This is how boron was determined after extracting it with methyl isobutyl ketone in the form of chelates with 2-ethylhexane-1,3-diol [393]. In the analysis of water samples [394] selenium was co-precipitated with iron hydroxide, the residue was washed and dried

274

(on a filter) and introduced into the flame-heated cavity. Indirect determination of As, Sb, Se and Te [395] involves chloroform extraction of these elements as diethyldithiocarbamates and measurement of molecular emission intensity of the S_2 molecule in the cavity. Several examples of the combination of concentration and MECA spectroscopy are available in ref. 396. Cases of combining preconcentration with other spectroscopic methods are also known. Thürauf and Assenmacher [397] have described a method for determining sulphur in "coal oil". It involves oxidation of sulphur to sulphur dioxide and subsequent determination by the dispersionless infrared (IR) absorption method. The sample was mineralized in a current of air or oxygen, the gases, after drying, were passed through silica gel cooled to $-70°C$, which adsorbed sulphur dioxide. After desorption by heating, the integral IR radiation absorption signal was measured. Differential thermolens calorimetry was employed to determine phosphorus in high-purity substances [398]. First, phosphorus was extracted with isobutanol as molybdophosphoric acid and then it was changed into an ion associate with auramine. The analytical signal was measured at the centre of a laser beam. A method has been developed for determining copper and gold in aqueous solutions after depositing them on a platinum cathode [399]. The deposited metals are determined by an electron probe from the X-ray lines of CuK_α and AuL_{α_1}.

Concentration methods are combined with α-spectrometry to determine actinides. For the determination of thorium in solutions, it was electrolytically deposited on a stainless-steel cathode which, after washing and drying, was subjected to α-spectrometry [400]. The procedure for simultaneous determination of Pu, Th and U in soft biological tissues [401] involves decomposition of sample (0.05–700 g) by heating it first with nitric acid and then with a mixture of nitric and sulphuric acids, adding periodically a drop each of nitric acid and hydrogen peroxide. Thereafter, the determinable elements were coprecipitated with iron hydroxide, the residue was dissolved in concentrated HCl, uranium and plutonium were extracted with 20% solution of triaurylamine in xylene. Then they were separated by back-extraction. With the same reagent, thorium left in the water phase was isolated from a medium ($4 M$ in HNO_3) and back-extracted with $10 M$ solution of HCl. In the final stage Pu, Th and U were electrolytically deposited on a platinum disc and their radioactivity was measured on an α-spectrometer with a surface-barrier Si detector.

References

1 Yu.A. Zolotov, in *XI Mendeleevsky S'ezd po obshchey i Prikladnoi Khimii: Ref. Dokladov i Soobshenii (XIth Mendeleev Congress on General and Applied Chemistry: Abstracts of Lectures and Papers)*, Nauka, Moscow, 1975, No. 5, p. 14.

2 Yu.A. Zolotov, *Ocherki Analit. Khimii (Essays in Analytical Chemistry)*, Khimiya, Moscow, 1977.

3 Yu.A. Zolotov, *Zh. Anal. Khim.*, 32 (1977) 2085.

4 Yu.A. Zolotov, *Analyst (London)*, 103 (1978) 56.

5 Yu.A. Zolotov, *Pure Appl. Chem.*, 50 (1978) 129.

6 N.M. Kuz'min, *Zh. Anal. Khim.*, 38 (1983) 2262.

7 N.M. Kuz'min, *Teoriya i Praktika Ekstraktsionnykh Metodov (Theory and Practice of Extraction Methods)*, Nauka, Moscow, 1985, p. 165.

8 Yu.A. Zolotov, *Tr. Kom. Anal. Khim. Akad. Nauk SSSR*, 15 (1965) 3.

9 Yu.A. Zolotov and N.M. Kuz'min, *Zh. Anal. Khim.*, 22 (1967) 773.

10 N.M. Kuz'min, Yu.A. Zolotov and Yu.A. Karbainov, *Tr. Kom. Anal. Khim. Akad. Nauk SSSR*, 17 (1969) 288.

11 Yu.A. Zolotov, *Ekstraktsiya Vnutrikompleksnykh Soedinenii (Extraction of Chelate Compounds)*, Nauka, Moscow, 1968.

12 H.A. Laitinen, *Anal. Chem.*, 48 (1976) 2049.

13 N.M. Kuz'min, V.S. Solomatin, A.N. Galaktionova and I.A. Kuzovlev, *Zh. Anal. Khim.*, 24 (1969) 725.

14 Sh.I. Peizulaev, in *Spektralnye Metody Analiticheskogo Kontrolya Veshchestv Vysokoi Chistoty s Predvaritelnym Kontsentrirovaniem (Atomic Emission Methods for Analytical Control of High-Purity Substances with Preliminary Concentration)*, D.Sc. Thesis, Lomonosov Moscow State University, Moscow, 1969.

15 Yu.A. Zolotov and N.M. Kuz'min, *Ekstraktsionnoe Kontsentrirovanie (Extraction Preconcentration)*, Khimiya, Moscow, 1971 (in Russian).

16 Yu.A. Zolotov, N.G. Vanifatova, T.A. Chanysheva and I.G. Yudelevich, *Zh. Anal. Khim.*, 32 (1977) 317.

17 N.R. Krivenkova, L.I. Pavlenko, B.Ya. Spivakov, I.A. Popova, T.S. Plotnikova, V.M. Shkinev, I.P. Kharlamov and Yu.A. Zolotov, *Zh. Anal. Khim.*, 31 (1976) 514.

18 L.I. Pavlenko, O.M. Petrukhin, Yu.A. Zolotov, A.V. Karyakin, G.M. Gavrilina and I.E. Tumanova, *Zh. Anal. Khim.*, 29 (1974) 933.

19 G.A. Vorobyeva, Yu.A. Zolotov, L.A. Izosenkova, A.V. Karyakin, L.I. Pavlenko, O.M. Petrukhin, I.V. Seryakova, L.V. Simonova and V.N. Shevchenko, *Zh. Anal. Khim.*, 29 (1974) 497.

20 N.M. Kuz'min, G.D. Popova, I.A. Kuzovlev and V.S. Solomatin, *Zh. Anal. Khim.*, 24 (1969) 899.

21 G.V. Myasoedova, G.I. Malofeeva, O.P. Shvoeva, E.V. Illarionova, S.B. Savvin and Yu.A. Zolotov, *Zh. Anal. Khim.*, 32 (1977) 645.

22 N.A. Rudnev, L.I. Pavlenko, G.I. Malofeeva and L.V. Simonova, *Zh. Anal. Khim.*, 24 (1969) 1223.

23 B. Nebesar, *Anal. Chim. Acta*, 39 (1967) 301.

24 B. Nebesar, *Anal. Chim. Acta*, 39 (1967) 309.

25 N.M. Kuz'min, I.A. Kuzovlev and L.S. Khorkina, *Zh. Anal. Khim.*, 27 (1972) 453.

26 N.M. Kuz'min, T.P. Dubrovina and O.M. Shemshuk, *Zh. Anal. Khim.*, 28 (1973) 364.

27 Kh.I. Zil'bershtein and M.P. Semov, *Zh. Anal. Khim.*, 34 (1979) 987.

28 O.G. Koch and G.A. Koch-Dedik, *Handbuch der Spurenanalyse*, Teil I, II, Springer Verlag, Berlin, Heidelberg, New York, 1974.

29 I.G. Yudelevich, L.M. Buyanova and I.R. Shelpakova, *Khimiko-spektralny Analiz Veshchestv Vysokoi Chistoty (Chemico-Atomic Emission Analysis of High-Purity Substances)*, Nauka, Novosibirsk, 1980.

30 F. Rohner, *Helv. Chim. Acta*, 21 (1938) 23.

31 N.M. Kuz'min and Yu.A. Zolotov, *Zavod. Lab.*, 38 (1972) 897.

32 G. Gorbach and F. Pohl, *Microchemie*, 36/37 (1951) 486.

33 G. Gorbach and F. Pohl, *Microchemie*, 38 (1951) 258.

34 G. Gorbach and F. Pohl, *Microchemie*, 38 (1951) 328.

35 G. Gorbach, *Mikrochim. Acta*, No. 2–3 (1955) 336.

36 F. Pohl, *Spectrochim. Acta*, 6 (1953) 19.

37 F. Pohl, *Fresenius' Z. Anal. Chem.*, 139 (1953) 423.

38 F. Pohl, *Fresenius' Z. Anal. Chem.*, 139 (1953) 241.

39 F. Pohl, *Fresenius' Z. Anal. Chem.*, 141 (1954) 81.

40 B.F. Scribner and H.J. Mullin, *J. Res. Natl. Bur. Stands.*, 37 (1946) 379.

41 S.L. Mandel'shtam, N.N. Semenov and Z.M. Turovtseva, *Zh. Anal. Khim.*, 11 (1956) 9.

42 A.N. Zaidel, N.I. Kalitievsky, L.V. Lipis, M.P. Chaika and Yu.I. Belyaev, *Zh. Anal. Khim.*, 11 (1956) 21.

43 S.L. Mandel'shtam and V.V. Nedler, *Optika Spektrosk.*, 10 (1961) 390.

44 A.N. Zaidel, N.I. Kalitievsky, L.V. Lipis and M.P. Chaika, *Emissionny Spektralny Analiz Atomnykh Materialov (Atomic Emission Analysis of Atomic Materials)*, Fizmatgiz, Moscow, 1960 (in Russian).

45 V.N. Muzgin and L.A. Gladysheva, *Zavod. Lab.*, 34 (1968) 1076.

46 O.F. Degtyareva and M.F. Ostrovskaya, *Zavod. Lab.*, 30 (1964) 174.

47 W. Geilmann, *Fresenius' Z. Anal. Chem.*, 160 (1958) 410.

48 G.I. Veres, *Zavod. Lab.*, 36 (1970) 1151.

49 Kh.I. Zil'bershtein and M.P. Semov, *Izv. Sib. Otd. Akad. Nauk SSSR, Ser. Khim. Nauk*, 9 (4) (1967) 72.

50 A.I. Staikov and N.F. Zakhariya, *Zh. Anal. Khim.*, 30 (1975) 1375.

51 A.G. Page, S.V. Godbole, K.H. Madraswala, M.J. Kulkarni, V.S. Mallapurkar and B.D. Joshi, *Spectrochim. Acta, Part B*, 39B (1984) 551.

52 V.Z. Krasil'shchik, E.I. Voropaeva and G.A. Shtenberg, *Zh. Anal. Khim.*, 38 (1983) 2082.

53 Kh.I. Zil'bershtein, N.I. Kalitievsky, A.N. Razumovsky and Yu.F. Fedorov, *Zavod. Lab.*, 28 (1962) 43.

54 N.K. Rudnevsky, L.N. Sokolova and S.G. Tsvetkov, *Tr. Khim. Khim. Tekhnol. (Gorky)*, No. 2 (1962) 341.

55 V.G. Pimenov, P.E. Gaivoronsky, V.N. Shimov and G.A. Maksimov, *Zh. Anal. Khim.*, 39 (1984) 1072.

56 V.G. Pimenov, P.E. Gaivoronsky and V.N. Shishov, in *Poluchenie i Analiz Chistykh Veshchestv (Obtaining and Analyzing Pure Substances)*, Gorky University Press, Gorky, 1983, p. 47.

57 V.G. Pimenov, A.N. Pronchatov, G.A. Maksimov, V.N. Shishov, E.M. Shcheplyagin and S.G. Krasnov, *Zh. Anal. Khim.*, 39 (1984) 1636.

58 T. Nakahara, *Prog. Anal. At. Spectrosc.*, 6 (1983) 163.

59 J.E. Goulter, in *Abstracts Papers of the Pittsburgh Conf. Anal. Chem. Appl. Spectrosc., Atlantic City, NJ, March 9–13, 1981*, Monroeville, PA, p. 129.

60 B. Pahlavanpour, M. Thompson and L. Thorne, *Analyst (London)*, 105 (1980) 756.
61 E. de Oliveira, J.W. McLaren and S.S. Berman, *Anal. Chem.*, 55 (1983) 2047.
62 P. Hulmston, *Anal. Chim. Acta*, 155 (1983) 247.
63 R. Röhl, H.-J. Hoffmann and W. Besler, *Fresenius' Z. Anal. Chem.*, 317 (1984) 872.
64 D. Marquardt, P. Lüderitzt, S. Leppin and J. Grosser, *Z. Chem.*, 24 (1984) 267.
65 T. Török, J. Mika and E. Gegus, *Emission Spectrochemical Analysis*, Akadèmiai Kiadô, Budapest, 1978.
66 A. Miyazaki, A. Kimura and Y. Umezaki, *Anal. Chim. Acta*, 127 (1981) 93.
67 F. Kurosawa, I. Tanaka, K. Sato and T. Otsuka, *Spectrochim. Acta, Part B*, 36B (1981) 723.
68 G.L. Bukhbinder, L.N. Shabanova and E.N. Gil'bert, *Zh. Anal. Khim.*, 39 (1984) 2120.
69 H. Tao, A. Miyazaki, K. Bansho and Y. Umezaki, *Anal. Chim. Acta*, 156 (1984) 159.
70 G.V. Myasoedova and S.B. Savvin, *Khelatoobrazuyushchie Sorbenty (Chelate-forming Sorbents)*, Nauka, Moscow, 1984 (in Russian).
71 O.A. Shiryaeva, L.N. Kolonina, I.N. Vladimirskaya, V.P. Baluda, V.G. Miskaryants and V.A. Shestakov, *Nauchn. Tr. Nauchno-Issled. Proekt. Inst. Redkometallichesk. Promyshlennosti (Giredmet, Moscow)*, 111 (1982) 55.
72 A.E. Watson and G.L. Moore, *S. Afr. J. Chem.*, 37 (1984) 81.
73 F.D. Pierce and H.R. Broun, *Anal. Lett.*, 10 (1977) 685.
74 R.J. Brown and W.R. Biggs, *Anal. Chem.*, 56 (1984) 646.
75 V.Z. Krasil'shchik, N.M. Kuz'min and E.Ya. Neiman, *Zh. Anal. Khim.*, 34 (1979) 2045.
76 M.S. Chupakhin, A.I. Sukhanovskaya, V.Z. Krasil'shchik, S.U. Kreingold, L.A. Demina, V.I. Bogomolov, E.V. Dobizha, M. Przhibyl, Z. Slovak, I. Borak and M. Smrzh, *Metody Analiza Chistykh Khimicheskikh Reaktivov (Methods of Analysis of Pure Reagents)*, Khimiya, Moscow, 1984.
77 V.Z. Krasil'shchik and T.G. Manova, *Zh. Anal. Khim.*, 30 (1975) 971.
78 V.Z. Krasil'shchik and G.A. Shteinberg, *Tr. Vses. Nauchno-Issled. Inst. Khim. Reakt. Osobo Chist. Khim. Veshchestv*, 40 (1978) 160.
79 V.Z. Krasil'shchik, T.G. Manova and N.I. Kuznetsova, *Zh. Anal. Khim.*, 32 (1977) 837.
80 V.Z. Krasil'shchik, T.G. Manova and N.I. Kuznetsova, *Zavod. Lab.*, 44 (1978) 1342.
81 K. Himeno, K. Yanagisawa, T. Yuki and Y. Nakamura, *Bunseki Kagaku*, 33 (1984) T43.
82 A. Sugimae and T. Mizoguchi, *Anal. Chim. Acta*, 144 (1982) 205.
83 N.M. Kus'min, V.S. Vlasov, V.Z. Krasil'shchik and V.G. Lambrev, *Zavod. Lab.*, 43 (1977) 1.
84 M.S. Cresser, *Solvent Extraction in Flame Spectroscopic Analysis*, Butterworths, London, 1978.
85 M.S. Cresser, *Prog. Anal. At. Spectrosc.*, 5 (1982) 35.
86 D.L. Tsalev, N.I. Tarasevich and I.P. Alimarin, *Zh. Anal. Khim.*, 28 (1973) 19.
87 H. Berndt and E. Jackwerth, *Spectrochim. Acta, Part B*, 30B (1975) 169.
88 B. Bailey and Lo-Fa-Chun, *Anal. Chem.*, 44 (1972) 1304.
89 H. Kowamura, G. Tanaka and W. Kutura, *Bull. Chem. Soc. Jpn.*, 43 (1970) 970.
90 A.V. Volynsky, B.Ya. Spivakov and Yu.A. Zolotov, *Talanta*, 31 (1984) 449.
91 J. Aggett and T.S. West, *Anal. Chim. Acta*, 57 (1971) 15.
92 L.N. Sukhoveeva, G.G. Butrimenko and B.Ya. Spivakov, *Zh. Anal. Khim.*, 35 (1980) 649.

93   A. Halász, K. Polyák and E. Gegus, *Mikrochim. Acta*, 11 (1981) 229.
94   I. Khavezov and D. Tsalev, *Atomno-Absorbtsionen Analiz (Atomic Absorption Analysis)*, Nauka i Izkustvo, Sofia, 1980.
95   H. Malissa and E. Schöffmann, *Mikrochim. Acta*, 1 (1955) 187.
96   U. Anders and D. Hailer, *Fresenius' Z. Anal. Chem.*, 278 (1976) 203.
97   A.I. Busev, V.M. Byrko, A.P. Tereshchenko, N.N. Novikova, V.P. Naidina and P.B. Terentyev, *Zh. Anal. Khim.*, 25 (1970) 665.
98   G.A. Vall, M.V. Usoltseva, I.V. Seryakova and Yu.A. Zolotov, *Zh. Anal. Khim.*, 31 (1976) 27.
99   V.P. Shaburova, I.G. Yudelevich, I.V. Seryakova and Yu.A. Zolotov, *Zh. Anal. Khim.*, 31 (1976) 255.
100 O.M. Petrukhin, Yu.A. Zolotov and L.A. Isosenkova, *Zh. Neorg. Khim.*, 16 (1971) 3285.
101 N.L. Fishkova and O.M. Petrukhin, *Zh. Anal. Khim.*, 28 (1973) 645.
102 V.A. Orlova, B.Ya. Spivakov, V.M. Shkinev, T.I. Kirillova, V.A. Ivanova, T.M. Malyutina and Yu.A. Zolotov, *Zh. Anal. Khim.*, 33 (1978) 91.
103 B.Ya. Spivakov, V.I. Lebedev, V.M. Shkinev, N.P. Krivenkova, T.S. Plotnikova, I.P. Kharlamov and Yu.A. Zolotov, *Zh. Anal. Khim.*, 31 (1976) 757.
104 Yu.A. Zolotov, O.A. Kiseleva, N.B. Shakhova and V.I. Lebedev, *Anal. Chim. Acta*, 79 (1975) 237.
105 G.A. Vall, L.P. Poddubnaya, N.G. Vanifatova, I.G. Yudelevich and Yu.A. Zolotov, *Zh. Anal. Khim.*, 35 (1980) 260.
106 E.A. Startseva, N.M. Popova, N.G. Vanifatova, T.A. Chanysheva, Yu.A. Zolotov and I.G. Yudelevich, in *Metody Vydeleniya i Opredeleniya Blagorodnykh Elementov (Methods of Isolation and Determination of Noble Elements)*, GEOKhI AN SSSR, Moscow, 1981, p. 59 (in Russian).
107 L.A. Terenteva, L.D. Afanaseva, G.K. Chalkova, N.G. Vanifatova, V.G. Torgov and Yu.A. Zolotov, *Zavod. Lab.*, 49 (1983) 25.
108 B.Ya. Spivakov, L.N. Sukhoveeva, K. Dittrikh, A.V. Karyakin and Yu.A. Zolotov, *Zh. Anal. Khim.*, 34 (1979) 1947.
109 L.N. Sukhoveeva, B.Ya. Spivakov, A.V. Karyakin and Yu.A. Zolotov, *Zh. Anal. Khim.*, 34 (1979) 693.
110 V.M. Shkinev, I. Khavezov, B.Ya. Spivakov, S. Mareva, E. Ruseva, Yu.A. Zolotov and N. Iordanov, *Zh. Anal. Khim.*, 36 (1981) 896.
111 N.A. Borshch, Yu.A. Zolotov, O.M. Petrukhin, V.N. Shevchenko, L.N. Kolonina and O.A. Shiryaeva, in *Metody Vydeleniya i Opredeleniya Blagorodnykh Elementov (Methods of Isolation and Determination of Noble Elements)*, GEOKhI AN SSSR, Moscow, 1981, p. 93 (in Russian).
112 Yu.A. Zolotov, G.A. Larikova, V.A. Bodnya, O.A. Efremova, S.L. Davydova, K.B. Yatsimirsky and A.G. Kolchinsky, *Dokl. Akad. Nauk SSSR*, 258 (1981) 889.
113 F.D. Pierce, M.J. Gortatowski, H.D. Mecham and R.S. Fraser, *Anal. Chem.*, 47 (1975) 1132.
114 B. Holynska, *Radiochem. Radioanal. Lett.*, 17 (1974) 313.
115 K. Nakano, T. Takada and T. Kakuta, *Jpn. Anal.*, 28 (1979) 325.
116 A. Isozaki, N. Soeda, T. Okutani and S. Utsumi, *J. Chem. Soc. Jpn., Chem. Ind. Chem.*, 4 (1979) 549.
117 T. Takada, H. Okano, T. Koide, K. Fujita and K. Nakano, *J. Chem. Soc. Jpn., Chem. Ind. Chem.*, No. 1 (1981) 13.

118 A. Isozaki, N. Soeda and S. Utsumi, *Bull. Chem. Soc. Jpn.*, 54 (1981) 1364.
119 A. Isozaki, T. Kawakami and S. Utsumi, *Jpn. Anal.*, 31 (1982) E311.
120 Z. Slovák, *Anal. Chim. Acta*, 110 (1979) 301.
121 Z. Slovák and B. Docekal, *Anal. Chim. Acta*, 117 (1980) 293.
122 Z. Slovák and H. Dočekalova, *Anal. Chim. Acta*, 115 (1980) 111.
123 I.V. Kubrakova, G.M. Varshall, E.M. Sedykh, G.V. Myasoedova, I.I. Antokolskaya and T.P. Shemarykina, *Zh. Anal. Khim.*, 38 (1983) 2205.
124 Yu.I. Belyaev, V.N. Oreshkin and G.L. Vnukovskaya, *Zh. Anal. Khim.*, 30 (1975) 503.
125 K. Bächmann, A. Möller, C. Spachidis and C. Zikos, *Fresenius' Z. Anal. Chem.*, 294 (1979) 337.
126 H. Heinrichs, *Fresenius' Z. Anal. Chem.*, 294 (1979) 345.
127 F. Dolinšek and J. Stupar, *Analyst (London)*, 98 (1973) 841.
128 J.D. Sheaffer, G. Mulvey and R.K. Skogerboe, *Anal. Chem.*, 50 (1978) 1239.
129 R.A. Nadkarni, *Anal. Chim. Acta*, 135 (1982) 363.
130 H. Bombach, B. Luft, E. Weinhold and F. Mohr, *Neue Hütte*, 29 (1984) 233.
131 K.S. Subramanian, *Fresenius' Z. Anal. Chem.*, 305 (1981) 382.
132 J.G. Crock and F.E. Lichte, *Anal. Chim. Acta*, 144 (1982) 223.
133 R.W. Ward and P.B. Stockwell, *J. Autom. Chem.*, 5 (1983) 193.
134 M.H. Arbab-Zavar and A.G. Howard, *Analyst (London)*, 105 (1980) 744.
135 B. Welz and M. Melcher, *At. Spectrosc.*, 5 (1984) 59.
136 S.S.M. Hassan, *Organic Analysis using Atomic Absorption Spectrometry*, Ellis Horwood, New York, Chichester, Brisbane, Toronto, 1984.
137 S.H. Kägler, *Erdoel Kohle, Erdgas, Petrochem. Brennst. Chem.*, 28 (1972) 232.
138 U. Lidums, *Chem. Ser.*, 2 (1972) 159.
139 W. Lund and B. Larson, *Anal. Chim. Acta*, 70 (1974) 299.
140 Y. Hoshino, T. Utsunomiya and K. Fukui, *J. Chem. Soc. Jpn., Chem. Ind. Chem.*, 6 (1977) 808.
141 C. Fairless and A. Bord, *Anal. Lett.*, 5 (1972) 433.
142 H. Heinrichs, *Fresenius' Z. Anal. Chem.*, 273 (1975) 197.
143 Y. Thomassen, B. Larsen, F. Langmyhr and W. Lund, *Anal. Chim. Acta*, 83 (1976) 103.
144 G. Voland, P. Tschöpel and G. Tölg, *Anal. Chim. Acta*, 90 (1977) 15.
145 S. Dogan and W. Haerdi, *Anal. Chim. Acta*, 76 (1975) 345.
146 C. Jincheng Chen, Guangzheng Ma, Jinqiu Ma and Baogui Zhang, *Fenxi Huaxue*, 11 (1982) 646.
147 B. Holen, R. Bye and W. Lund, *Anal. Chim. Acta*, 130 (1981) 257.
148 V. Hudnik, S. Gomišček and B. Gorenc, *Anal. Chim. Acta*, 98 (1978) 39.
149 J.A. Nichols, A. John and R. Woodriff, *J. Assoc. Off. Anal. Chem.*, 63 (1980) 500.
150 A. Mizuike, *Enrichment Techniques for Inorganic Trace Analysis*, Springer-Verlag, Berlin, Heidelberg, New York, 1983.
151 M. Hiraide and A. Mizuike, *Pure Appl. Chem.*, 54 (1982) 1555.
152 S. Nakashima and M. Yagi, *Anal. Chim. Acta*, 147 (1983) 213.
153 V.I. Rigin and A.O. Eremina, *Zh. Anal. Khim.*, 39 (1984) 510.
154 W.J. Price, *Analytical Atomic Absorption Spectrometry*, Heyden and Son, London, New York, Rheine, 1974.
155 G.F. Kirbright and M. Sargent, *Atomic Absorption and Fluorescence Spectrometry*, Academic Press, New York, 1974.
156 N. Omenetto and J.D. Winefordner, *At. Fluoresc. Spectrom., Prog. Anal. At. Spectrosc.*, 2 (1–2) (1979) 183.

157 V.A. Razumov, *Zh. Anal. Khim.*, 32 (1977) 596.

158 A.N. Zaidel, *Atomno-Fluorestsentny Analiz (Atomic Fluorescence Analysis)*, Khimiya, Leningrad, 1983 (in Russian).

159 L. Ebdon and J.R. Wilkinson, *Anal. Chim. Acta*, 136 (1982) 191.

160 K. Braun, W. Slavin and A. Walsh, *Spectrochim. Acta, Part B*, 37B (1982) 721.

161 V.I. Rigin, *Zh. Anal. Khim.*, 38 (1983) 1060.

162 V.I. Rigin and I.V. Rigina, *Zh. Anal. Khim.*, 34 (1979) 1121.

163 V.I. Rigin, *Zh. Anal. Khim.*, 34 (1979) 261.

164 M. Jones, G.F. Kirbright, L. Ranson and T.S. West, *Anal. Chim. Acta*, 63 (1973) 210.

165 V.M. Shkinev, B.Ya. Spivakov, G.A. Vorob'eva and Yu.A. Zolotov, *Anal. Chim. Acta*, 167 (1985) 145.

166 P.P. Kish, B.Ya. Spivakov, V.V. Roman and Yu.A. Zolotov, *Zh. Anal. Khim.*, 32 (1977) 1942.

167 I.A. Blyum, F.P. Kalupina and T.I. Tsenskaya, *Zh. Anal. Khim.*, 29 (1974) 1572.

168 R. Belcher, *Talanta*, 15 (1968) 357.

169 V.A. Nazarenko and G.V. Flyantikova, *Zh. Anal. Khim.*, 32 (1977) 1217.

170 M. Satake, *Anal. Chim. Acta*, 92 (1977) 423.

171 A. Wasey, R.K. Bansal, B.K. Pure and M. Satake, *Analyst (London)*, 109 (1984) 601.

172 H.S. Sumiyoshi and K. Nakahara, *Talanta*, 24 (1977) 763.

173 A. Sanz-Medel, Blanco D. Gomis, E. Fuente and S. Arribas Jimeno, *Talanta*, 31 (1984) 515.

174 M. Takagi, K. Nakamura, Y. Sanui and K. Ueno, *Anal. Chim. Acta*, 126 (1981) 185.

175 K. Nakamura, H. Nishida, M. Takagi and K. Ueno, *Anal. Chim. Acta*, 139 (1982) 219.

176 Yu.A. Zolotov, V.P. Ionov, N.V. Nizyeva and A.A. Formanovsky, *Dokl. Akad. Nauk SSSR*, 277 (1984) 1145.

177 E.I. Morosanova, Yu.G. Bundal, E.D. Matveeva, A.A. Formanovsky and Yu.A. Zolotov, in *Second All-Union Conference on Chemistry of Macrocycles, Odessa, November 20–22, 1984*, Abstracts of Lectures and Reports, Odessa, 1984, p. 169 (in Russian).

178 E. Upor, M. Mohai and Gy. Novák, *Photometric Methods in Inorganic Trace Analysis*, Akadèmiai Kiadô, Budapest, 1985.

179 T. Kiriyama and R. Kuroda, *Fresenius' Z. Anal. Chem.*, 288 (1977) 354.

180 U. Nevoral, *Collect. Czech. Chem. Commun.*, 43 (1978) 2274.

181 K. Yoshimura, H. Waki and S. Ohashi, *Talanta*, 23 (1976) 449.

182 Yi. Ruishi, Wen Li and Feng Yugi, *Fenxi Huaxue*, 12 (1984) 215.

183 K. Yoshimura and H. Waki, *Talanta*, 32 (1985) 345.

184 Z.G. Szabó and I. Thege Konkoly, *Proc. Anal. Div. Chem. Soc.*, 15 (1978) 45.

185 A.B. Blank and L.P. Eksperiandova, *Zavod. Lab.*, 45 (1979) 7.

186 S. Taguchi, E. Ito-oka and K. Goto, *Bunseki Kagaku*, 33 (1984) 453.

187 M. Aoyama, T. Hobo and S. Suzuki, *Anal. Chim. Acta*, 141 (1982) 427.

188 A.N. Smagunova and E.N. Bazykina, *Zh. Anal. Khim.*, 40 (1985) 773.

189 N.F. Losev and A.N. Smagunova, *Osnovy Rentgenospektralnogo Fluorestsentnogo Analiza (Fundamentals of X-Ray Fluorescence Analysis)*, Khimiya, Moscow, 1982 (in Russian).

190 D.E. Leyden and W. Wegscheider, *Anal. Chem.*, 53 (1981) A1059.

191 V.P. Afonin, *Zh. Anal. Khim.*, 35 (1980) 2428.

192 F.I. Lobanov, *Zavod. Lab.*, 47 (1981) 1.

193 F.I. Lobanov, I.M. Yanovskaya and N.V. Makarov, *Usp. Khim.*, 52 (1983) 854.

194 R. van Grieken, *Anal. Chim. Acta*, 143 (1982) 3.
195 P. Marijanović, J. Makjanić and V. Valković, *J. Radioanal. Nucl. Chem.*, 81 (1984) 353.
196 R. Iessuè and C. Tarali, *Rass. Chim.*, 30 (1978) 75.
197 H.R. Linder, H.D. Seltner and B. Schreiber, *Anal. Chem.*, 50 (1978) 896.
198 E. Bruninx, A. van Eenbergen and A. Schouten, *Anal. Chim. Acta*, 109 (1979) 419.
199 J.E. Kessler and J.M. Mitchell, *Anal. Chem.*, 50 (1978) 1644.
200 Y. Saiton, A. Yoneda, Y. Maeda and T. Azumi, *Bunseki Kagaku*, 33 (1984) 412.
201 A.J. Pik, A.J. Cameron and J.M. Eckert, *Anal. Chim. Acta*, 110 (1979) 66.
202 R. Panayappan, D.L. Venezky, J.V. Glifrich and L.S. Birks, *Anal. Chem.*, 50 (1978) 1125.
203 M. Marshall and J.A. Page, *Spectrochim. Acta*, 33B (1979) 795.
204 B.H. Vassos, R.F. Hirsch and H. Letterman, *Anal. Chem.*, 45 (1973) 792.
205 B. Vanderborght, J. Verbeeck and R. van Grieken, *Bull. Soc. Chim. Belg.*, 86 (1977) 23.
206 J. Verbeeck, B. Vanderborght, R. van Grieken and G. Ex, *Anal. Chim. Acta*, 128 (1981) 207.
207 K. Lorber and K. Müller, *Mikrochim. Acta*, 4–5 (1976) 375.
208 C.J. Toussaint, G. Aina and F. Bo, *Anal. Chim. Acta*, 88 (1977) 193.
209 D.E. Leyden, R.E. Channel and C.W. Blount, *Anal. Chem.*, 44 (1972) 607.
210 R. van Grieken, C.M. Bresseleers and B.M. Vanderborght, *Anal. Chem.*, 49 (1977) 1326.
211 F. Clanet and R. Deloncle, *Anal. Chim. Acta*, 117 (1980) 343.
212 F. Clanet, R. Deloncle and G. Popoff, *Analusis*, 9 (1981) 276.
213 H. Kingston and P.A. Pella, *Anal. Chem.*, 53 (1981) 223.
214 G.D. Brykina, A.V. Stefanov, G.A. Okuneva, V.A. Alekseeva and Yu.S. Nikitin, *Zh. Anal. Khim.*, 39 (1984) 1750.
215 V.A. Shestakov, G.I. Malofeeva, O.M. Petrukhin, E.V. Marcheva, N.K. Esenova, Yu.I. Murinov, Yu.E. Nikitin and Yu.A. Zolotov, *Zh. Anal. Khim.*, 38 (1983) 2131.
216 V.A. Shestakov, G.I. Malofeeva, O.M. Petrukhin, O.A. Shiryaeva, E.V. Marcheva, Yu.I. Murinov, Yu.E. Nikitin and Yu.A. Zolotov, *Zh. Anal. Khim.*, 39 (1984) 311.
217 A. Disam, P. Tschöpel and G. Tölg, *Fresenius' Z. Anal. Chem.*, 295 (1979) 97.
218 D.L. Taylor and H. Zeitlin, *Anal. Chim. Acta*, 64 (1973) 139.
219 A. Kawase, S. Nakamura and N. Fudagawa, *Bunseki Kagaku*, 30 (1981) 229.
220 K. Iwasaki and K. Tanaka, *Jpn. Anal.*, 24 (1974) 619.
221 M. Murata, M. Omatsu and S. Mushimoto, *X-Ray Spectrom.*, 13 (1984) 83.
222 V.V. Berdikov, O.I. Grigoryev and B.S. Iokhin, *J. Radioanal. Chem.*, 68 (1982) 181.
223 K. Kato and M. Murano, *Jpn. Anal.*, 23 (1974) 1292.
224 R.C. Bearse, D.A. Close, J.J. Malanefy and C.J. Umbarger, *Anal. Chem.*, 46 (1974) 499.
225 T.G. Dzubay and R.K. Stevens, *Environ. Sci. Technol.*, 9 (1975) 663.
226 S. Tosiuki and S. Kontiro, *Jpn. Anal.*, 18 (1969) 1032.
227 A.J. Ahearn (Editor), *Trace Analysis by Mass Spectrometry*, Academic Press, New York, London, 1972.
228 M.S. Chupakhin, G.I. Ramendik and O.I. Kryuchkova, *Analiticheskie Vozmozhnosti Iskrovoi Mass-Spektrometrii (Analytic Possibilities of Spark-Source Mass Spectrometry)*, Atomizdat, Moscow, 1972.
229 V.M. Gladskoi, G.A. Ivanova, I.A. Kuzovlev and N.M. Kuz'min, *Zh. Anal. Khim.*, 26 (1971) 1087.

230 M. Murozumi, S. Nakamura, T. Igarashi and H. Tsubota, *J. Chem. Soc. Jpn., Chem. Ind. Chem.*, No. 4 (1978) 565.
231 Yu.A. Zolotov, N.V. Shakhova, O.I. Kryuchkova, S.I. Gronskaya, B.Ya. Spivakov, G.I. Ramendik and V.N. Gushchin, *Zh. Anal. Khim.*, 33 (1978) 1253.
232 O.M. Petrukhin, Yu.A. Zolotov, V.N. Shevchenko, O.I. Kryuchkova, S.I. Gronskaya, V.N. Gushchin, G.I. Ramendik, V.V. Dunina and E.G. Rukhadze, *Zh. Anal. Khim.*, 34 (1979) 334.
233 B.W. Vanderborght and R.E. van Grieken, *Talanta*, 27 (1980) 417.
234 F.W.E. Strelow and P.F.S. Jackson, *Anal. Chem.*, 46 (1974) 1481.
235 K.J.R. Rosman, J.R. de Laeter and A. Chegwidden, *Talanta*, 29 (1982) 274.
236 K.G. Heumann, *Trends Anal. Chem.*, 1 (1982) 357.
237 I.R. Shelpakova, A.I. Saprykin, T.A. Chanysheva and I.G. Yudelevich, *Zh. Anal. Khim.*, 38 (1983) 581.
238 T.A. Chanysheva, I.R. Shelpakova, A.I. Saprykin, L.M. Yankovskaya and I.G. Yudelevich, *Zh. Anal. Khim.*, 38 (1983) 979.
239 S.K. Aggarwal, F. Adams and E. Adriaenssens, *Fresenius' Z. Anal. Chem.*, 318 (1984) 402.
240 R. Alvares, P.J. Paulsen and D.E. Kelleher, *U.S. Dep. Commer. Nat. Bur. Stand. Spec. Publ.*, No. 492 (1977) 185.
241 J.W. Arden and N.H. Gale, *Anal. Chem.*, 46 (1974) 687.
242 J. Puymbroeck and R. Gijbels, *Bull. Soc. Chim. Belg.*, 87 (1978) 803.
243 J.R. Wasson, *Anal. Chem.*, 54 (1982) 121R.
244 J.R. Wasson and J.E. Salinas, *Anal. Chem.*, 52 (1980) 50R.
245 J.R. Wasson and P.J. Corvan, *Anal. Chem.*, 50 (1978) 92R.
246 Yu.A. Zolotov, in *Euroanalysis III, Reviews on Analytical Chemistry*, Applied Science Publishers, London, 1979, p. 26.
247 V.Yu. Nagy and M.V. Evstiferov, in *Teoriya i Praktika Ekstraktsionnykh Metodov (Theory and Practice of Extraction Methods)*, Nauka, Moscow, 1985, p. 210.
248 Yu.A. Zolotov, O.M. Petrukhin, V.Yu. Nagy and L.B. Volodarskii, *Anal. Chim. Acta*, 115 (1980) 1.
249 D. Yamamoto, T. Fukumoto and N. Ikawa, *Bull. Chem. Soc. Jpn.*, 45 (1972) 1403.
250 P.M. Solozhenkin, G.G. Sidorenko, N.G. Klassen and F.A. Shvengler, *Zh. Anal. Khim.*, 30 (1975) 2219.
251 P.M. Solozhenkin, G.G. Sidorenko, G.M. Larin, L.M. Shalyukhina and I.A. Glukhov, *Zh. Anal. Khim.*, 34 (1979) 808.
252 P.M. Solozhenkin, V.S. Pupkov, S.V. Ysova, M.B. Saidova and M.M. Yunusov, *Zh. Anal. Khim.*, 39 (1984) 2165.
253 A.R. Forrester, J.M. Hay and R.H. Thompson, *Organic Chemistry of Stable Free Radicals*, Academic Press, London, 1968.
254 J.K. Kochi (Editor), *Free Radicals*, Wiley, New York, 1973.
255 E.G. Rozantsev and V.D. Sholle, *Organicheskaya Khim. Svobodnykh Radikalov (Organic Chemistry of Free Radicals)*, Khimiya, Moscow, 1979 (in Russian).
256 A.L. Buchachenko and A.M. Vasserman, *Stabilnye Radikaly, Elektronnoe Stroenie, Reaktsionnaya Sposobnost' i Primenenie (Stable Radicals: Electronic Structure, Reactivity and Application)*, Khimiya, Moscow, 1973.
257 E.G. Rozantsev, *Svobodnye Iminokisilnye Radikaly (Free Iminoacid Radicals)*, Khimiya, Moscow, 1970.
258 L.B. Volodarsky, I.A. Grigor'ev and R.Z. Sagreev, in L.J. Berliner and J. Reuben

(Editors), *Biological Magnetic Resonance*, Vol. 2, Plenum Press, New York, 1980, p. 169.

259   J.F.W. Keana, *Chem. Rev.*, 78 (1978) 37.

260   V.N. Parmon, A.I. Kokorin and G.M. Zhadomirov, *Stabilnye Biradikaly (Stable Biradicals)*, Nauka, Moscow, 1980 (in Russian).

261   L.N. Skripnichenko, A.B. Shapiro, E.G. Rozantsev and L.B. Volodarskii, *Izv. Akad. Nauk. SSSR, Ser. Khim. Nauk*, No. 1 (1982) 109.

262   S.V. Larionov, V.I. Ovcharenko, V.N. Kirichenko and V.K. Mokhosoeva, *Izv. Sib. Otd. Akad. Nauk SSSR, Ser. Khim. Nauk*, No. 1 (1982) 14.

263   A. Rassat and P. Rey, *Bull. Soc. Chim. Fr.*, No. 3 (1967) 815.

264   A.A. Rzaev, A.A. Medzhidov and Kh.S. Mamedov, *Zh. Inorg. Khim.*, 25 (1980) 1277.

265   A.A. Medzhidov, L.N. Kirichenko and G.I. Likhtenshtein, *Izv. Sib. Otd. Akad. Nauk SSSR, Ser. Khim. Nauk*, No. 3 (1969) 698.

266   Yu.A. Zolotov, V.A. Bodnya, M.P. Kelareva, E.I. Morosanova, L.B. Volodarskii and V.A. Reznikov, *Zh. Anal. Khim.*, 37 (1982) 981.

267   D. Jahr, K.E. Schwarzhans, D. Nöthe and P.K. Burkert, *Z. Naturforsch. B*, 26B (1971) 1210.

268   L.N. Kirichenko and A.A. Medzhidov, *Izv. Sib. Otd. Akad. Nauk SSSR, Ser. Khim. Nauk*, No. 12 (1969) 2849.

269   D. Jahr, K.H. Rebhan, K.E. Schwarzhans and J. Wiedemann, *Z. Naturforsch. B*, 23B (1973) 55.

270   Yu.A. Zolotov, O.M. Petrukhin, N.A. Kurdyukova, V.V. Zhukov and I.N. Marov, *Zh. Anal. Khim.*, 33 (1978) 1307.

271   P.M. Solozhenkin, N.G. Klassen and F.A. Shvengler, *Zh. Anal. Khim.*, 32 (1977) 2080.

272   P.M. Solozhenkin, N.G. Klassen and G.G. Sidorenko, *Avtorskoe Svidetel'stvo, U.S.S.R. Pat.*, 448023 (1971); *Byulleten Izobretenii*, No. 40 (1974).

273   P.M. Solozhenkin and G.G. Sidorenko, *Kontrol Soderzhaniya Metallov Metodom EPR v Rudnvkh Pulpakh i Rastvorakh (Controlling Metal Content in Ore Pulps and Solutions by the ESR Method)*, TsNIITE Tsvetnoi Metallurgii, Moscow, 1973.

274   V.Yu. Nagy, O.M. Petrukhin, Yu.A. Zolotov and L.B. Volodarskii, *Izv. Sib. Otd. Akad. Nauk SSSR, Ser. Khim. Nauk*, No. 9 (1978) 2186.

275   Yu.A. Zolotov, V.A. Bodnya, M.P. Kelareva, E.I. Morosanova, L.B. Volodarskii and V.A. Reznikov, *Zh. Anal. Khim.*, 37 (1982) 981.

276   M.P. Kelareva, T.A. Gromova, V.A. Bodnya, L.B. Volodarskii, V.A. Reznikov and Yu.A. Zolotov, *Zh. Anal. Khim.*, 37 (1982) 1563.

277   I.M. Rottshafer, R.I. Boczkowski and H.B. Mark, *Talanta*, 19 (1972) 163.

278   G.G. Glukhov, *Trudy Nauchno-Issled. Inst. Yad. Fiz., Elektron. Avtom. Tomsk. Politekh. Inst. S.M. Kirova*, No. 7 (1977) 36.

279   H.A. Das, *Pure Appl. Chem.*, 54 (1982) 755.

280   R.A. Kuznetsov and M.M. Ustanova, *Zh. Anal. Khim.*, 40 (1985) 965.

281   V.G. Lambrev, *Ekstraktsiya Vhutrikompleksnykh Soedinenii Elementov s 1-Fenil-3-metil-4-benzoilpirazolonom-5 i 8-Oksikhinolinom i ye Ispolzovanie v Radioaktivat-sionnom Analize (Extraction of Chelate Compounds of Elements with 1-Phenyl-3-Methyl-4-Benzoylpyrazolone-5 and 8-Hydroxyquinoline and its Application in Radioactivation Analysis)*, Cand. Sc. Thesis, GEOKhI AN SSSR, Moscow, 1966.

282   G.G. Glukhov, L.A. Larionova and E.N. Gilbert, *Izv. Sib. Otd. Akad. Nauk SSSR, Ser. Khim. Nauk*, 7 (3) (1973) 87.

283   R.A. Kulmatov, A.A. Kist and E.S. Gureev, *Dokl. Akad. Nauk UzSSR*, No. 9 (1976) 25.

284 R.A. Kulmatov, A.A. Kist and V.D. Garbuzov, *Dokl. Akad. Nauk UzSSR*, No. 12 (1976) 23.
285 H. Akaiwa, H. Kawamoto and N. Nakata, *J. Radioanal. Chem.*, 36 (1977) 59.
286 S. Sava, L. Zikovsky and J. Boisvert, *J. Radioanal. Chem.*, 57 (1980) 23.
287 S.A. Harrison, P.D. La Fleure and W.H. Zoller, *U.S. Dep. Commer. Nat. Bur. Stand. Spec. Publ.*, No. 492 (1977) 148.
288 K.H. Lieser, W. Calmano, E. Heuss and V. Neitzert, *J. Radioanal. Chem.*, 37 (1977) 717.
289 V.G. Lambrev, M.M. Ivanova, V.A. Koftyuk and N.M. Kuz'min, *Zh. Anal. Khim.*, 38 (1983) 138.
290 B.Ya. Spivakov and Yu.A. Zolotov, *Zh. Anal. Khim.*, 25 (1970) 616.
291 O.V. Stepanets, B.Ya. Spivakov, Yu.V. Yakovlev, V.P. Kolotov and B.V. Savelev, *Zh. Anal. Khim.*, 31 (1976) 440.
292 V.P. Kolotov, V.M. Shkinev, Yu.V. Yakovlev and B.Ya. Spivakov, *Zh. Anal. Khim.*, 34 (1979) 2176.
293 E.N. Gilbert, G.V. Veriovkin, B.N. Botchkaryov, A.A. Godovikova, V.A. Mikhailov and V.Ya. Zhavoronkov, *J. Radioanal. Chem.*, 20 (1974) 253.
294 G.V. Veriovkin, E.N. Gilbert, A.M. Nemirovsky and E.G. Obrazovsky, *Zh. Anal. Khim.*, 36 (1981) 1073.
295 E.N. Gilbert, G.V. Veriovkin, V.I. Semenov and V.A. Mikhailov, *J. Radioanal. Chem.*, 38 (1977) 229.
296 E.N. Gilbert, G.V. Veriovkin, V.A. Mikhailov and N.A. Korol, *Zh. Anal. Khim.*, 35 (1980) 2300.
297 E.N. Gilbert, G.V. Veriovkin and V.A. Mikhailov, *Izv. Sib. Otd. Akad. Nauk SSSR, Ser. Khim. Nauk*, 10 (3) (1975) 102.
298 E.N. Gilbert, G.V. Veriovkin and V.A. Mikhailov, *J. Radioanal. Chem.*, 59 (1980) 381.
299 T. Braun and G. Ghersini (Editors), *Extraction Chromatography*, Akadémiai Kiadó, Budapest, 1975.
300 I. Akaza, T. Tajima and T. Kiba, *Bull. Chem. Soc. Jpn.*, 46 (1973) 1199.
301 E.P. Horwitz, C.A.A. Bloomquist and W.H. Dolphin, *J. Chromatogr. Sci.*, 15 (1977) 41.
302 A.M.G. Figueiredo and L.T. Atalla, *Publ. IEA*, 553 (1979) 18.
303 S.K. Kalinin and G.A. Yakovleva, *Zh. Anal. Khim.*, 33 (1978) 1995.
304 V.K. Markov, A.Ya. Yablochkin, M.I. Krapivin and V.I. Nagein, *Radiokhimiya*, 18 (1976) 751.
305 N.I. Gusev and E.I. Balashova, *Radiokhimiya*, 20 (1978) 883.
306 E.N. Gilbert, G.V. Veriovkin, V.A. Yakhina and E.S. Gureev, *Zh. Anal. Khim.*, 35 (1980) 656.
307 A. Kunbazarov, M.V. Rudomino, A.M. Sorochan, N.M. Dyatlova and M.M. Senyavin, *Zh. Prikl. Khim.*, 47 (1974) 2452.
308 K.V. Chmutov, P.P. Nazarov, E.A. Chuveleva and O.V. Kharitonov, *Radiokhimiya*, 19 (1977) 431.
309 A.I. Kalinin, *Zh. Anal. Khim.*, 32 (1977) 21.
310 M.M. Ivanova, I.P. Ogloblina, S.A. Genel, V.V. Mitina, A.I. Kalinin and V.G. Lambrev, *Zh. Anal. Khim.*, 32 (1977) 1066.
311 J.S. Fritz, *Pure Appl. Chem.*, 49 (1977) 1547.
312 V.G. Lambrev, I.P. Ogloblina, M.M. Ivanova, V.V. Mitina and S.A. Genel, *Zh. Anal. Khim.*, 35 (1980) 270.

313 J.M. Peters and G. Del Fiore, *J. Chromatogr.*, 108 (1975) 415.
314 L.I. Guseva and G.S. Tikhomirova, *Radiokhimiya*, 19 (1977) 188.
315 H.A. van der Sloot, G.D. Wals, C.A. Weers and H.A. Das, *Anal. Chem.*, 52 (1980) 112.
316 P. Schramel, *Mikrochim. Acta*, No. 2–3 (1978) 287.
317 G. Török, L.G. Nagy and J. Ruip, in *Abstract Papers of the Pittsburgh Conf. Anal. Chem. Appl. Spectrosc., Atlantic City, NJ, 1980*, Pittsburgh, PA, 1980, p. 703.
318 J.L. Fasching and D.P. Stout, in *Abstract Papers presented at the Pittsburgh Conf. and Expo. Anal. Chem. and Appl. Spectrosc., Atlantic City, NJ, March 8–13*, 1982, p. 282.
319 Kh.Z. Brainina, *Inversionnaya Voltamperometriya Tverdykh Faz (Stripping Voltametry of Solid Phases)*, Khimiya, Moscow, 1972.
320 F. Vydra, K. Štulík and E. Yuláková, *Rozpouštecí Polarografie a Voltametrie*, SNTL-Nakladetelststvi Technické Literatury, Prague, 1977.
321 Kh.Z. Brainina and E.Ya. Neiman, *Tverdofaznye Reaktsii v Elektroanaliticheskoi Khim. (Solid Phase Reactions in Electroanalytical Chemistry)*, Khimiya, Moscow, 1982 (in Russian).
322 P.T. Kissinger and V.R. Heineman (Editors), *Laboratory Techniques in Electroanalytical Chemistry*, Marcel Dekker, New York, Basel, 1984.
323 E.Ya. Neiman, *Zh. Anal. Khim.*, 29 (1974) 438.
324 A.A. Kaplin and V.M. Pichugina, *Zh. Anal. Khim.*, 39 (1984) 664.
325 Kh.Z. Brainina, L.S. Fokina and N.D. Fedorova, *Zh. Anal. Khim.*, 33 (1978) 225.
326 G. Gillain, G. Duyckaerts and A. Disteche, *Anal. Chim. Acta*, 106 (1979) 23.
327 V. Kemula, *Zh. Anal. Khim.*, 22 (1967) 562.
328 E.Ya. Neiman and L.N. Trukhacheva, *Zavod. Lab.*, 38 (1972) 1058.
329 S.B. Adeloju, A.M. Bond and M.L. Noble, *Anal. Chim. Acta*, 161 (1984) 303.
330 M. Nobufumi, *Jpn. Anal.*, 26 (1977) 224.
331 G.K. Budnikov and N.A. Ulakhovich, *Usp. Khim.*, 49 (1980) 147.
332 G.K. Budnikov, T.V. Troepol'skaya and N.A. Ulakhovich, *Elektrokhimiya Khelatov Metallov v Nevodnykh Sredakh (Electrochemistry of Metal Chelates in Non-aqueous Media)*, Khimiya, Moscow, 1980.
333 S.R. Rajagopalan, *Bull. Mater. Sci.*, 5 (1983) 317.
334 Yu.A. Karbainov and A.G. Stromberg, *Zh. Anal. Khim.*, 20 (1965) 769.
335 A.G. Stromberg, Yu.A. Karbainov and S.N. Karbainova, *Zavod. Lab.*, 36 (1970) 257.
336 Yu.A. Karbainov and G.N. Sutyagina, *Zh. Fiz. Khim.*, 47 (1973) 1885.
337 Yu.A. Karbainov and G.N. Sutyagina, *Zh. Fiz. Khim.*, 47 (1973) 2099.
338 Yu.A. Karbainov, *Izv. Tomsk. Politekh. Inst.*, 164 (1967) 228.
339 V.F. Toropova, G.K. Budnikov, N.A. Ulakhovich and E.P. Medyantseva, *J. Electroanal. Chem.*, 144 (1983) 1.
340 J. Labuda amd R. Dančíková, *Chem. Zvesti*, 37 (1983) 733.
341 Y. Nagaosa and K. Kobayashi, *Talanta*, 31 (1984) 593.
342 N.A. Ulakhovich, N.E. Naida, T.S. Gorbunova, I.R. Akhmetzyanova, A.P. Sturis and G.K. Budnikov, *Zavod. Lab.*, 50 (1984) 3.
343 T.V. Nghi and F. Vydra, *Collect. Czech. Chem. Commun.*, 40 (1975) 1485.
344 Y. Nagaosa and T. Sana, *Anal. Lett.*, A17 (1984) 243.
345 N.A. Ulakhovich, G.K. Budnikov, I.V. Postnova and N.K. Shakurova, *Zavod. Lab.*, 46 (1980) 587.
346 V.F. Toropova, Yu.N. Polyakov, G.N. Zhdanova and A.R. Garifzyanov, *Zh. Anal. Khim.*, 39 (1984) 1238.

347 N.A. Ulakhovich, G.K. Budnikov and G.A. Kutyrev, *Zh. Anal. Khim.*, 38 (1983) 671.
348 V.F. Toropova, Yu.N. Polyakov and G.N. Zhdanova, *Zh. Anal. Khim.*, 38 (1983) 238.
349 V.N. Maistrenko, Yu.I. Murinov, L.B. Reznik, F.A. Amirkhanova and Yu.E. Nikitin, *Zh. Anal. Khim.*, 37 (1982) 52.
350 V.N. Maistrenko, Yu.I. Murinov, L.B. Reznik, F.A. Amirkhanova, L.I. Yufereva and Yu.E. Nikitin, *Zh. Anal. Khim.*, 39 (1984) 272.
351 Y. Nagaosa and T. Yamada, *Talanta*, 31 (1984) 371.
352 V.N. Maistrenko, F.A. Amirkhanova, L.B. Reznik and V.V. Yunak, in *Electrochemical Methods of Analysis, Abstracts of the 2nd All-Union Conference on Electrochemical Methods of Analysis, Tomsk, June 4–6, 1985*, Part I, Tomsk, 1985, p. 149.
353 V.N. Maistrenko, Yu.I. Murinov, L.B. Reznik, F.A. Amirkhanova and Yu.E. Nikitin, *Zh. Anal. Khim.*, 40 (1985) 272.
354 A.B. Blank and L.P. Eksperiandova, *Zh. Anal. Khim.*, 37 (1982) 1749.
355 A.R. Guadalupe and H.D. Abruña, *Anal. Chem.*, 57 (1985) 142.
356 J.A. Cox and P.J. Kulesza, *J. Electroanal. Chem.*, 159 (1983) 337.
357 K. Izutsu, T. Nakamura and T. Oku, *J. Chem. Soc. Jpn., Chem. Ind. Chem.*, 10 (1980) 1656.
358 H. Müller, M. Otto and I. Werner, *Katalytische Methoden in der Spurenanalyse*, Akademie-Verlag, Leipzig, 1980.
359 S.U. Kreingold, *Catalysis in Analysis of Reagents and High-Purity Substances*, Khimiya, Moscow, 1983 (in Russian).
360 A.M. Bulgakova, G.Ya. Antipova and L.A. Egorova, *Avtorskoe Svidetel'stvo, U.S.S.R. Pat.*, 257840 (1969); *Byulleten' Izobretenii*, 36 (1969).
361 M. Otto and H. Müller, *Talanta*, 24 (1977) 15.
362 H. Müller, H. Schuried and G. Werner, *Talanta*, 21 (1974) 581.
363 M. Otto and H. Müller, *Anal. Chim. Acta*, 90 (1977) 159.
364 M. Otto, P.R. Bontchev and H. Müller, *Mikrochim. Acta (Wien)*, 1/3–4 (1977) 193.
365 S.U. Kreingold and E.M. Yutal', *Khim. Reakt. Osobo Chist. Vesh.*, 41 (1979) 39.
366 S.U. Kreingold, E.D. Shigina and E.V. Loginova, *Zh. Anal. Khim.*, 38 (1983) 1397.
367 G. Schwedt, *Chromatographic Methods in Inorganic Analysis*, Dr. A. Huthig Verlag, Heidelberg, Basel, New York, 1981.
368 *Adv. Chromatogr.*, 23 (1984).
369 C.J. Cower and A.J. Derose, *The Analysis of Gases by Chromatography*, Pergamon Press, New York, 1983.
370 Yu.S. Drugov, A.B. Belikov, G.A. Dyakova and V.M. Tul'chinsky, *Metody Analiza Zagryaznenii Vozdukha (Methods of Analysis of Air Contaminations)*, Khimiya, Moscow, 1984 (in Russian).
371 Yu.S. Drugov, *Zh. Anal. Khim.*, 40 (1985) 585.
372 D.N. Sokolov, *Gazovaya Khromatografiya Letuchikh Kompleksov Metallov (Gas Chromatography of Volatile Complexes of Metals)*, Nauka, Moscow, 1981.
373 K. Bächmann, *Talanta*, 29 (1982) 1.
374 J.M. O'Laughlin, *J. Liq. Chromatogr.*, 7 (1984) 127.
375 R.E. Majors, H.G. Barth and C.H. Lochmüler, *Anal. Chem.*, 56 (1984) 300R.
376 P.R. Haddad and A.L. Heckenberg, *J. Chromatogr.*, 300 (1984) 357.
377 A.R. Timerbaev, O.M. Petrukhin and Yu.A. Zolotov, *Zh. Anal. Khim.*, 36 (1981) 1160.
378 V.P. Mikhailenko, I.P. Sereda and A.N. Korol, *Zh. Anal. Khim.*, 34 (1979) 2260.
379 M.Ya. Bykhovsky and A.Yu. Braude, *Zh. Anal. Khim.*, 38 (1983) 2236.
380 N.K. Rudnevsky, V.T. Demarin and L.V. Sklemina, *Zh. Prikl. Spektrosk.*, 38 (1983) 61.

381 K. Jinno and C. Fujimoto, *J. Jpn. Chem.*, (1983) suppl. No. 138, p. 115.
382 R. Frigieri and R. Ferraroli, *Cron. Chim.*, No. 70 (1982) 3.
383 I.S. Krull, *Trends Anal. Chem.*, 3 (1984) 76.
384 V.G. Berezkin, *Khimicheskie Metody v Gazovoi Khromatografii (Chemical Methods in Gas Chromatography)*, Khimiya, Moscow, 1980 (in Russian).
385 S. Dillio and I. Sutikno, *J. Chromatogr.*, 298 (1984) 21.
386 V.I. Rigin, *Zh. Anal. Khim.*, 39 (1984) 648.
387 G. Schwedt, *Fresenius' Z. Anal. Chem.*, 295 (1979) 382.
388 Y. Shibata, M. Morita and K. Fuwa, *Anal. Chem.*, 56 (1984) 1527.
389 R.M. Cassidy and S. Elchuk, *J. Chromatogr. Sci.*, 18 (1980) 217.
390 M.A. Alawi, *Fresenius' Z. Anal. Chem.*, 313 (1982) 239.
391 M. Czuha and W.M. Riggs, *Anal. Chem.*, 47 (1975) 1836.
392 J.S. Brinen and J.E. McClure, *J. Electron Spectrosc. Relat. Phenom.*, 4 (1974) 243.
393 A. Ghonain, *Proc. Soc. Anal. Chem.*, 11 (1974) 138.
394 Th.A. Kouimtzis, M.C. Sofoniou and I.N. Papadoyannis, *Anal. Chim. Acta*, 123 (1981) 315.
395 A. Safavi and A. Townshend, *Anal. Chim. Acta*, 164 (1984) 77.
396 S.L. Bogdanski and A. Townshend, *Wiss. Z. Karl-Marx-Univ. Leipzig, Math.-Naturwiss Reihe*, 28 (1979) 377.
397 W. Thürauf and H. Assenmacher, *Fresenius' Z. Anal. Chem.*, 307 (1981) 265.
398 V.I. Grishko, I.G. Yudelevich and V.P. Grishko, *Zh. Anal. Khim.*, 39 (1984) 1813.
399 R. Bock, E. Zimmer and G. Weichbrodt, *Fresenius' Z. Anal. Chem.*, 293 (1968) 377.
400 O. Frindik, *Fresenius' Z. Anal. Chem.*, 318 (1984) 45.
401 N.P. Singh, C.J. Ziramerman, L.L. Lewis and W.E. McDonald, *Radiochim. Acta*, 35 (1984) 219.

Chapter 4

Preconcentration in Analysis of Various Substances

For the convenience of this discussion, let us arbitrarily divide the main materials analysed into several groups within each of which the sample preparation technique including preconcentration is similar: (1) environmental samples; (2) mineral raw material; (3) metals and alloys, inorganic materials including pure substances; (4) organic substances and biological samples.

4.1. Environmental samples

4.1.1 DISTINGUISHING FEATURES

Pollution of the air caused by the development of industry and urbanization, and the associated disturbances in ecological equilibria are causing more and more problems for society. Incomplete and irrational use of natural resources and of industrial products, harmful gases discharged by vehicles and power stations and imperfect techniques are the causes of pollution. The biosphere is contaminated by industrial and agricultural waste waters, domestic sewage, gaseous and aerosol wastes into the atmosphere. Often wastes entering into the biosphere are in excess of its natural potential for self-cleaning. Appearance of industrial products and wastes in the biosphere results in an increase in the content of toxic elements (As, Cd, Hg, Pb, Se and others) in soils, natural waters, the lower layers of air, flora and fauna. Certain inorganic compounds change the organoleptic properties of water or make it unfit for drinking and industrial use. A list of typical inorganic substances detected in waste water samples of different industries is available in the handbook *Harmful Inorganic Compounds in Industrial Waste Waters* [1].

The chemical composition of natural, waste (to say nothing of drinking) waters, and also of air is strictly controlled. In the USSR, the sanitary inspectorate has ascertained limiting concentrations for waters for more than 900 inorganic and organic substances; the fishing department determines the quality of water from about a hundred indexes. Such control is very complicated and expensive. It is therefore accomplished selectively; main indexes are checked with due consideration for specific properties of the contaminations in the given area.

Analysts are expected to give reliable analytical information about environmental samples, in particular information to identify harmful trace elements, to determine their content, forms in which they are present in air, atmospheric precipitation, soils, inland and sea waters, and bottom sediments. It is needed not only for determining the extent to which the biosphere is contaminated, but also to find the contamination sources and to localize them, to estimate the efficiency of purification techniques and of the methods of eliminating wastes, and to develop a waste-free technology.

In the U.S.S.R., the limiting allowable concentrations (LAC) of components are the main criteria by which the quality of the atmospheric air and water is determined. The LAC tables [2] list the main requirements placed upon the quality of water and air: the content of the components to be determined varies in the range $100-1000\,\mu g/l$ (for very toxic components: $1-2\,\mu g/l$) and $5-100\,\mu g/m^3$ (often $1-5\,\mu g/m^3$ and even less), respectively. Environment controllers demand that methods should be worked out for determining As, B, Ba, Be, Bi, Br, Cd, Co, Cr, Cs, Cu, Hg, I, Li, Mn, Mo, Ni, Pb, Sb, Se, Sn, Sr, W, U, Zn and other elements at a concentration level of $10^{-7}-10^{-10}\,\%$, and also for ascertaining the forms in which they occur. Taking into consideration the specific features of the samples to be studied and the fact that environmental samples are usually analysed for trace elements by atomic emission and atomic absorption spectrometry, spectrophotometry and fluorimetry, voltammetry, and kinetic and activation methods, one can easily understand the need for preconcentration of trace elements.

The determination of trace elements in environmental samples is a very complicated problem. First, a representative sample has to be selected in the field in changeable conditions; in so doing care should be taken that its chemical composition is not affected. The trace elements may be present in environmental samples in different states. From the viewpoint of the phase state, the trace elements may be present in the dissolved form as suspensions and colloids. The dissolved compounds are noted for their large variety: hydrolysed forms, complex compounds, etc. Depending on temperature, the presence or absence of oxidizing agents including dissolved oxygen, and solar radiation, one form may change into another because most often, water and air are non-equilibrium systems. Therefore, for determining total content of a component it (the component) should be changed into a form convenient for analysis, for example, into one oxidation state.

Zyrin and Zvonareva (see ref. 3, p. 3) have arranged the environmental samples in an order according to the complication of the analytical cycle: air, natural waters, plants, soils. In the analysis of these samples the main difficulty consists of obtaining the sample in a form suitable for analysis. Most often the sample is first obtained as solution or dry residue; at this stage it is important that the determinable elements are not lost or gained (from outside). Evaporation, drying and mineralization of the sample are also carried out at this stage; concentration also takes place in these operations. This procedure is not simple always and involves various steps.

The use of preconcentration in analysis of environmental samples has been considered in many publications [3–26]. The preconcentration methods preferred for analysis of different samples are discussed elsewhere.

Let us now consider preconcentration in analysis of certain environmental samples.

4.1.2 NATURAL AND WASTE WATERS

One would think that water is a comparatively simple substance for carrying out concentration. The trace elements present in it can be easily isolated from the matrix by several techniques: distillation of matrix, separation of trace elements by solvent extraction or sorption methods. However, as mentioned earlier, trace elements may be present in water in diverse forms; therefore, measures must be taken to obtain forms suitable for attaining an effect. The task becomes complicated when saline and often waste water samples are to be analyzed: water is "contaminated" with salts, organic substances and other components which create additional difficulties in isolating the trace elements. Thus, if water contains suspended matter, it is either first dissolved and then concentrated, or filtered out or centrifuged. In the latter case, it is first mixed with the concentrate obtained and then analyzed. The

TABLE 4.1

Examples of analysis of natural and waste water samples with preliminary concentration of trace elements

Samples analysed	Elements to be determined	Sample preparation features	Method of concentration	Method of determination	Detection limit	Ref.
River water	Al, Ba, Ca, Cd, Co, Cr, Cu, Fe, K, Mg, Mn, Mo, Na, Ni, Pb, Sn, Sr, V, Zn	Addition of conc. HNO_3	Volatilization	AES–ICP	0.03–4.8 μg/ml	27
Water, precipitation and fish tissues	As, Se	Addition of $K_2S_2O_8$ and HCl	Volatilization and subsequent generation of hydrides	AES–ICP	0.02–0.03 μg/ml	28
Natural water	As, Sb, Se	Addition of $K_2S_2O_8$ and HCl	Volatilization and subsequent generation of hydrides	AES–ICP	–	29
Waste water of caustic soda production	$Hg(II)$, Hg_2Cl_2, methyl and ethyl mercury	Addition of HNO_3 and $K_2Cr_2O_7$	Extraction of $Hg(II)$ with 0.1% diethyldithio-carbamate in CCl_4; extraction of organic compounds of Hg with benzene from HCl containing urea	NAA	–	30
Borehole water	K	Filtration	Extraction of potassium dimethyldibenzo-18-crown-6 in the	Photometry	5 μg/ml	31

Sample	Element	Separation/preconcentration I	Separation/preconcentration II	Method	Detection limit	Ref.
River water	Cu(II)	Boiling in the presence of $KMnO_4$ and HNO_3; addition of hydroxylamine hydrochloride	Extraction of Cu(II) with n-heptane (1 M in caproic acid and 0.5 M in benzylamine) in the presence of picrate ion	AAS	0.3 µg/l	32
Quarry and mine water	Be (as suspension and in ionic form)	Filtration through membrane filter; photochemical oxidation of beryllium complexes with humic acids	Extraction of beryllium with $5 \cdot 10^{-3}$ M trifluoroacetyl-acetone in carbon disulphide and subsequent gas chromatographic isolation	AFS	0.5 pg/ml	33
Sea water	Cu(II), Mn(II)	—	Sorption of 8-hydroxyquinolinates on silica gel coated with Bondapak Porasil B	AES–ICP	0.02 µg/ml	34
Sea water	Sb(III), Sb(V), Se(IV), Se(VI)	—	Sorption of pyrrolidine dithiocarbamates on silica gel coated with Bondapak Porasil B	ETAAS	7–50 µg/l	35

(Continued on p. 294)

References pp. 340–349

TABLE 4.1 (*continued*)

Samples analysed	Elements to be determined	Sample preparation features	Method of concentration	Method of determination	Detection limit	Ref.
	Au, Cd, Ir, Pd, Pu, Tc, Zn	Addition of complexing agent	Sorption on microporous anion exchanger JRA-900	ETAAS or MS	–	36
Sea and tap water	Cu, Pb	–	Sorption on 2-mercaptoben-zothiazole coated with the aid of collodion	Flame AAS	0.04–0.1 g/ml	37
River water, plant ash and biological samples	Cd, Cu(II), In, Ni, Pb, Tl(III)	Boiling H_2SO_4 with addition of $K_2S_2O_8$ or evaporating samples with HNO_3 and $HClO_4$	Sorption of diethyldithio-carbamates on polymeric thio ether	Flame AAS	–	38
Sea water	Cd, Co, Cu, Fe, Ni, Pb, V, Zn	–	Sorption on complex-forming sorbent Chelex-100	SSMS	n ng/ml	39
Water	Cd, Cu(II), Fe(III), Mn(II), Ni, Pb, Zn	UV radiation by mercury lamp in the presence of H_2O_2	Sorption on complex-forming cellulose sorbent Hyphan	Flame AAS	–	40
	F	–	Sorption on activated coal impregnated with $ZrO(NO_3)_2$	–	–	41

Sample	Elements	Pretreatment	Separation	Technique	Detection limit	Ref.
Natural water	U	UV radiation by quartz lamp, filtration	Co-precipitation with dibenzyldithio-carbamate of iron(III)	XRFS	0.4 μg/ml	42
Natural water	Ag, Cr, Cu, Fe, Hg, Mo, Ni, Pb, Pd, Sn, V, Zn, Zr	Filtration, addition of $K_2S_2O_8$ and heating to boiling (50 min)	Co-precipitation with pyrrolidinedithio-carbamate of cadmium or cobalt	XRFS	—	43
	Ag, Bi, Cd, Co, Cu, Fe, Mo, Ni, Pb, Sn, V, Zn	—	Co-precipitation with dibenzyldithio-carbamate of silver or nickel	AES–ICP	—	44
Sea water	Cd(II), Co(II), Cr(III), Cu(II), Mn(II), U(VI), Zn	—	Co-precipitation with 1-(2-pyridylazo)-2-naphthyl alcohol	NAA	5 ng–0.3 μg	45
Natural water	As, Cd, Co, Cu, Hg, Mo, Sb, Sn, Te, Ti, U, V, W	—	Flotation with combined collector of Fe(III) and pyrrolidinedithio-carbamate of ammonia	NAA	—	46

sample preparation stage is very important for obtaining reliable results of analysis.

With the main methods of preconcentrating trace elements in analysis of natural and waste waters are grouped various volatilization, extraction and sorption methods, particularly those which employ ion-exchange membranes and complex-forming sorbents, and coprecipitation. Electrochemical methods, precipitation and flotation are rarely applied. Some examples of utilizing preconcentration of trace elements in analysis of natural and waste waters are given in Table 4.1.

Volatilization methods are in general use because in most cases the matrix and the trace elements boil at significantly different temperatures. For determining Ag, Al, Cd, Co, Cr, Cu, Fe, Mn, Ni, Pb, Sb and V in the radioactive liquid flows of nuclear power stations, Pickford and Rossi [47] have suggested a technique involving successive evaporation of five portions of water, of total volume up to 0.5 ml, directly in a graphite furnace. The dry residue is then heated for evaporation and atomization. Thanks to better reproducibility of conditions under which an aliquot of water is evaporated in a furnace, the precision of analysis is better than with manual sampling.

Arsenic, a readily volatile element, is not easily determined. Often it is reduced to arsine and determined by the Marsh–Guttsait method, or preferably by AES–ICP [28, 29], AAS [48–50] or AFS [51]. In a similar manner other hydride-forming elements are determined in natural and waste water samples [52]. We shall also mention a method of isolating boron in aqueous media used in nuclear power engineering [53]. On adding sulphuric acid and methanol, volatile methyl borate forms, which, with an air current, is introduced into the capillary of the pulverizer of a flame emission spectrometer.

The extraction method is also important. As early as 1953, Pohl [54] suggested the preconcentration of Ag, Al, Au, Bi, Cd, Co, Cr, Cu, Fe, Ga, Hg, In, Mn, Mo, Ni, Pb, Pd, Sn, Ti, Tl, Th, V, Zn and rare-earth elements in analysis of natural waters by extracting the trace elements as chelates with 8-hydroxyquinoline, dithizone and diethyldithiocarbamate at stepwise varying values of pH (3–5–7–9) and using chloroform as solvent. The concentrate obtained after volatilization of the solvent was analyzed by the atomic emission method. The limits of detection were 10^{-7}–10^{-9}%.

Among the extraction systems used in analysis of natural waters including sea, waste and other waters, the system of ammonium pyrrolidinedithiocarbamate (APDTC, II) and methyl isobutyl ketone

(MIBK) is widely used. The APDTC system was proposed in refs. 55 and 56. This is one of the most stable dithiocarbamates; it is also stable in acid media and forms chelates with at least thirty elements. The procedures for separating trace elements with this reagent are discussed in ref. 57. For the determination of Co, Cr, Cu, Fe, Mn, Ni and Zn in sea water, Morris [58] extracted them as pyrrolidinedithiocarbamates with MIBK. The extract was mixed with fine cellulose powder. The pulp formed was dried, ground, pressed into a tablet, and analyzed by the XRF method. For this purpose other dithiocarbamates may be used as well as dithizone, 8-hydroxyquinoline, β-diketones and other reagents.

Extraction chromatography is also a convenient technique. Sturis et al. [59] used Teflon powder as carrier of benzene; organic reagent (2-isopropyl-8-mercaptoquinoline) was introduced directly into the solution being analyzed. Trace amounts of Cd, Co, Cu, Hg and Zn were quantitatively concentrated at pH 5–10. The trace elements were eluted from the column with $6 M$ HCl or benzene. The concentration coefficient attained values of 80–125.

Yano et al. [60] studied the conditions of extraction and back-extraction of mercury from sea waters in columns packed with dithizone or thenoyltrifluoroacetone gels. The gels were obtained by impregnating polystyrene (2% divinylbenzene) beads with chlorobenzene solutions of dithizonate of zinc or thenoyltrifluoroacetone. Mercury is quantitatively separated with dithizone gel at pH 0.3–2 and is easily eluted with $1 M$ solution of hydrobromic acid or $8 M$ HCl. At pH $\geqslant 4$ mercury is concentrated with thenoyltrifluoroacetone gel; elution is done with a mixture of 40 ml of $0.05 M$ sulphuric acid and 10 ml 10% solution of potassium bromide. Along with mercury, Ag, Cu and Pd are also extracted under conditions recommended by these authors.

For indirect atomic emission spectral determination of phosphorus (referred to molybdenum and antimony) in river and sea water samples [61] the extracts of reduced molybdo–antimonous heteropoly phosphoric acid in diisobutylketone were introduced into an inductively coupled plasma. The volume of the sample analyzed was 500 ml and that of diisobutylketone was 5 ml. Only arsenic(V) hinders the determination of phosphorus; it is therefore first reduced to arsenic(III). A detection limit of 0.3–6.0 ng of phosphorus per ml can be achieved.

In the analysis of natural and waste waters wide use is made of sorption methods which ensure concentration coefficients of $1 \cdot 10^4$. Sorption can be carried out in static as well as in dynamic conditions, sorbent being used in the column or membrane, or as suspension.

Organic and inorganic ion exchangers and complex-forming sorbents are also employed. Columns packed with sorbents are conveniently used for sampling and preconcentration under field conditions.

The techniques based on absorption of chelates are also finding application. For example, Vanderborght and Van Grieken [62] converted Cd, Co, Cr, Cu, Dy, Eu, Fe, Hf, Hg, Lu, Mn, Ni, Pb, Re, Sm, Tb, Yb and Zn into 8-hydroxyquinolinates at pH 8 and adsorbed the formed chelates with activated charcoal. The trace elements were adsorbed quantitatively; the concentration coefficient was $1 \cdot 10^4$. Different methods were used for determining trace elements in tap, soil, surface, and sea water samples including spectrophotometry, AAS, XRFS, NAA and the gamma activation method.

Sorption filters are even more convenient (see Chapter 2). We shall illustrate this by considering an example [63]. A cellulose-based filter, cellulose being coated with $-N(CH_2-CH_2-NH_2)_2$ groups, quantitatively adsorbs twelve trace elements (mainly heavy metals) contained in drinking water as suspensions and in the ionic form. After drying, the filter was analyzed by the XRF method and the trace elements were determined with detection limits of $n \cdot 10^{-7}\%$.

Higher concentration coefficients were attained on performing exchange reactions in static conditions. Koval'chuk $et\ al.$ [64] used metatitanic acid as sorbent for the trace elements present in sea water samples, and as collector in atomic emission analysis of the obtained concentrate. A sorbent (200 mg) was added to 2 l of sample (pH 8) and thoroughly mixed; the mixture was allowed to stand, the residue was separated and calcinated at 350–400°C. The concentrate obtained was mixed with graphite powder and sodium chloride, and analysed. Al, Bi, Cd, Co, Cr, Cu, Fe, Ga, Ge, In, Mn, Mo, Ni, Pb, Sn, W, V and Zn were quantitatively sorbed. The limits of detection were 10^{-7}–$10^{-8}\%$, the concentration coefficient being $1 \cdot 10^4$.

In the analysis of water samples the electrochemical methods of preconcentration are characterized by their high efficiency in a number of cases. They allow flexible control of preconcentration conditions, high concentration coefficients and automatic control. Very often, electrochemical concentration is an integral part of stripping voltammetry [22, 65, 66]. This method was used, for instance, for simultaneous determination of traces of Cd, Cu, Pb and Zn in underground and spring waters [67]. For determining $n \cdot 10^{-6}\%$ Co, Cr, Cu, Hg, Ni and Zn in fresh water [68] the trace elements were electrochemically deposited on pyrolytic graphite; the analysis was completed using the XRF method.

298

Other methods of preconcentration are also used for analysis of natural waters. Co-precipitation of Cd, Hg and Pb from sea water samples on copper sulphide is of interest [69]. Immediately after sampling the water was filtered through a membrane filter and the filtrate was then treated with sulphuric acid and potassium permanganate. Elimination of filtration raised the determination results of trace elements (especially of mercury) by 1.5–4 times. This was because the total content of ionic and colloidal forms and suspended particles was determined. Sulphuric acid and potassium permanganate apparently destroy the colloidal forms and hinder sorption of trace elements on the vessel walls. After concentration, the trace elements were determined by AAS. Depending on the sample volume (2–20 l) the determination was carried out in the following ranges of concentration: mercury, 0.001–0.01; lead, 0.025–0.25; cadmium, 0.005–0.05 μg/l. In analysis of sea water for selenium(IV), Nakashima [70] co-precipitated the trace element with iron(III) hydroxide and then accomplished flotation of the residue in the presence of sodium lauryl sulphonate. The separated residue was dissolved in hydrochloric acid and an aliquot of the obtained solution was introduced into the hydride generator containing NaBH$_4$. The hydride formed with nitrogen was transported to the flame atomizer; determination was completed by AAS.

Flotation was used for preconcentration of nitrite ions in river, rain and well water [71]. To a 1-ml sample of water, n-amino benzoylsulphoamide, N-(1-naphthyl)-ethylenediamine and sodium lauryl sulphonate were added as surfactants. The ion associate formed, in which the cationic part is also an azo dye obtained upon reaction of nitrite ions with the first two reagents, and the anion part is a surfactant, was floated by passing nitrogen. The foam removed was dissolved in propyl alcohol and, after adding a definite amount of HCl, determined photometrically.

In gas chromatographic (GC) determinations of nitrate, sulphate and phosphate ions in natural water samples [72], these anions were preconcentrated by freezing them (after the addition of a silver salt). The silver compounds so formed were transformed by n-butyl iodide into butyl esters, which are suitable for GC analysis. For the determination of chloride, bromide, iodide, cyanide, thiocyanate and nitrate ions, Faigle and Klockow [73] transformed these ions into highly volatile compounds by treating the water sample with tetrabutyl ammonium sulphate, freezing out the compounds formed with acetone and dry ice, and subsequent alkoxylation with a solution of decyl ester of methanesulphonic acid in acetonitrile. The compounds obtained [72, 73] were separated at different temperatures in glass columns. A flame ionization detector was used. The detection limits were found to be 0.2–6 μmol/l.

4.1.3 AIR

Filtration is usually employed for determining trace elements in solid particles present in air (see Chapter 2). In analysis of atmospheric aerosols, air is pumped through Whatman, Millipore or Taio filters, glass fibre, graphite materials or filter cloth with different pore sizes. Sometimes large volumes of the sample are taken. Thus, for determining the radioactive isotope ^{55}Fe in the atmosphere, Winkler [74] pumped 10 000–15 000 m^3 air through Microsorban filters. As methods of analysis of concentrated particles, X-ray fluorescence spectrometry and NAA methods [75, 76] are widely used. Thus, O'Connor et al. [77] have determined simultaneously up to forty elements collected on one filter. Systematic neutron activation determination of 34 elements was carried out in atmospheric aerosol samples collected in different parts of Belgium at different times of a year [78]. The samples were taken by pumping about 400 m^3 air through a Whatman cellulose filter for 24 h. Other instrumental methods were also employed.

Trace elements determined	Method of analysis	Ref.
Ag, Ba, Ca, Co, Cr, Cu, Fe, Mg, Mn, Na, Ti, V and Zn	AES–ICP	79
22 elements	AES–ICP	80
Al, Ca, Cu, Fe, Mn, Pb, Si, Ti and Zn	AES–ICP	81
Se and Te	ETAAS	82
Cl$^-$, NO$_3^-$ and SO$_4^{2-}$	Ion chromatography	83

Atmospheric dust can also be isolated from air by electrostatic precipitation. Torsi et al. [84] have described an arrangement used to isolate dust for subsequent determination of lead by the ETAAS method. They applied high voltage to the graphite furnace tube placed in the luminous flux of a hollow-cathode tube and the thin tungsten wire fixed near the oven along its axis, and passed a current of air. As a result, the charged dust particles settled on the surface of the atomizer. The wire was then removed and lead was determined. Isolation of dust particles from air can also be effected by the impact precipitation method [85].

The trace elements present in air either in the gaseous or vapour form are concentrated by absorption or adsorption [11, 24, 86, 87]. Thus, mercury and its compounds can be concentrated from air by different methods [87]: solid particles are isolated on fine-pore filters; mercury chlorides, on a 3% SE-30 coated on HCl-treated Chromosorb W; methyl mercury(II) salts, on Chromosorb treated with an alkali solution; mercury vapours, on silver-plated glass beads. Besides these, a technique involving absorption of mercury vapours from air on a gold wire/plate or on a gold-plated carrier is often used; this forms the basis of the methods employed for atomic fluorescent determination of mercury in air [88].

Dumas and Bond [89] have developed a simple method for sorption of traces of arsine from air in a tube filled with Chromosorb 102. After desorption, arsine is determined by the GC method.

Absorption concentration yields better results if use is made of an absorber with a porous plate [90] ensuring a large contact surface for phases. Particularly good results are obtained when the absorbing solution reacts with the adsorbed forms of trace elements as, for instance, in absorption of mercury vapours with a solution of potassium iodide and elemental iodine.

For concentration of tetraalkyl compounds of lead from air (sample volume $\leqslant 10\,l$) use was made of filtration together with freezing in a trap filled with glass beads [91]. Lead may be determined by a suitable method, for instance by a spectrophotometric or atomic absorption method.

4.1.4 SOILS AND BOTTOM SEDIMENTS

For obtaining exhaustive information about soils, it is necessary to analyse them for about 40 macro- and trace elements. It is important to know what trace elements (useful and harmful) are deficient, or are present in excess, in the soil. Besides the trace elements (B, Ca, Cu, I, Mn, Mo and Zn) whose biological activity does not give rise to any doubts, Cr, Ni and V have to be determined; As, Be, Br, Cd, F, Hg, Sb, Se, Tl and other elements are also determined to check soil pollution.

In agricultural chemistry one determines in soils the exchange or mobile forms of trace elements that affect crop yields. For this the sample is not completely decomposed; the ions to be determined are removed from the surface by mixing the soil with acids and buffer solutions. The extracts obtained are fit for analysis; they do not require special additional treatment [92–96].

If it is necessary to determine the total content of trace elements in soils, the sample is dried, fused and/or decomposed with mixtures of concentrated acids: nitric, hydrochloric and hydrofluoric acids. At this stage of analysis the trace elements are concentrated by removal of water, organic substances and silicon. Often the soil samples are dry mineralized at 450–500°C for 6–8 h [97, 98].

After the sample has been dissolved, different methods may be used to concentrate the trace elements. Sorption and solvent extraction methods are quite often employed for the purpose; electrochemical methods and co-precipitation are rarely used. The stage preceding preconcentration is the most complicated stage in the analysis. This is evident from the following example. For the determination of Ag, Bi, Co, Cu, Mn, Mo, Ni, Pb, Sn and V in bottom sediments, 1 g air-dried sample was ground, heated several times with concentrated sulphuric acid until no more sulphur dioxide appeared [99]. The sample was again moistened with sulphuric acid, ammonium fluoride and hydrochloric acid were added, and the solution was slightly clarified. The acidity of the solution was adjusted to 6 M in HCl. Iron, which interferes with the determination, was extracted with a mixture of isoamylacetate and methyl isobutyl ketone (2:1). Besides iron, the extract contained Mo and Sn. Fe, Mo and Sn were back-extracted in water slightly acidified with hydrochloric acid, and iron was precipitated with ammonia. The solution obtained after the removal of iron was mixed with another iron-free solution, iron being removed by extraction, and only then the trace elements were concentrated by extracting their diethyldithiocarbamates with chloroform.

The time needed for analysis of soils and bottom sediments is reduced by employing autoclave decomposition [100–104]. For the AES–ICP determination of molybdenum in soils and rocks [105] the sample was treated with hydrochloric acid in the presence of bromine in a thick-walled tube having a pressure-tight Teflon cover. Molybdenum was extracted from 6 M HCl solution with methyl isobutyl ketone. Rapid determination of Cd, Cr, Cu, Fe, Hg, Mn, Ni, Pb and Zn in river sediments [106] involves decomposition of the sample in a closed Teflon vessel with a mixture of hydrofluoric, nitric and hydrochloric acids at 140°C. Decomposition takes about 1 h.

302

4.2. Mineral raw material

Different mineral raw materials, *e.g.* rocks, ores, minerals and non-metallic mineral resources, in particular civil engineering materials and salts, are some of the most complex, important and widely found materials to be analysed. The search for, and assessment and development of, mineral deposits and their mining and treatment are based to a large extent on the analysis results of mineral raw materials. Such analyses are of great importance particularly for geological surveys and for the branches of industry utilizing mineral raw materials, *e.g.* ferrous and non-ferrous metallurgy, chemical and construction materials industries.

Determination of trace elements is one of the main tasks of the laboratories specializing in analysis of mineral raw materials. From the viewpoint of geologists, the elements whose concentration does not exceed 0.01% are usually considered as trace elements. Quantitative trace analysis of mineral raw materials is a complicated task due to the varied compositions of the samples, forms of macro- and trace elements, and the need for different standard (reference) samples.

In mineral raw materials the ratio of total concentrations of accompanying components to the concentration of the determinable trace component can vary between 10^4 and 10^9. This, as well as the non-homogeneity of distribution of trace elements being determined in the sample, allows one to conclude that preconcentration methods are necessary for analysis (although, in a number of cases, Be, Cs, Cu, Li, Rb, Sc, Sn, Ta and some other elements can be determined without separating them). Besides, it may be required to mine and process progressively poorer ores. It is necessary to work out procedures ensuring, for a conditional weighed amount of 1 g, the determination of trace elements in the range 10–$100\,\mu g$ for fluorine to 10^{-4}–$10^{-5}\,\mu g$ for gold, beryllium and bismuth.

Separation of trace elements is the main technique of concentration used in the analysis of mineral raw materials: the matrix is very complex so that it is worthwhile to remove it. It is true that partial separation of the main components takes place on decomposing silicate and carbonate rocks; carbon dioxide and silicon tetrahalide are removed from them.

Preconcentration of trace elements has long been used in the analysis of mineral raw materials. For information on concentration of trace elements by precipitation, distillation, electrolysis, separation and extraction, the reader is referred to the classic manual on inorganic analysis [107]. Several new publications, in particular [108–113], have appeared which describe the use of instrumental methods instead of classic chemical methods of analysis; the methods of concentration are also discussed therein.

Using as an example the laboratories of the U.S.S.R. Geological Survey, engaged in the determination of trace elements, Blyum and Zolotov [114] have shown that, of the preconcentration techniques, extraction, co-precipitation, sorption and volatilization are most extensively employed in the U.S.S.R.

We shall consider two examples of preconcentration of trace elements in the analysis of carbonate rocks. Dorokhova *et al.* [115] have developed an atomic emission method for determining Be, Bi, Co, Cr, Ga, La, Mn, Mo, Nb, Ni, Pb, Sb, Sc, Sn, Ti, V, Y, Yb, Zn and Zr in carbonate and carbonate–silicate rocks. The method involves the use of complex-forming sorbents based on aminopolystyrenepolyarsenazo and polypyridylazoresorcinol. Silica is first removed as silicon tetrafluoride if its content in the rock exceeds 21%. Conditions have also been found for group concentration of trace elements in the presence of 0.5–1 g calcium. The concentrate was ashed at 450–500°C. Under the determined conditions, a concentration coefficient of 10–20 was attained in the analysis of pure carbonates. Preliminary treatment of silicate–carbonate rocks in the chamber used for volatilizing silicon tetrafluoride makes it possible to raise the concentration coefficient up to 40. The detection limits were 10^{-5}–10^{-6}%. In the analysis of limestone, Kawabuchi *et al.* [116] concentrated Fe(III), Cd, Co, Cu, Ni, Pb, Zn, Cr(III) and Mn(II) using ion exchangers of the iminodiacetic type. The trace elements were desorbed from the column with mineral acid solutions; the determination was completed with the atomic absorption method.

Determination of noble, rare-earth and radioactive metals in mineral raw materials is not possible without the application of preconcentration techniques.

Preconcentration of noble metals (gold, silver, platinum-group elements) is very essential. Fire melting, discussed in Chapter 2, is especially used for preconcentration of these metals. Extraction methods of concentrating noble metals are also finding extensive application. Examples illustrating the use of these methods may be found in Chapters 2 and 3 of this book and in other monographs and reviews [117–124]. Some examples given in Table 4.2 corroborate the significance of extraction concentration of noble and other metals in the analysis of mineral raw materials.

Sorption methods are often used for concentration of noble metals [119, 121, 124, 136]. Ion exchangers, complex-forming sorbents and modified polymers are more often used for concentration of gold, silver and platinum-group metals. Many complex-forming sorbents are very active towards noble metals, the reason being the large tendency of noble metals to form complexes; the sorption selectivity increases when concentration is accomplished from highly acidic solutions. A complex-forming sorbent with the 3(5)-methylpyrazole group (PVB-MP) exhibits unique properties towards noble metals including Ir, Rh and Ru; it enables their group concentration to be accomplished in the presence of copper, nickel and other elements [137]. Recently, a new complex-forming sorbent, Polyorgs XI–H [138], has been suggested. It is a fibrous material. In the process of formation its fibres are filled with a finely dispersed polymer sorbent exhibiting weakly basic and complex-forming properties. Polyorgs XI–H has a highly developed surface and good kinetic characteristics. From acid solutions, it extracts Au(III), Pd(II), Pt(II) and Cr(VI); from neutral solutions, Ag, Au(III), Cu(II), Ni, Zn, Al, Co, Cd, Fe(III) and V(IV). Complex-forming sorbents are conveniently used in combination with XRFS and NAA methods, for the concentrate obtained can be immediately used for analysis. Of interest is the procedure for introducing directly into an inductively coupled plasma the water suspension of the sorbent Monivex containing concentrated Au, Ir, Pd, Pt, Rh and Ru [139].

Co-precipitation and, less frequently, precipitation are also employed for preconcentration of noble metals. Sulphides of certain metals (Bi, Cu, Hg, Sb) are used as collectors for concentration of noble metals [136]; thioacetamide may be employed as a reagent. In that case, platinum-group metal sulphides are quickly formed at relatively low temperatures and in moderately concentrated (in H_2SO_4) solutions [140, 141]. Inorganic collectors like finely dispersed precipitates of elementary sulphur, selenium, tellurium as well as Al, Fe(III), Be, Mg, Mn(IV), La, Ti or Zn hydroxides and others are also used for the purpose. Trace amounts of noble metals are also successfully isolated by using organic co-precipitants, e.g. n-dimethylaminobenzyledene rhodanine, dithizone, thiobenzamide, α-benzyldioxime, thionalide, thiooxine, etc. Such organic collectors are not always sufficiently selective at high concentrations of non-noble metals.

Davydova et al. [142] concentrated platinum-group metals using a

TABLE 4.2

Some methods of analysis of mineral raw materials, based on extraction of trace elements

Material analysed	Elements to be determined	Extractant	Aqueous phase	Method of analysing concentrate	Detection limit	Ref.
Limestone, garnet, diorite, pegmatoidal rock, CT-1 (trap brilliant), CG-1 (albetized garnet), CGD-1 (modified gabbro-diorite)	Ag	0.05 M diphenylthiourea in chloroform	0.5–3 M HCl	AAS (flame and ETA)	0.1 μg/ml or 0.5 ng	125
Albitophrite	Hg	Astraphloxine FF, benzene	1–2.5 M H_2SO_4	Photometry	$n \cdot 10^{-4}\%$	126
Ores	Ir, Pd, Pt, Rh	Di-2-ethyl-hexyldithiophosphoric acid	1–4 M HCl	AES	–	127
Quartz, limestone, diverse eruptions of rocks and tectonites	Ag, Au	0.1 M di-n-butylsulphide in toluene (for Au) and 0.05 M N,N'-diphenyl-thiourea in chloroform (for Ag)	1.0–1.5 M HCl	AES	Ag 0.01 g Au 0.005 g	128
Minerals, rocks	Ir, Pd, Pt, Ru	N,N'-Hexamethyl-ene-N'-phenyl-thiourea, chloroform	6 M HCl	SMS	$1.6 \cdot 10^{-6}\%$	129

Sample	Element(s)	Extraction reagent	Acid medium	Method	Detection limit	Ref.
Ores, pyrites, ashes of coals and river sediments	In	Methyl isobutyl ketone	$ca.$ 2 M NH$_4$I	ETAAS	—	130
Chromium ores	V	Solution of N-benzoyl-N-phenylhydroxylamine in toluene	Dilute H$_2$SO$_4$ + HCl	Flame AAS	—	131
Rocks	Mo	0.2% α-benzoinoxime in chloroform	—	ETAAS	0.1 g/ton	132
Geological materials	Ag, As, Bi, Cd, Cu, Mo, Pb, Sb, Zn	10% Aliquat 336 in methyl isobutyl ketone	3.5–4.0 M HCl containing about 6% KI	Flame AAS	0.1–5 μg	133
Cu–Ni sulphide ores	Ag, Au, Pd, Pt, Rh, Ru	Mixture of 0.05 M di-2-ethylhexyldithiophosphoric acid and 0.25 M n-octylamine in toluene	2–3 M HCl	ETAAS	$1.5 \cdot 10^{-2}$– $1 \cdot 10^{-4}$ g/ton	134
Silicate rocks	Ce, Eu, Gd, La, Lu, Nd, Sm, Tb, Tm, Yb	Mixture of 0.5 M solution of thenoyltrifluoroacetone and 0.1 M solution of 1,10-phenanthroline	0.02 M HCl	NAA	—	135

combined scheme: fire melting in nickel sulphide and subsequent selective sorption of noble metals with the chelate-forming sorbent PVB-MT-20T or co-precipitation with thiourea. This method of concentration in combination with the kinetic method of determination enabled rocks containing 10^{-6}–$10^{-7}\%$ platinum-group metals to be successfully analysed.

Rare-earth elements are found in different rocks: limestone, granite, schists, granodiorites, etc. Their concentration in rocks is usually in the range of 10^{-4} to $10^{-3}\%$, sometimes $10^{-2}\%$. Precipitation as mixed oxalates is the classic method of isolating all rare-earth elements from silicate rocks [108]. After ignition to mixed oxides, weighing is done. Determination may be done by a photometric method, for example with arsenazo III [143]. In this case, finely ground silicate rock was decomposed with a mixture of hydrofluoric, nitric and sulphuric acids. From the solution obtained, the rare-earth elements were separated from Fe, Th, U and Y and other metals interfering with subsequent determination by precipitating the hydroxides with ammonia followed by the precipitation of oxalates (calcium was used as co-precipitant). The precipitates of oxalates were ignited in a muffle furnace at 500°C; the dry residue was dissolved in hydrochloric acid in the presence of hydrogen peroxide. After the addition of arsenazo III solution, photometric evaluation was done at pH 1.8–2.0. For separating rare-earth elements from the solutions obtained after treating the geological materials [144], use was made of the cation-exchanger Dowex 50W-X8 and 0.8–$6.0\,M$ HCl solutions as eluents.

On the whole, in the analysis of raw materials use is made of diverse preconcentration techniques, but preference is given to solvent extraction, sorption methods and co-precipitation.

4.3. Metals, alloys and other inorganic substances and materials, including high-purity substances

4.3.1 GENERAL

In this section we shall discuss the possibilities of concentration methods as applied to a large group of, what one would imagine to be, different materials, that is inorganic substances with relatively small concentrations of the matrix elements. Generally, these are compounds of one, two or three elements. With these are grouped the materials of ferrous metallurgy (steels, cast irons, refractory materials), non-ferrous metallurgy (more than seventy metals, alloys and compounds of metals, semiconducting materials), and of the chemical industry (acids and alkalis, soda, inorganic chemical reagents and others). The interest in the determination of impurities in such substances and materials is growing all the time. This is due to the creation and development of the nuclear industry, which required pure uranium, thorium, beryllium, sodium, zirconium, niobium, graphite, etc. More pure substances were needed by electrical engineering and electronics: silicon, germanium, sulphur, selenium, gallium, indium, arsenic, antimony, cadmium in the elemental form and as compounds, in particular $A^{II}B^{VI}$ and $A^{III}B^{V}$, and several related materials. Pure compounds of rubidium and rare-earth elements are needed for manufacturing lasers. Fibre optics demanded extra-pure glasses. The demands on the purity of chemical reagents and components used in the manufacture of high-temperature and chemically stable steels and alloys have appreciably increased.

Trace impurities can vary to a large extent the properties of metals, alloys and inorganic materials. For example, the carbon content affects the properties of iron, but high-purity iron does not corrode unlike technical grade iron. The plasticity of titanium, zirconium or tungsten increases with the purity of the metals. Beryllium is very brittle at ordinary temperatures, but becomes very plastic after its purification by the zone melting method. Impurities affect the electrical, magnetic and other properties of many metals and their compounds.

Besides trace impurities, an analyst has to determine the alloying traces specially introduced for imparting particular properties to the substance. Often it is required to determine not only the total content of trace elements in the bulk sample, but also their distribution over the surface, at depth and in the volume. Development of new grades of steels and alloys including high-temperature and refractory, expansion of the range of semiconducting materials, inorganic reagents and preparations, and increased demands on their purity complicate the analytical control of the trace component composition. Thus, industry places heavy demands on high-quality steels in respect to content of non-ferrous metals; often they have to be determined at a content of 10^{-4}–10^{-6}%. Now 10^{-5}–10^{-7}% and sometimes 10^{-8}–10^{-10}% impurities can be determined.

Different methods of preconcentration and their various combinations with the determination methods are employed in the analysis of metals and alloys, inorganic materials and pure substances. In the

U.S.S.R., wide use is made of atomic emission methods; extensive use is also made of extraction–photometric methods, stripping voltammetry, the chemical activation method, and the combination of concentration with atomic absorption spectrophotometry. The atomic emission methods enable multi-elemental analysis to be carried out with detection limits of 10^{-5}–10^{-8}%.

In the U.S.S.R., the use of different methods of preconcentration for analysis of pure substances was first reviewed in two books [145, 146], which critically assessed the possibilities of preconcentration, in general, and certain techniques in particular. Most of the discussed procedures with some modifications are still in use. This refers primarily to the analysis of silicon, germanium, selenium, sulphur, iodine and chemical reagents.

More stringent requirements are placed on the purity of laboratory air, reagents, vessels and devices when pure substances are analysed [9, 101, 117, 147–152]. Preference is given to preconcentration techniques that do not demand reagents, e.g. volatilization, sublimation and controlled crystallization. But each concentration technique has its own problems and restricted field of application; this is the reason why different techniques are used for the analysis of different substances.

An analyst generally starts his work after a pure substance has been obtained by a producer who makes use of different purification methods similar in many respects to preconcentration and separation techniques. For instance, crystallization techniques are effectively used for obtaining and purification of different poly- and monocrystals; sorption methods are employed for purification of silicon, germanium and titanium halides, silanes, phosphorus, arsenic, water and acids; distillation is used to obtain volatile metal halides, organic solvents, acids, water and some metals (cadmium, mercury, zinc). It must be noted that the analytical method of preconcentration should, as a rule, be different from the method that was used to obtain or purify the substance. For this reason, analysts are sometimes forced to make use of less effective methods.

All elements are present in each substance or material, but their concentrations differ significantly. However, eight elements (Al, Ca, Fe, K, Mg, Na, O and Si), which account for 98.6% of the Earth's crust, are very likely to be present in the material analysed. It is quite difficult to free the substance of these elements and, equally, to keep them in the purified form. It is very difficult to lower the detection limit of these elements. Karpov and Alimarin [153] have correctly mentioned that a

310

series of chemical, physico-chemical and mainly physical methods of analysis have been worked out at present with detection limits at a level of 10^{-6}–$10^{-10}\%$, but it is not possible to determine the elements that are widely distributed in nature if their concentration is less than 10^{-6}–$10^{-7}\%$.

Pure substances are costly, of great variety and are manufactured in small lots. Therefore, analysts are sometimes compelled to work with small amounts of the sample containing insignificant amounts of impurities.

The effects associated with the preparation of vessels and tare, the selection of main and auxiliary equipment, reagents and solutions and reference samples, and with the sterile conditions of analysis are more critically assessed if the concentrations are $10^{-7}\%$ or less (in the case of widely distributed elements, $10^{-5}\%$ or less).

Constant struggle to lower the detection limits, the elimination or strict account of impurities or losses which occur when the sample reacts with the environment (microclimate of the laboratory, reagents, solutions, vessel material and researcher), and specific attention to the metrology of analysis are typical of the theory and practice of analytical chemistry of pure substances. The struggle to lower the detection limits for trace elements is always important, but often leads to a sacrifice of precision.

Tölg's opinion and experience [101, 154–157] in eliminating and lowering the errors caused by preconcentration are of interest. He, like other specialists engaged in the analysis of high-purity substances (see ref. 117), mentions important sources of such errors; on the one hand, the impurities introduced by the laboratory air, reagents and apparatus, and, on the other hand, the losses due to adsorption on the vessel's walls, volatilization and incomplete separation of trace components. For attaining low limits of detection of widespread, as well as of less abundant, elements which are often present in the laboratory, e.g. Cu, Hg, Pb and Zn, it is necessary that the air in the laboratory should be dust-free, that use should be made of purified reagents and apparatus which does not give off impurities into the atmosphere, and that vessels made of inert materials be used, particularly those of synthetic quartz or Teflon. As mentioned by Tölg [154, 156], elimination of dust from the air of the laboratory makes it possible to lower the detection limits of aluminium, iron, silicon and other widespread elements by at least two orders of magnitude.

Tölg suggests a close combination of decomposition, concentration and determination, in both time and space, preferring to use a single closed vessel. This was illustrated by determining nanogram amounts of carbon and sulphur in highly pure iron, copper, noble and refractory metals by fusing the sample in a high-frequency field created in an atmosphere of pure oxygen. The isolated CO_2 and SO_2 are determined by the conductometric method. Combination of decomposition, separation and determination is useful for the separation of macro- and trace elements having different volatilities. An example of this is the separation of silicon, arsenic, osmium, boron and tin as SiF_4, As_2O_3, OsO_4, BF_3, and $SnBr_4$, respectively, from heavy metals.

Another method of eliminating systematic errors, according to Tölg, involves the use of micro- and ultramicro-techniques. It enables analysts to work with very small weighed amounts of the sample in a sterile evacuated volume with a minimum surface area of inert material that comes into contact with the sample. Tölg has approved in his laboratory different micro-techniques of preconcentration of trace elements including volatile substances. These are co-precipitation, electrodeposition, volatilization, sublimation, extraction, adsorption in thin layers of silver halides, ion exchange and fire melting.

The detection limits of widely found elements are determined from the result of a blank experiment, which gradually varies even within one laboratory. A blank experiment is therefore indispensable in an analysis; the sequence of this experiment is dictated by definite conditions. A strict procedure for the blank experiment is often very important for obtaining correct results. Such an experiment must simulate the analysis procedure of the sample for all parameters (time, temperature, equipment, reagents and solutions, vessel, etc.) and thereby take into account the possible contaminations. A blank experiment that is correctly carried out makes it possible to eliminate systematic errors connected with foreign impurities.

It is usually performed without the sample being analysed, *i.e.* by running the used reagents through the steps of the analysis, often not through all. The results obtained are not always correct because they do not take into consideration the sources of possible contaminations step-by-step. Moreover, the matrix residues in the actual concentrate can affect the determination conditions while, in the concentrate of the blank epxeriment conducted in the absence of the sample, similar effects are not produced.

An original technique for conducting a blank experiment has been worked out by Blank [158] for determining trace elements. A single result of analysis is computed from the results of two determinations using two different weighed amounts of the sample. A similar technique was used by Zolotovitskaya and Yakovenko [159]. Blank discussed this technique by considering extraction photometric determinations. We used an unusual blank experiment for atomic emission determination of a number of trace impurities in phosphoric anhydride [160]. The procedure involved extraction of trace impurities as 8-hydroxyquinolinates with a mixture of isoamyl alcohol and carbon tetrachloride from the solution obtained after hydrolysis of the sample and neutralization with ammonium hydroxide until the pH reached 8. The extract was washed, evaporated on a graphite collector, mineralized with nitric acid, and analysed. The blank experiment was performed in two variants. In the first, commonly used variant, appropriate volumes of all the utilized reagents and solutions, water, ammonium hydroxide, extractant and nitric acid were evaporated on a graphite collector. In the second variant, 0.1 part of the sample being analysed was passed through all stages of analysis. The difference from the analysis of the unknown sample consisted in the fact that the 0.9 amount of ammonium hydroxide used to neutralize the tested sample was evaporated on the concentrate obtained after evaporating the extract; nitric acid and others were evaporated thereafter. After atomic emission analysis of the concentrates of the samples and the blank experiment, a set of simple equations was solved:

$$c + x = a \tag{4.1}$$

$$0.1c + x = b \tag{4.2}$$

Here, c stands for the true content of the element impurity in the sample being analysed in μg; x is the true content of the element impurity in the blank sample in μg; a and b are the measured contents of the element impurity in the analysed and blank samples, respectively, in μg.

Hence

$$c = 1.1(a - b) \tag{4.3}$$

The following amounts of copper were found in the concentrate of the blank experiment carried out in two variants: first variant, $0.03\,\mu$g; second variant, $0.2\,\mu$g. The contents of aluminium and iron agreed satisfactorily. The detection limit of copper, which was $1.5 \cdot 10^{-6}$ and $9 \cdot 10^{-6}\%$ respectively in the first and second variants, was estimated with consideration for the weighed amount and the results and fluctuations of the blank experiment. Of course, the blank experiment carried out with 0.1 part of the analysed sample took a more strict account of the impurities introduced from outside (from the vessel material and the air of the laboratory). Without this experiment we would have, first, decreased the detection limit of copper by six times and second, systematically overestimated by about 0.2 μg the results of copper determination in absolute units.

Now we shall consider the use of different methods of preconcentration in the analysis of metals, alloys and inorganic substances and materials including high purity substances and materials. This is reviewed many times [9, 100, 104, 109, 117–120, 123, 124, 136, 145, 146, 151–153, 161–166]. Here preconcentration is effected mainly by volatilization and related methods, extraction, electrolytic precipitation and sorption techniques. Numerous examples characterizing the possibilities of various methods of preconcentration used in the analysis of pure substances are discussed in Chapters 2 and 3. Therefore, we shall mention only a few works which characterize the possibilities of preconcentration.

4.3.2 VOLATILIZATION METHODS

These methods are indispensable in the analysis of diverse substances including acids, volatile halides and substances that can be readily changed into a highly volatile state.

Matrix distillation is the simplest form of preconcentration used in the analysis of mineral acids. It is therefore widely used in the analysis of hydrofluoric, hydrochloric, hydrobromic, hydroiodic and nitric acids. The use of this technique in the analysis of phosphoric, perchloric and sulphuric acids does not yield good results due to high boiling points and possible decomposition of the matrix [167]. In matrix distillation As, B, Cr, Ge, Hg, Re, Ru, Sb(III) and Se may be lost, particularly if complete evaporation is done. It is a convenient method for determining Ag, Al, Ba, Be, Ca, Cd, Co, Cs, Cu, Fe, Ga, Hf, In, K, Li, Mg, Mn, Na and also rare-earth elements, determination being carried out by atomic emission spectroscopy; the last drops of the sample, that is of the concentrate of impurities, can be evaporated directly on the end face of a graphite electrode. Other methods of preconcentration (precipitation and co-precipitation of trace impurities as chelates and solvent extraction) are also employed for the purpose. These methods are, however, less effective in determining widespread elements due to

314

the appearance of blank correction, and also due to the concentration of a small number of elements.

Volatilization of matrix elements by chemical transformations is used for analysis of various substances that are simple in macroelement composition (Table 4.3). Table 4.4 [168] also illustrates the possibilities of matrix volatilization in a reagent gas current in analysing different metals, inorganic materials and high-purity substances. Of course, it is not a universal method, particularly for the determination of volatile elements. However, several techniques have been worked out on this basis and they satisfy the practical needs. These are of

TABLE 4.3

Methods for atomic emission analysis of high-purity substances which involve volatilization of matrix from solutions after chemical transformations [164]

Analysed sample	Compound to be volatilized	Determinable elements	Reagent	Detection limit (%)
As, As_2O_3	AsX_3 (X = Br, Cl, F, I)	Al, Bi, Ca, Cr, Cu, Fe, Mg, Mn, Ni, P, Pb, Sb, Si	HCl, Br_2	$10^{-5}-10^{-6}$
B	BF_3	Mn, Mo, Ta, W	HF	$10^{-5}-10^{-6}$
B	$B(OCH_3)_3$	Al, As, Cu, Fe, Mg, Mo, Na, P, Pb, Si	CH_3OH	$10^{-4}-10^{-5}$
Cr	CrO_2Cl_2	Ag, Al, Bi, Ca, Cd, Co, Cu, Fe, Ga, In, Mg, Mn, Mo, Ni, Pb, Sb, Ti, V	HCl, $HClO_4$	$n \cdot 10^{-4}$
GaAs	AsX_3 (X = Br, Cl, F, I)	Ag, Al, Bi, Cr, Cu, Mg, Mn, Ni, Pb	HCl, Br_2	$10^{-5}-10^{-7}$
Ge	$GeCl_4$	Ag, Al, Au, Ba, Bi, Ca, Cd, Co, Cr, Cu, Fe, In, Mg, Mn, Mo, Pb, Sb, Sn, Ti, Tl, V	HCl, HNO_3	$10^{-5}-10^{-6}$
GeO_2	$GeCl_4$	Ag, Al, Be, Bi, Cd, Fe, Ga, In, Mg, Mn, Ni, Pb	HCl	$10^{-5}-10^{-6}$
InAs	$AsBr_3$, $InBr_3$	Ag, Al, Be, Bi, Ca, Cd, Co, Cr, Cu, Mg, Mn, Ni, Pb, Zn	HBr, Br_2	$n \cdot 10^{-7}$
Se	$SeBr_4$	Cd, Cu, Fe, Pb, Te, Tl	HBr	$10^{-7}-10^{-9}$
Si	SiF_4	Ag, Al, Bi, Ca, Cd, Co, Cr, Cu, Fe, Mg, Mn, Mo, Ni, Pb, Sb, Sn, Ti, V	HF, HNO_3	$10^{-4}-10^{-5}$

TABLE 4.4

Conditions for volatilization of certain matrices in atmospheres of chemically active gases

Element or compound	Compound or be volatilized	Reaction gas	Volatilization temperature (°C)
As	As_2O_3	Air (26.7 kPa)	224
S	SO_2	Air or oxygen	250–270
Se	SeO_2	Nitrogen oxides	315
Os	OsO_4	Air	700–800
As, Sb	Chlorides	Chlorine	No heating
Sn	$SnCl_4$	Chlorine	60–70
Ti	$TiCl_4$	Chlorine diluted wtih CO_2 gas	110–120
Al	$AlCl_3$	Hydrogen chloride	200
Ga	$GaCl_3$	Chlorine	< 200
V_2O_5	$VOCl_3$, VCl_4	Hydrogen chloride	300
Zr	$ZrCl_4$	Chlorine	350–370
MoO_3, WO_3	Chlorides and oxychlorides	Hydrogen chloride	600
TiO_2	$TiCl_4$	CCl_4 vapours diluted with CO_2 gas	550–650
SiO_2	SiF_4	Vapours of hydrofluoric acid	160
Si	SiF_4	Vapours of hydrofluoric and nitric acids	160–220
TiO_2	TiF_4, $TiOF_2$	Hydrogen fluoride	350
Ni	$Ni(CO)_4$	Carbon monoxide (14.4 MPa)	200

importance for analytical control in obtaining As, Bi, Ni, S, Sb, Se, Si and Sn compounds. For improving the metrological characteristics it is necessary that the matrix should be volatilized in liquid- and gas-phase autoclaves.

For eliminating systematic errors during decomposition of sample, Tölg [154–157] suggests that:

(1) all substances including those which are difficult to decompose should be completely decomposed;

(2) the determinable trace elements including volatile ones (Hg, Se, As, etc.) must be left in the apparatus for chemical decomposition;

(3) the residues obtained after decomposition must be completely dissolved with minimum amounts of readily purified reagents in the vessel used for decomposition;

(4) the amount of contaminants getting into the sample during de-

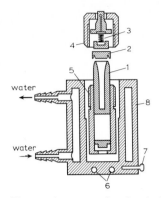

Fig. 4.1. Apparatus for chemical decomposition of the inorganic and organic matrix by nitric and hydrofluoric acid vapours. 1 = Teflon vessel; 2 = Teflon cover; 3 = spring; 4 = threaded cover; 5 = high-quality steel cylinder; 6 = Teflon tubes for water cooling; 7 = heating elements each of 200 W; 8 = metallic block.

composition should be a minimum;

(5) the interaction of decomposed material with the walls of the vessel should be reduced to a minimum, and the decomposition temperature must be kept as low as possible;

(6) the vessel used for decomposition must be used, where possible, for other operations in the course of analysis.

Conforming to these requirements, an apparatus has been designed for chemical decomposition [156] (Fig. 4.1). It consists of a Teflon vessel with a cover (volume, 10 ml; height, 80 mm; diameter, 24 mm; wall thickness, 6 mm) which is put in a high quality steel cylinder (wall thickness, 8 mm). A spring tightly presses the cover against the vessel. As a result, a safety valve is formed, through which gas at a very high pressure is passed; the gas comes out through the channels in the steel cylinder. The reaction vessel inserted in an aluminium or brass cylinder is electrically heated. The decomposition temperatures must not exceed 170°C, otherwise the vessel may deform. Decomposition temperature and time are assigned depending on the nature of the matrix. After the decomposition of the sample, the vessel is immediately cooled with water. Under such conditions, 1 g of quartz decomposes in 2.5 h with 4 ml 40% hydrofluoric acid or 0.5 ml 70% perchloric acid. When metals are dissolved with acids, the hydrogen formed diffuses out through the Teflon joint and, as a result, the pressure does not increase abruptly. The vessel used for decomposition of sample is so designed that excess acid, after the decomposition has completed, can be evaporated in the heating block by using the transporting gas flow. When the desired acidity is attained in the vessel, trace elements are extracted by adding an organic reagent and a solvent. For determining extremely low contents of iron, the steel cylinder must be gold-plated. The author indicates that even nanogram amounts of readily volatile elements, *e.g* As, Hg and Se, quantitatively remain in the decomposition system.

Combination of decomposition of sample and volatilization of matrix elements in a single closed vessel has been employed earlier also (see, for example, ref. 169). Thus, Brodskaya *et al.* [170] developed an atomic emission method of analysis of high-purity antimony for Ag, Al, Bi, Cd, Co, Cr, Cu, Fe, Mg, Mn, Ni, Pb and Zn, the detection limit being 10^{-5}–$10^{-6}\%$. The procedure involved preliminary conversion of the sample into volatile trichloride by reacting it with gaseous hydrogen chloride at 300°C, and volatilization of the matrix. Powdery metallic antimony (1 g) was taken in a 70 mm long and 10 mm wide quartz boat, at the bottom of which was placed 50 ml spectrally pure graphite collector. The boat with the sample was placed in a quartz furnace; after that, dry HCl was passed through the furnace. The impurities were adsorbed by the graphite powder and the obtained concentrate was analysed by AES.

4.3.3 SOLVENT EXTRACTION

Solvent extraction is extensively used for preconcentration of trace impurities in the analysis of certain substances.

We shall dwell on the use of this method in analysis of rare and scattered elements. Analysis of rare-earth elements and their compounds is a difficult task; very specific requirements are placed on their purity. Besides usual elements, trace amounts of impurities of other rare-earth elements have also to be determined. AES enables trace impurities to be determined with a detection limit of 10^{-5}–$10^{-6}\%$; chemical activation analysis enables the determination to be carried out at a level of 10^{-5}–$10^{-7}\%$. Vakulenko *et al.* [171] have developed a rapid atomic emission method of analysis of high-purity Er, Lu, Tm, Y and Yb compounds for Ce, La, Nd and Pr after the extraction separation of a large part of the matrix element, separation being carried out from chloride solutions (0.5 M in HCl) with 1 M solution of di(2-ethylhexyl)phosphoric acid in CCl_4. Trace impurities of light rare-earth elements were co-precipitated as oxalates from the aqueous phase; for this, the matrix element (left in the solution) was used as collector. The

318

precipitate, after filtration and calcination at 900°C, was subjected to AES. The detection limit was 10^{-4}–10^{-3}%.

A method has been suggested for concentration of Al, Bi, Co, Cu, Fe, Mn and other elements from solutions obtained after decomposition of high-purity oxides of rare-earth elements [172]. Trace impurities were extracted as 8-hydroxyquinolinates and 8-mercaptoquinolinates with chloroform, and the extract, after evaporation on graphite collector, was analysed by AES. In analysing rare and scattered elements and their compounds, Mo Shengjun and Zharg Zhuoyong [173] first extracted cobalt and cadmium as pyrrolidinedithiocarbamates with methyl isobutyl ketone and then analysed the extract by the ETAAS method.

Sometimes it is necessary to determine trace impurities at a level of 10^{-5}–10^{-6}% in pure refractory metals, *e.g.* in Nb, Ta, Ti and V. With this aim in view, Notkina *et al.* [174] extracted the trace impurities as diethyldithiocarbamates using chloroform or carbon tetrachloride. The metals (Nb, Ta or Ti) to be analysed were dissolved in a mixture of nitric and hydrofluoric acids, excess acid was evaporated, solution of ammonium tartrate and ammonium hydroxide was added, and the pH was fixed at 6–6.5. Under these conditions, quantitative separation of Ag, Au, Bi, Cd, Co, Cu, Fe, In, Mn, Ni, Pb, Se, Tl and Zn was accomplished. When analysing vanadium, the weighed amount was dissolved in nitric acid, the solution was evaporated and the residue was calcinated at 400°C; the obtained vanadium pentoxide was dissolved in ammonia, pH was fixed at 8, and Ag, Au, Bi, Cu, Fe, Cd, Co, In, Mn, Ni, Pb, Tl and Zn were extracted. The extracts were evaporated on a graphite collector. The analysis was completed with atomic emission spectrometry.

Yudelevich *et al.* [175] carried out atomic emission analysis of high-purity rhenium for Al, As, Ba, Be, Bi, Ca, Cd, Co, Cr, Fe, In, Mg, Mn, Ni, Pb, Sb, Si, Te, Tl and Zn, the detection limit being 10^{-4}–10^{-7}%. The sample was dissolved in hydrogen peroxide, diluted with water and the matrix element was extracted with a solution of tri-*n*-octylamine oxide in toluene. The aqueous phase, after evaporation on graphite collector, was subjected to AES. Yudelevich *et al.* [164] have proposed several methods for analysis of metals and high-purity substances which involve extraction of matrix elements and subsequent determination of trace impurities in an aqueous phase by the atomic emission method. Thus, 2,2'-di-chlorodiethyl ether was used to extract matrix in the analysis of Au, Fe, Ga, In, Sb, Tl, GaSb, InSb and InAs; di-2-ethylhexylphosphoric acid, in the analysis of In and InSb; and salts of

TABLE 4.5

Extraction–atomic emission analysis of uranium, thorium, platinum and their compounds (activation source = ICP)

Sample to be analysed	Elements to be determined	Extractant	Composition of aqueous phase	Detection limit ($\mu g/ml$)	Ref.
U and U_3O_8	B, Cd	0.2 M trioctylphosphine oxide in hexane	HNO_3	0.02–0.05	176
U, UF_6 and others	Al, B, Ba, Be, Bi, Ca, Cd, Co, Cr, Cu, Dy, Eu, Fe, Gd, Li, Mg, Mn, Mo, Na, Nb, Ni, Pb, Ru, Sb, Si, Sm, Sn, Sr, Ta, Ti, W, Zn, Zr	Mixture of tri-(2-ethylhexyl)phosphate and hexane (3:1)	10% HNO_3 in the presence of 100 $\mu g/ml$ F^-	–	177
U	Dy, Eu, Gd, Sm, Th	20% trioctylamine in xylene	1 M HCl	0.007–0.07	178
ThO_2	13 Rare-earth elements	Mixture of di-(2-ethylhexyl)phosphoric acid and tributyl phosphate in an organic solvent	–	–	179
Pu and its oxides	Al, B, Be, Ca, Cr, Cu, Fe, Ga, K, Mg, Mn, Mo, Ni, Pb, Si, Sn, Ti	N,N-Diethylcarbamyl-methylene phosphonate in xylene	HCl + HNO_3	–	180
Plutonium nitrate	Al, Fe, Li, Na, Ni, P, Pb, Ta, Ti, Zr	20% tri-n-octylammonia in xylene	4 M HNO_3	–	181

tertiary ammonium bases and organic sulphides, in the analysis of Ag, Au and Pd.

Extraction concentration is extensively used in analysis of actinides and their compounds for various impurities. Several other techniques involving combination of matrix isolation and AES–ICP are given in Table 4.5.

For analysis of pure substances, a number of procedures have been worked out by us. These involve combination of extraction of macro- and trace elements with various methods of determination (Table 4.6). We shall explain this table further.

There are several reasons for preconcentration of trace impurities in analysis of compounds of alkali and alkaline-earth metals. First, the widespread analytical methods (spectrophotometry, voltammetry, atomic emission spectrometry, especially gravimetry and titrimetry) do not enable small absolute amounts of impurities to be determined in small samples. Second, the matrix element often hinders the determination even if the technique ensures a low detection limit. Particularly, in neutron activation determination of trace impurities in sodium chloride, in which both the main elements are readily activated, it is very difficult to carry out analysis due to induced radioactivity. As the matrix metals are moderately extracted compared to most of the element impurities, the determinable trace components are best separated by solvent extraction while the macro-components remain in the aqueous phase. They are often extracted as chelates using sodium diethyldithiocarbamate, 8-hydroxyquinoline, 1-phenyl-3-methyl-4-benzoylpyrazolone-5 (PMBP) and their mixtures. This ensures simultaneous separation of a large group of element impurities.

Proceeding from the concept of coordinatively saturated and unsaturated (with respect to the central atom) chelates and of the effect of solvent on their extraction [221–223], we used a mixture of organic solvents (oxygenated polar and non-oxygenated non-polar solvents). This helped to increase the number of determinable element impurities that could be separated in one stage. Moreover, the use of heavy organic solvents (chloroform or carbon tetrachloride) made it possible to transfer the extract into the lower part of the separating funnel and to separate it readily. This allowed for simultaneous determination of up to 13–15 trace impurities [186–189]. Much better results were obtained on using (for the analysis of alkali metal halides) the extraction system containing sodium diethyldithiocarbamate, cupferron, trioctylphosphine oxide and chloroform: twenty elements were separated in one stage [190].

TABLE 4.6

Some methods of analysis of metals and high-purity substances, including extraction concentration

Materials analysed	Elements determined	Aqueous phase characteristics	Organic solvent, extractant	Final method	Detection limit (%)	Ref.
Water	Al, Bi, Co, Cu, Fe, Ga, In, Mn, Ni, Pb, Zn	pH 5.5–6.0	Chloroform + isoamyl alcohol, 8-hydroxyquinoline + sodium diethyldithiocarbamate (DDTC)	AES	10^{-7}–10^{-9}	182, 183
Hydroiodic acid	Co, Cr, Cu, Mn, Ni, Pb	pH 5.5–6.0	Chloroform + isoamyl alcohol, 8-hydroxyquinoline	AES	10^{-7}–10^{-8}	184
Metallic sodium	Al, Be, Bi, Co, Cu, Fe, Mn, Ni, V, Zn	pH 7.5–8.0	CCl_4 + ethyl alcohol + 8-hydroxyquinoline + 1-phenyl-3-methyl-4-benzoylpyrazolone-5 (PMBP)	AES	10^{-5}–10^{-7}	185
Sodium hydroxide	Al, Ca, Co, Cr, Cu, Fe, Mg, Mn, Mo, Ni, Pb, Sn, Ti, V	pH 8.0 and 2.5	Chloroform + isoamyl alcohol, 8-hydroxyquinoline + PMBP	AES	10^{-5}–10^{-7}	186
KCl, NaCl, KBr, NaBr, NaI, KI, $CaCl_2$	Al, Au, Bi, Cd, Co, Cu, Cr, Fe, In, Mn, Ni, Pb, Zn	pH 5.5–6.0	Chloroform + isoamyl alcohol, 8-hydroxyquinoline + DDTC	AES	10^{-7}–10^{-9}	187, 188

Matrix	Elements	Medium	Reagent	Method	Range	Ref.
KOH, NaOH, KNO$_3$, NaNO$_3$	Ag, Al, Au, Bi, Cd, Co, Cu, Ga, In, Mn, Ni, Pb, Ti, Tl, Zn	pH 5.5–6.0	Chloroform + isoamyl alcohol, 8-hydroxyquinoline + DDTC	AES	10^{-6}–10^{-8}	189
Alkali metals halides	Ag, Al, As, Bi, Ca, Co, Cr, Cu, Fe, Cd, Hg, Mg, Mn, Ni, Pb, Sb, Sn, Ti, Zn	pH 8–9	Chloroform, trioctylphosphine oxide, DDTC, Cupferron	AES	10^{-5}–10^{-7}	190
Halides of Ba, Cd, Co, Cs, K, Li, Ni, Rb, Sr	Ag, Cu, Tl	HCl, HBr, and HI solutions	Chloroform solution of diphenylthiourea	AAS	10^{-5}–10^{-7}	191
Co, Cu, Fe, Ni	Ir, Pt, Rh	1–4 M HCl	Di-2-ethylhexyldithio-phosphoric acid in acetone	AES	—	192
Silver	Al, As, Ba, Be, Bi, Ca, Cd, Co, Cr, Cu, Fe, Ga, In, Mg, Mn, Ni, Pb, Sn, Te, Tl, V, Zn, Zr	1 M HNO$_3$	O-Isopropyl-N-ethylthio-carbamate in chloroform	AES	10^{-6}–10^{-8}	193
Silver	Al, As, Ba, Be, Bi, Ca, Cd, Co, Cr, Cu, Fe, Ga, In, Mg, Mn, Ni, Pb, Sb, Sn, Te, Tl, V, Zn	0.9 M HNO$_3$	O-Isopropyl-N-methyl-thiocarbamate in chloroform	AES	10^{-5}–10^{-8}	194
Calcium oxide	Al, Co, Cu, Fe, Ga, Mn, Ni, Pb, Zn	pH 5.5–6.0	Chloroform + isoamyl alcohol, 8-hydroxyquinoline	AES	10^{-6}–10^{-8}	195

(Continued on p. 324)

324

TABLE 4.6 (*continued*)

Materials analysed	Elements determined	Aqueous phase characteristics	Organic solvent, extractant	Final method	Detection limit (%)	Ref.
Zn, ZnO, ZnSO$_4$	Cd	pH 7	Iodide ions, 1,4-dimethyl-1,2,4-triazolenes-(3-azo-4)-dimethyl-aniline, benzene + cyclohexane	Spectro-photometry	–	196
Cadmium sulphide	Al, Co, Cr, Fe, Mg, Mn, Ni, Ti, Zn	2.8–3.0 M HI in the presence of HCl	Diethyl ether	AES	10^{-5}–10^{-7}	197
Cadmium chloride and metallic cadmium	Al, Ca, Co, Cr, Fe, Ga, Mg, Mn, Ni, Sr, Ti, Zn	3.3 M HI in the presence of HCl	Diethyl ether	AES	10^{-4}–10^{-7}	198, 199
Glass B$_2$O$_3$·CaO	Al, Bi, Co, Cu, Fe, Ga, Ni	pH 4.5–5.0	CCl$_4$ + isoamyl alcohol, 8-hydroxyquinoline	AES	10^{-5}–10^{-6}	200
Gallium chloride	Al, Bi, Ca, Co, Cr, Cu, Fe, Mg, Mn, Ni, Pb, Ti, Zn	9 M HCl	Butyl acetate	AES	10^{-4}–10^{-7}	201
Trimethyl gallium etherate	Al, Cu, Mn, Ni, Pb, Zn	9 M HCl	Butyl acetate	AES	10^{-5}–10^{-6}	202
Metallic indium, nickel chloride, aluminium	Au	3 M in H$_2$SO$_4$ 1 M in CCl$_3$COOH	Chloroform, 8-mercaptoquinolinate of zinc	Spectro-photometry	10^{-5}–10^{-6}	203

Titanium tetrachloride	Cu, Fe, Mo, V, Zn	HCl, HCl + NH_4I solutions	Solution of trioctylamine in toluene	AAS	10^{-5}–10^{-6}	204
Toluene, m-xylene	Ag, Al, Bi, Cr, Cu, Fe, Ga, Mg, Mn, Ni, Pb, Ti, Zn	6 M HCl	Toluene or m-xylene, 8-hydroxyquinoline	AES	10^{-6}–10^{-9}	205
Vanadium	P(V)	0.6–0.8 M in HCl and 0.1–0.3 M in Na_2MoO_4	Solution of dinonyl tin dinitrate in a mixture of chloroform, n-butyl alcohol and ethyl acetate	Spectrophotometry	$5 \cdot 10^{-6}$	206
	As(V)	1–1.2 M HCl			$1 \cdot 10^{-5}$	
Phosphorus pentoxide	Ag, Al, Bi, Ca, Co, Cu, Fe, Ga, Mn, Ni, Pb, Ti, Zn	pH 8	CCl_4 + isoamyl alcohol, 8-hydroxyquinoline + PMBP	AES	10^{-5}–10^{-7}	166
Phosphorus trichloride	Al, Bi, Co, Cu, Fe, Ga, Mn, Ni, Pb, Zn	pH 4	CCl_4 + isoamyl alcohol, 8-hydroxyquinoline + DDTC	AES	10^{-5}–10^{-7}	207

(Continued on p. 326)

TABLE 4.6 (*continued*)

Materials analysed	Elements determined	Aqueous phase characteristics	Organic solvent, extractant	Final method	Detection limit (%)	Ref.
Gaseous mixture of phosphine and hydrogen	Se	pH 2–3 and 7–8	Toluene, 3,3′-diamino-benzidine	Fluorimetry	$5 \cdot 10^{-5}$	208
	Te	5.5 M H_2SO_4 + KBr	Benzene, butylrhoda-mine C	Spectrophotometry	$8 \cdot 10^{-4}$	
Metallic arsenic	Al, Bi, Co, Cu, Fe, Mn, Ni, Pb, Ti	pH 5.5–6.0	CCl_4 + isoamyl alcohol, 8-hydroxy-quinoline + DDTC	AES	10^{-5}–10^{-7}	209
	Au, Cr, Cu, Fe, Ga, In, Mn, Ni, Zn	12.5 M H_2SO_4	Benzene	NAA	10^{-4}–10^{-7}	210
Gaseous mixture of arsine and hydrogen	Ag, Al, Bi, Co, Cr, Cu, Fe, In, Mn, Ni, Pb, Sn, Ti, Zn	pH 5.5–6.0	CCl_4 + isoamyl alcohol, 8-hydroxyqui-noline	AES	10^{-5}–10^{-7}	211
Glass $As_2O_3 \cdot CaO$	Ag, Al, Cu, Fe, Ga, In, Mn, Ni, Pb, Ti, Tl, Zn	pH 4	CCl_4 + isoamyl alcohol, 8-hydroxyqui-noline + DDTC	AES	10^{-5}–10^{-6}	212

		conc. HCl	m-Xylene	AES		
Arsenic trichloride	Ag, Al, Au, Ba, Bi, Ca, Co, Cr, Cu, Fe, Ga, In, Mg, Mn, Ni, Sb, Te, Zn	conc. HCl	m-Xylene	AES	10^{-10}	213
Epitaxial structures of gallium arsenide	Ni	$GaBr_3$ + 10 M HCl	n-Decyl alcohol	AAS	$5 \cdot 10^{-7}$	214
	Ag, Al, Bi, Co, Cr, Cu, Fe, In, Mg, Mn, Ni, Pb, Sn, Zn	$GaBr_3$ + 10 M and 5 M HCl	n-Decyl alcohol	AES	10^{-4}–10^{-6}	215
	Te	$GaBr_3$ + 6 M HCl	n-Decyl alcohol	Fluorimetry	$5 \cdot 10^{-4}$	216
Antimony trichloride	Ag, Al, Bi, Ca, Co, Cr, Cu, Fe, Mg, Mn, Ni, Pb, Tl, Zn	9 M HCl	Diethyl ether (Fe) and butyl acetate (other elements)	AES	10^{-4}–10^{-7}	217
Alloy steels	Bi, Cd, Cu, Pb, Sb, Sn, Zn	1 M in NH_4I and 2.3 M in H_2SO_4	Methyl isobutyl ketone	AES	10^{-5}–10^{-7}	218
Steels	Bi, Cd, Cu, Sb, Zn	1 M in KI(NH_4I) and 2.3 M H_2SO_4	Solution of trioctylammonium sulphate in o-xylene or methyl isobutyl ketone	AAS	10^{-5}–10^{-7}	219
Stainless steel, sulphuric acid	As, Bi, Cd, Cu, Pb, Sb, Sn, Zn	1 M in NH_4I and 2.3 M in H_2SO_4	Methyl isobutyl ketone	SSMS	10^{-5}–10^{-7}	220

References pp. 340–349

Dinonyl and dioctyl tin dinitrates are highly selective extractants for multicharge anions: AsO_4^{3-}, PO_4^{3-}, SeO_4^{2-}, SeO_3^{2-}, WO_4^{2-}, MoO_4^{2-}, and SO_4^{2-} [206, 224, 225]. An extraction–spectrophotometric method involving the use of dinonyl tin dinitrate has been worked out for determining P and As in metallic vanadium. O-Isopropyl-N-ethyl-thiocarbamate and its analogue with a methyl group are highly selective extractants for silver. These have been used for extraction of macro-elements in the analysis of metallic silver for a large group of element impurities [193, 194].

Extraction chromatography was successfully used in combination with AES (a hollow cathode being the source of excitation) and AAS and also with fluorimetry for analysing epitaxial structures of gallium arsenide [214–216]. The authors succeeded in separating (and concentrating) trace and macro elements with very similar properties; detection limits of 10^{-4}–$10^{-6}\%$ were attained by using milligram amounts of weighed samples.

Arsine was passed through concentrated nitric acid placed in quartz bottles to analyse it for Ag, Al, Bi, Co, Cr, Cu, Fe, In, Mn, Ni, Pb, Sn, Ti and Zn by the atomic emission method [211]. In so doing, arsine was oxidized to arsenic acid. The solution was then evaporated to dryness, the dry residue was dissolved in deionized water, neutralized with ammonia to pH 5.5–6.0, and the trace impurities were extracted with $0.1\,M$ solution of 8-hydroxyquinoline in a mixture of carbon tetrachloride and isoamyl alcohol (2:1). The extract after evaporation on a graphite collector was subjected to AES. In aliquots of the solution, selenium and tellurium were determined by the fluorimetric method after the sorption of arsine. In an almost similar manner, selenium and tellurium present in a gaseous mixture of phosphine and hydrogen were determined [208].

4.3.4 OTHER METHODS OF PRECONCENTRATION

Among other methods of preconcentration, a mention must be made of sorption and co-precipitation methods. Jackwerth et al. [226] have developed AA and XRF methods for determining trace amounts of Ag, Cd, Co, Cu, In, Ni, Pb, Tl and Zn in powdery tungsten. A weighed amount of 5 g was dissolved in 30% hydrogen peroxide, sodium diethyldithiocarbamate was added to the solution obtained, and the chelates formed were adsorbed under static conditions with activated charcoal. After that, the trace impurities of the metals being determined were desorbed with nitric acid, and the solution obtained was analysed. The detection limits were 10^{-4}–10^{-6}%. A similar technique has been used for analysis of alkali and alkaline-earth elements [227], silver and thallium nitrate [228], metallic zinc and its nitrate [229] and manganese [230].

Ion exchangers are also employed for concentration of trace impurities in the analysis of materials of the considered group. Thus, a sample of carbon steel, analysed for Ce, La and Pr, after dissolving in nitric acid and suitable treatment, was passed through a column packed with the anion exchanger Bio-Rad 1X10 in the chloride form [231]. The sorbent was separated out, dried, mixed with cellulose, and pressed into a tablet. The tablet was analysed by the XRF method which ensured a detection limit of 10^{-2}–10^{-3}%. Ryabukhin et al. [232] studied the ion-exchange behaviour of yttrium, containing the trace impurities Ce, La, Nd and Pr, under dynamic conditions. In the column packed with the cation exchanger KU-2-8 a concentration coefficient of 1.5–2.0 orders was attained by using 0.25 M citric acid as eluent. They recommend completing the determination by atomic emission spectrometry.

Co-precipitation of trace impurities with partial precipitation of the matrix element was employed, for instance, by Slyusareva and Kondratyeva [233]. From the hydrochloric acid solution obtained after the decomposition of a lanthanum sample (pH 6–6.5), 1.5–2% lanthanum was precipitated as hydroxide. A solution of tannin and quinine was added for more complete co-precipitation of the trace impurities Al, Hf, Mo, Nb, Ta, Ti, V, W and Zr. The hydroxide precipitate was filtered out, calcinated, and analysed by AES. For a weighed amount of 5 g, the detection limit was 10^{-4}–10^{-6}% and the concentration coefficient was 50–70. In the analysis of pure anodic copper and other copper-containing substances for Bi, Fe, Pb, Sb, Se, Sn, Te and As [234] the trace impurities were also co-precipitated with lanthanum hydroxide in ammonia solution at pH 9–10; after dissolution, the concentrate was analysed by AAS.

Table 4.7 contains examples of using various preconcentration techniques in analysis of metals, alloys, inorganic materials and high-purity substances.

4.4. Organic substances and biological samples

4.4.1 DISTINGUISHING FEATURES

Determination of trace elements in organic substances (natural and industrial) is a common task of analytical practice. Such determina-

TABLE 4.7

Some methods of analysis of metals, alloys, inorganic materials and high-purity substances

Sample analysed	Elements determined	Method of concentration	Method of determination	Detection limit	Ref.
Magnesium and its nitrate	Ag, Cu, Fe, Hg, In, Mn, Pb, Zn	Sorption of trace elements on activated charcoal	AAS	0.014–13 ng/g	235
Tungsten	Ca, Cu, Mg, Mn, Ni, Pb, Zn	Sorption of trace elements on complex-forming sorbent Wofatit MC-50	AAS	–	236
Phosphoric acid	Ca, Dy, Eu, Fe, Pb	Sorption of trace elements on ion exchanger (matrix = polyacrylamide) and hydrated ZrO_2	AAS	–	237
Ammonium fluoride and hydrofluoric acid	Co, Cu, Fe, Mn, Ni, Zn	Co-precipitation with potassium fluoride	AAS	–	238
Lead	Bi, Sb, Sn	Co-precipitation with iron(III) hydroxide	AAS	0.5–5 µg/g	239
Steel	V	Co-precipitation with 1-phenyl-3-methyl-4-capronylpyrazolone-5	XRFS	2 µg	240
High-purity platinum	Ca, Co, Cr, Fe, Mg, Mn, Mo, Ni, Ti, V	Electrolytic separation of matrix	AES	10^{-4}–$10^{-5}\%$	241
Nickel-based alloys	Se	Electrolytic isolation of trace element	AAS	–	242
PH_3, AsH_3 and SiH_4 of extra	Al, B, Bi, Co, Cr, Cu, Fe, In, Mn, Ni,	Washing-out suspended particles with water	Photometry (B) and AES	1 ng/ml	243

tions are essential for different fields of the national economy. Trace elements, *e.g.* Co, Cu, Mn, Mo, Zn and others, vital for plant and animal life are receiving ever increasing attention. There are many physiologically active, rare and scattered elements whose low content in plants can be the cause of mineral deficiency in fodder and food products, and can bring down the crop yield and the production of milk and meat. Large contents of molybdenum in fodder and food products can cause disease (endemic podagra) in animal and man. Presence of non-ferrous metals in plants serves as a biogeochemical indicator in prospecting ore deposits. It is also important to know the content of toxic elements, in particular of As, Be, Bi, Cd, Hg, Pb, Sb, Se, Sn and Te. The development of the industry of organic synthesis, particularly on the basis of crude petroleum, natural gas and coal, the increase in the manufacture of polymers and polymer-based materials, organometallic compounds, biologically active preparations and pharmaceuticals, the demands of the synthetic organic chemistry and the petroleum refining industries drastically widened the range of organic substances. The content of harmful trace elements or of strictly stipulated impurities has also to be controlled in most organic substances.

Of the methods employed for concentration of trace elements in the analysis of organic substances and biological samples, mention must be made of the following:

(1) dry mineralization, including ashing in air and in an atmosphere of oxygen or other reaction gases, thermal decomposition or pyrolysis, plasma destruction;

(2) wet mineralization by nitric, hydrochloric, sulphuric or other acids or their mixtures, including mineralization in the presence of hydrogen peroxide or other oxidizing agents;

(3) extraction, sorption, electrolytic isolation and others.

Dry and wet ashing are the main methods of obtaining concentrates of impurities when substances of organic nature are analysed. It is true that these are usually not considered as a technique for concentration of trace elements. Specific features of ashing and its application in analysis of organic substances and biological samples are discussed in a number of publications [3, 100, 244–247]. We have discussed this method in Chapter 2.

Besides preconcentration of trace elements, ashing has at least two more advantages: it ensures destruction of many compounds of trace elements and frees them for subsequent analytical determination, and it decreases significantly the demand for reference samples and allows their unification. Ashing should of course be carried out without losses of the trace elements being determined, or these losses must be controlled at least.

Ashing is carried out at moderate temperatures, say at temperatures not exceeding 400–500°C, for even drying of plant samples to the air-dried (20–30°C), or absolutely dry (100–120°C) state, is attended with complete or partial loss of volatile compounds of Br, Hg, I, Se, etc. [6]. Decomposition with a mixture of sulphuric and nitric acid or with a mixture of nitric, sulphuric and perchloric acid is a classic example of wet ashing of biological samples.

4.4.2 DRY ASHING

Dry ashing is used in combination with different methods of analysis. Ashing in a platinum crucible and in the presence of potassium or magnesium sulphonate, if the sample is liquid, has been used for atomic absorption determination of As, Ba, Be, Ca, Cd, Co, Cr, Cu, Fe, K, Mg, Mn, Mo, Na, Ni, Pb, Sb, Si, Sn, Ti, V and Zn in petroleum products [248]. An almost similar procedure was employed in the atomic absorption and atomic emission flame-photometric analysis of residual fuel oil (mazut) and petroleum coke, respectively, for Ba, Ca, K, Na, Ni and Sr; the detection limits were 10^{-5}–$10^{-7}\%$ [249]. The application of dry as well as wet ashing in the analysis of petroleums and petroleum products, including motor fuels and lubricants, is discussed in detail by Kyuregyan [246].

In the analysis of bark and leaves for nickel, Oemar [250] restricted himself to ashing of the sample by gradually raising the temperature to 550°C. Ashing continued for 2 h. The residue, after dissolving in concentrated hydrochloric or nitric acid, was diluted with water, pulverised into a flame, and nickel was determined by AAS. The atomic emission analysis of Lavsan for Co, Sb and Ti [251] is also based on preliminary ashing of the sample at 400–450°C; the ashing takes 5–6 h. A weighed sample was taken so as to obtain 10–15 g ash. Thus, if an ash content is equal to 0.05%, 20–30 g sample should be taken.

A method has been worked out for decomposing organic samples in oxygen plasma created in a microwave field [156, 252]. A quartz tube 140–160 mm long and 13 mm in diameter having a cooling finger, was placed in a microwave cavity resonator. Purified oxygen was supplied under pressure through a fine quartz nozzle to the substance placed at the bottom of the test tube. Oxygen was activated by a microwave generator (2450 MHz) of variable power (maximum 200 W). The sample, by coming in contact with the oxygen plasma, gradually burnt from the top to the bottom, and the volatile elements collected on the cooling finger. The residue obtained after burning was dissolved in pure acid (HNO_3, HCl, or their mixture); the solution was then heated in the field of the microwave generator. The condensate formed on the walls dissolved when the liquid was boiled with a reflux condenser. The solution obtained contained all of the trace elements and was used for further analysis. The authors regard it to be a convenient method for determining As, Be, Cd, Pb, Sb, Se, Te and other volatile trace elements in various organic substances.

Volodina *et al.* [253], who have been working in a similar field, have developed a method for effective plasma destruction of organic compounds at low temperatures (100–200°C). In these conditions, the reaction products volatile at high temperatures remain quantitatively in the residue obtained after ashing. This allows an increase in the number of elements being determined from one weighed amount. Depending on the particular analytical task, different plasma-forming gases (oxygen, ammonia, argon) can be used. The method has been employed for analysis of organo–boron compounds and organometallic compounds, polymers containing halides and organic-based metal-containing catalysts. Volodina *et al.* [253] determined in these substances not the trace elements, but the elements that are the main components of the substance: boron, tellurium, halides and metals. However, this method is also suitable for determining trace elements. Buyanovskaya *et al.* [254] also studied the behaviour of boron-containing organic compounds in a low-temperature plasma created by a high-frequency discharge. Low-temperature (about 100°C) decomposition of the sample and quantitative isolation of boron were accomplished in 10–15 min. Similar studies were made on organic compounds containing Co, Fe, Ni and Zn [255]. Here also, the metals being determined were concentrated as stable oxides in the ash residue using a low-temperature oxygen plasma created by a high-frequency discharge.

For concentration of trace elements in the analysis of blood and human tissues by the X-ray spectrometry method with proton excitation [256] use has been made of low-temperature ashing in oxygen plasma (150°C) generated by a high-frequency electromagnetic field. Four different methods of decomposing biological samples, including normal ashing and ashing in oxygen plasma at 100°C, have been used for obtaining graphite-based electrodes for group determination of As, Ba, Cr, Cu, Fe, Mn, Rb, Sr, Ti and Zn by spark source MS [257].

In most cases, wet ashing is preferred to dry ashing as the possibility of losing highly volatile forms of determinable trace elements is reduced. However, the characteristics of this technique may deteriorate owing to the use of corrosive reagents. The reagents (usually nitric, hydrochloric, sulphuric, and other organic acids) must be pure and the material of the vessel in which the sample is subjected to ashing must be inert and should not sorb the trace elements to be determined. Fused quartz, Teflon, glassy carbon and noble metals comply with these requirements to a large extent and are therefore in general use. Numerous examples of applying ashing to analysis of organic materials and biological samples are available in refs. 100, 244–247.

Often it is advantageous to carry out wet ashing in a closed vessel under pressure so as to decrease to the minimum the loss of highly volatile elements and the possibility of introducing impurities. The use of autoclaves enables the materials to be decomposed by reagents, which at atmospheric pressure dissolve with difficulty or do not dissolve at all. Besides, the consumption of reagents decreases and decomposition speeds up.

The device shown in Fig. 4.1 has been used for wet mineralization (under pressure) of various biological samples by nitric and hydrofluor-

TABLE 4.8

Conditions for mineralization of biological samples in a polytetrafluoroethylene autoclave at 170°C

Substance analysed	Weighed amount (mg)	Reagent	Volume of reagent (ml)	Duration of decomposition (h)
Smoked meat	300	65% HNO_3	0.75	1.5
Meat	400	Highly concentrated HNO_3	0.75	2
Salted pork fat	100	Highly concentrated HNO_3	0.5	4
Blood	200	Highly concentrated HNO_3	0.5	3
Cellulose	200	65% HNO_3	0.5	1
Leaves	300	65% HNO_3	1	2
Slime	100	Highly concentrated HNO_3	0.5	3
Slime	100	40% HF	0.25	3

ic acids [156]. A weighed amount (100–400 g) of the sample was placed in a polytetrafluoroethylene vessel (volume 10 ml) and ashed under the conditions listed in Table 4.8. Temperature and decomposition time were varied depending on the type of sample analysed. After decomposition the unreacted excess acid was removed by volatilization, the latter being carried out in the same vessel. This may also be used for subsequent extraction concentration of trace elements.

There are data which reveal that glassy carbon is superior in inertness and stability to Teflon [258]. The use of glassy carbon enables ashing to be done with nitric, perchloric and hydrofluoric acid at temperatures up to 250°C, whereas the maximum allowable temperature for Teflon is 170°C. Besides, this material does not allow such volatile elements as mercury to penetrate through the vessel walls. Fig. 4.2 shows a device used for decomposition of organic and inorganic substances under pressure. Ashing in this device makes it possible to concentrate volatile elements like As, Hg, Sb, Se and Sn, which are of interest for toxicological analysis and environmental control.

Table 4.9 contains characteristics of wet ashing for certain biological materials. Using the method of radioactive indicators it has been shown, by considering the examples of As, Hg and Se, that $\geqslant 97\%$ determinable trace elements remain in the concentrate obtained after ashing.

The work by Voellkopf and Grobenski [259] is an example of using

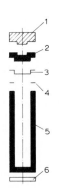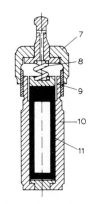

Fig. 4.2. Device for decomposing sample under pressure. 1 = Upper cover made of graphitized Teflon; 2 = lower cover made of glassy carbon; 3 = Teflon packing; 4 = Teflon packing ring; 5 = cell made of glassy carbon; 6 = washer; 7 = threaded cover; 8 = safety valve spring; 9 = plate (compression); 10 = steel bomb; 11 = cell made of glassy carbon (assembled).

TABLE 4.9

Characteristics of wet autoclave mineralization for some biological materials

Material	Weighed amount (mg)	Conc. acid, 0.5 ml	Heating time (h) 80°C	140°C	200°C	Heating period at 200°C (h)
Pig bones	200	HNO$_3$	–	–	0.5	1
Dry milk	100	HNO$_3$	–	–	1	2
Human hair	50	HNO$_3$	1	0.5	0.5	2
Suet	100	HNO$_3$	–	–	0.5	1
Plants and hay	60	HNO$_3$ + 0.2 ml HF	1	1	0.5	0.5
Liver	200	HNO$_3$	0.5	0.5	0.5	2
Serum	200	HNO$_3$	0.5	0.5	0.5	2.5
Protein concentrate	50	HNO$_3$	1	0.5	0.5	2.5
Conifer needles	50	HNO$_3$	0.5	0.5	0.5	0.5

wet autoclave (bomb) mineralization in the analysis of substances belonging to this group. Dry milk, human hair, urine and other biological materials were mineralized in (1:1) nitric acid for 30–60 min. Thereafter, As, Cd, Cr, Pb and Se were determined by the ETAAS method.

4.4.4 OTHER METHODS OF PRECONCENTRATION

The analysis of organic materials and biological substances without mineralization is of interest.

Analysing edible fats and butters for trace amounts of copper and zinc, Jacob and Klevay [260] boiled the sample (50 g) for 2 h in a flask fitted with a reflux condenser, together with 55 ml solution obtained by mixing 525 ml concentrated HCl, 500 ml water and 0.11 g ethylenediaminetetraacetic acid. In so doing, the trace components went over to the aqueous phase which, after necessary sample treatment, was analysed by the atomic absorption method. Aneva and Ivancheva [261] have described a procedure for determining arsenic and lead in crude oils which involves their extraction with a 1% solution of nitric acid as iodides that are formed upon adding toluene-dissolved iodine into the solution sample. The extract obtained was analysed by the ETAAS method. For determining diethyl lead in urine [262], hydrochloric acid, sodium chloride and methyl isobutyl ketone were added to the sample. After mixing and centrifugal separation, the aqueous layer was separated; to it was added the methanol solution of glyoxal-bis-(2-

336

hydroxyanil) and lead was extracted with methyl isobutyl ketone at pH 9.8. After evaporating the extract, lead was determined by the ETAAS method. Determination of thallium in urine by the same method [263] is based on its extraction as diethyldithiocarbamate with toluene. A procedure for serial rapid determination of radioisotopes of iodine in milk has been suggested [264]. The milk sample (5 l) is mixed with 250 ml of 37% solution of formaldehyde and passed through a polyurethane foam cylinder previously impregnated with 0.1 M solution of iodine in 10% solution of Alamin 336 in toluene. The sorbent was then analysed by γ-spectrometry.

For concentration of trace elements from liquid biological samples sorption methods are used. An ion-exchange membrane is suitable for concentration of trace elements from solutions because, after drying, it can be directly used as radiator in XRFS. That is how milk was analysed: the membrane was immersed into the sample for 4–7 days [265]. Spectrophotometric determination of uranium with arsenazo III in urine [266] also involves preliminary sorption separation of trace element.

Often simple distillation is used in analysis of motor fuels for trace elements [246]. But it is unlikely to yield correct results because the fractions which are mixed to obtain motor oils are volatilized from petroleum and petroleum-refining products; single volatilization cannot be more effective than multiple rectification.

4.4.5 COMBINING PRECONCENTRATION TECHNIQUES

The coefficients of concentration attained in mineralizing biological samples, charcoals, shales, low-quality oils and certain organic materials are not very high. That is the reason why mineralization is used in combination with other preconcentration techniques when it is necessary to appreciably lower the detection limits of trace elements.

In blood serum, Al, Co, Cr, Mn, Mo, Ni and V were determined by the ETAAS method in the following manner [267]. The sample was frozen at $-20°C$ and sublimation drying was carried out. The dried sample was then ashed in a muffle furnace by keeping it (the sample) for 1 h at 100, 150 and 250°C, and for all night at 480°C. The residue was dissolved in (1:20) nitric acid and the solution was put into an electrothermal spray gun.

Extensive use is made of mineralization in combination with hydride-formation (Table 4.10).

TABLE 4.10

Examples of combining mineralization with formation of hydrides

Substance analysed	Element determined	Method of determination	Ref.
Biological and water samples	As	AAS	268
Biological tissues of marine origin	As, Sb, Se	AES–ICP	269
Biological tissues of marine origin	As, Hg, Se (Hg in the form of element)	AAS	270
Blood serum	As, Se	AES–ICP	271
Blood, serum and blood plasma	Se	AAS	272
Vegetable and animal tissues, hair and human nails, blood serum and others	Se	AAS	273
Marine organisms	Sn	AAS	274

Many examples can be given of the combination of mineralization and extraction.

Analysing vegetable materials for Co, Mn, Mo, Ni, Pb, V and Zn, Stracheim et al. [275] ashed an air-dried sample of 10 g at 430–450°C for 16–18 h. The dry residue was dissolved in hydrochloric acid. The precipitates of silicates were filtered off, treated with a mixture of hydrofluoric and sulphuric acids, dissolved in hydrochloric acid, and the solution was mixed with the filtrate. After adding sulphosalicylic acid, the pH was set at 4.8 using ammonium hydroxide; the trace impurities were precipitated as pyrrolidinedithiocarbamates, chloroform was added until the precipitate dissolved completely, the mixture was shaken and the organic layer was singled out. The extract was mixed with graphite collector, evaporated to dryness, calcinated at 450°C for 30 min, and analysed by atomic emission spectrometry. For the determination of lead in plants by the ETAAS method [276] the dried and ground material was dissolved successively in nitric and hydrochloric acid, the solution was evaporated and the residue was dissolved in water. The trace element was then extracted as pyrrolidinedithiocarbamate with chloroform.

Destruction of the sample analysed also precedes the determination of mercury in biological materials (tissues, lymph, blood, urea, other secretions, plants, food products). For the determination of mercury in blood and samples of homogenized inner human organs, 10 g of the sample was ashed with a mixture of concentrated sulphuric and nitric acid in the presence of potassium permanganate [277]. After the removal of nitrates, EDTA and hydrazine sulphate (masking of interfering elements) was added to the aliquot of the transparent solution and mercury was extracted with chloroform solution of dithizone. The extract was then subjected to spectrophotometric determination. Kish *et al.* [278] have developed an extraction photometric method for determining mercury in urea. This analysis is necessary to reveal mercury intoxication and for giving proper treatment in the case of mercury poisoning. The method involves extraction with benzene and photometry of the coloured ionic associate formed by the anionic mercury iodide complex with crystal violet cation. The latter, unlike dithizone, is stable; its aqueous solutions necessary for determination can be stored for several months. Besides, this reagent is more specific to mercury: 10^4–10^6-fold (with respect to mercury) amounts of Al, Cd, Co, Cr, Cu, Mg, Mn, Ni, Pb, Sn and a number of other elements do not hinder the determination. The detection limit of mercury is about $n \cdot 10^{-7} \%$.

A neutron activation procedure has been worked out for determining Au, Cd, Cu, Hg, Fe, Mo and Zn in biological samples (liver tissues, kidneys, plants) after wet ashing of the sample, subsequent isolation and separation of trace elements by the extraction–chromatographic method in the system of di-2-ethylhexylphosphoric acid or tri-n-octylamine and hydrochloric acid solution [279]. Teflon powder is used as carrier of the immobile phase.

Use is made also of mineralization in combination with sorption methods. In analysing plants for Be, Hf, Nb, Ta and Zr, Vanaeva *et al.* [280] mineralized the sample at 500°C and, after dissolving the dry residue and adjusting the pH to 4, sorbed Hf, Nb, Ta and Zr with 50 mg pyrogallol formaldehyde resin under static conditions and then Be at pH 8.5–8.7. The concentrate, after mixing with graphite collector, was analysed by AES. In analysing blood serum or animal tissues for Cd, Co, Fe, Mo, V and Zn, Mianzhi and Barnes [271] neutralized the sample after wet autoclave mineralization and isolated Cd, Co, Mo, V (pH 4.5) and Cu, Fe and Zn (pH 8) in a column packed with a complex-forming sorbent having dithiocarbamate groups. The sorbent was decomposed with H_2O_2 and HNO_3; the solution obtained was analysed by the AES–ICP method.

For determining mercury in natural and powder milk, peas, potatoes and fish tissues, Rigin [281] has proposed the following technique. Homogenized sample is decomposed in an autoclave (at 85°C) for 2 h with a mixture of equal volumes of concentrated hydrochloric, sulphuric and nitric acid in the presence of hydrogen peroxide. Mercury is then electrolytically separated and determined by the AFS method.

References

1 Ya. M. Grushko, *Vrednye Neorganicheskie Soedineniya v Promyshlennykh Stochnykh Vodakh (Harmful Inorganic Compounds in Industrial Waste Waters)*, Khimiya, Leningrad, 1979 (in Russian).
2 G.P. Bespamyatnov, K.K. Bogushevskaya, A.V. Bespamyatnova, Yu. A. Krotov, L.A. Zelenskaya, V.F. Plekhotkin and G.G. Smirnov, *Predelno Dopustimye Kontsentratsii Vrednykh Veshchestv v Vozdukhe i Vode (Limiting Allowable Concentrations of Harmful Substances in Air and Water)*, Khimiya, Leningrad, 2nd ed., 1975 (in Russian).
3 *Metody Opredeleniya Mikroelementov v Prirodnykh Obektakh (Methods of Determining Trace Elements in Natural Samples)*, Vol. III, Nauka, Moscow, 1976 (in Russian).
4 *Sovremennye Metody Opredeleniya Mikroelementov, Tezisy Dokladov Vsesoyuznogo Simpoziuma (Modern Methods of Determining Trace Elements, Abstracts of Reports made at the All-Union Symposium)*, Izd. Kishinev University, Kishinev, 1977 (in Russian).
5 *Vtoroe Vsesoyuznoe Soveshchanie po Analizu Prirodnykh i Stochnykh Vod, Tezisy Dokladov (2nd All-Union Conference on Analysis of Natural and Waste Waters, Abstracts of Lectures)*, Nauka, Moscow, 1977 (in Russian).
6 A.V. Karyakin and I.F. Gribovskaya, *Emissionny Spektralny analiz Obektov Biosfery (Atomic Emission Analysis of Substances of the Biosphere)*, Khimiya, Moscow, 1979 (in Russian).
7 L.A. Kul'chinsky, I.T. Goronovsky, A.M. Koganovsky and M.A. Shevchenko, *Spravochnik po Svoistvam, Metodam Analiza i Ochistke Vod, v Dvukh Chastyakh (Handbook on the Properties, Methods of Analysis, and Purification of Water, in two volumes)*, Naukova Dumka, Kiev, 1980 (in Russian).
8 *Metody Analiza Obektov Okruzhayushchei sredy, Tezisy dokladov Vsesoyuznoi konferentsii (Methods of Analysis of Environmental Samples, Abstracts of Reports made at the All-Union Conference, Moscow, September 27–29, 1983)*, Nauka, Moscow, 1983 (in Russian).
9 A. Mizuike, *Enrichment Techniques for Inorganic Trace Analysis*, Springer-Verlag, Berlin, Heidelberg, New York, 1983.
10 Yu. Yu. Lurye, *Analiticheskaya Khimiya Promyshlennykh Stochnykh Vod (Analytical Chemistry of Industrial Waste Waters)*, Khimiya, Moscow, 1984 (in Russian).
11 Yu. S. Drugov, A.B. Belikov, G.A. Dyakova and V.M. Tul'chinsky, *Metody analiza Zagryaznenii Vozdukha (Methods of Analysing Air Pollution)*, Khimiya, Moscow, 1984 (in Russian).
12 L.W. Canter, *River Water Quality Monitoring*, Wiley, New York, 1985.
13 Y.C. Lamb, *Water Quality and its Control*, Wiley, New York, 1985.

14 J.C. Méranger, K.S. Subramanian and C.H. Langford, *Rev. Anal. Chem.*, 5 (1980) 29.
15 W. Slavin, *At. Spectrosc.*, 1 (1980) 66.
16 R.A. Isaac, *J. Assoc. Off. Anal. Chem.*, 63 (1980) 788.
17 D.E. Leyden and W. Wegscheider, *Anal. Chem.*, 53 (1981) A1059.
18 H.W. Nürnberg, *Fresenius' Z. Anal. Chem.*, 316 (1983) 557.
19 G.V. Myasoedova, N.I. Shcherbinina and S.B. Savvin, *Z. Anal. Khim.*, 38 (1983) 1503.
20 Y. Wada, *Bull. Soc. Sea Water Sci., Jpn.*, 36 (1983) 291.
21 Yu. S. Drugov, *Z. Vses. Khim. Obs.*, 28 (1983) 80.
22 H.W. Nürnberg, *Anal. Chim. Acta*, 164 (1984) 1.
23 A. Alian and B. Sansoni, *J. Radioanal. Nucl. Chem.*, 89 (1985) 191.
24 Yu. S. Drugov, *Zh. Anal. Khim.*, 40 (1985) 585.
25 M.J. Fishman and D.E. Erdmann, *Anal. Chem.*, 51 (1979) 317.
26 J.C. van Loon (Editor), *Chemical Analysis of Inorganic Constituents of Water*, CRC Press, Boca Raton, FL, 1982.
27 P.D. Goulden and D.H.J. Anthony, *Anal. Chem.*, 54 (1982) 1678.
28 P.D. Goulden, D.H.J. Anthony and K.D. Austen, *Anal. Chem.*, 53 (1981) 2027.
29 D.D. Nygaard and J.H. Lowry, *Anal. Chem.*, 54 (1982) 803.
30 J.-G. Lo and J.-Y. Yang, *J. Radioanal. Nucl. Chem. Lett.*, 94 (1985) 311.
31 A.L. Shabanov, G.A. Babaev, A.B. Elchiev and Yu. G. Mamedova, *Zh. Anal. Khim.*, 39 (1984) 1621.
32 T.A. Onishchenko, I.V. Pyatnitsky, Yu.K. Onishchenko, V.V. Sukhan and V.N. Kashpor, *Zh. Anal. Khim.*, 40 (1985) 1595.
33 V.I. Rigin, *Zh. Anal. Khim.*, 39 (1984) 807.
34 H. Watanabe, K. Goto, S. Taguchi, J.W. McLaren, S.S. Berman and D.S. Russell, *Anal. Chem.*, 53 (1981) 738.
35 R.E. Sturgeon, S.N. Willie and S.S. Berman, *Anal. Chem.*, 57 (1985) 6.
36 M. Koide, D.S. Lee and M.O. Stallard, *Anal. Chem.*, 56 (1984) 1956.
37 K. Terada, K. Matsumoto and T. Inaba, *Anal. Chim. Acta*, 170 (1985) 225.
38 I.I. Nazarenko, I.V. Kislova, L.I. Kashina and G.I. Malofeeva, *Zh. Anal. Khim.*, 40 (1985) 2129.
39 A.P. Mykytiuk, D.S. Russell and R.E. Sturgeon, *Anal. Chem.*, 52 (1980) 1281.
40 P. Burba and P.G. Willmer, *Fresenius' Z. Anal. Chem.*, 311 (1982) 222.
41 H. Hashitani, M. Okumura and K. Fujinaga, *Fresenius' Z. Anal. Chem.*, 320 (1985) 773.
42 G.S. Caravajal, K.I. Mahan and D.E. Leyden, *Anal. Chim. Acta*, 135 (1982) 205.
43 T. Tisue, C. Seils and R.T. Keel, *Anal. Chem.*, 57 (1985) 82.
44 C.L. Smith, J.M. Motooka and W.R. Willson, *Anal. Lett.*, 17 (1984) 1715.
45 H. Bem and D.E. Ryan, *Anal. Chim. Acta*, 166 (1984) 189.
46 Feng Xi and D.E. Ryan, *Anal. Chim. Acta*, 162 (1984) 47.
47 C.J. Pickford and G. Rossi, *Analyst (London)*, 97 (1972) 647.
48 B.B. Mesman and T.C. Thomas, *Anal. Lett.*, 8 (1975) 449.
49 B.J.A. Haring, W. van Delft and C.M. Bom, *Fresenius' Z. Anal. Chem.*, 310 (1982) 217.
50 C.T. Tye, S.J. Haswell, P. O'Neill and K.C. Bancroft, *Anal. Chim. Acta*, 169 (1985) 195.
51 V.I. Rigin, *Zh. Anal. Khim.*, 33 (1978) 1966.
52 T. Nakahara, *Prog. Anal. At. Spectrosc.*, 6 (1983) 163.
53 D.D. Siemer, *Anal. Chem.*, 54 (1982) 1321.
54 F.A. Pohl, *Fresenius' Z. Anal. Chem.*, 139 (1953) 241.

55 H. Malissa and E. Schoffmann, *Mikrochim. Acta*, No. 1 (1955) 187.
56 H. Malissa and S. Gomišček, *Fresenius' Z. Anal. Chem.*, 169 (1959) 402.
57 C.A. Watson, *Ammonium Pyrrolidine Dithiocarbamate*, Hopkin and Williams, London, 1971.
58 A.W. Morris, *Anal. Chim. Acta*, 42 (1968) 397.
59 A.P. Sturis, D.K. Apsite, A.K. Sturis, O.E. Veveris, V.F. Rone and Yu. A. Bankovsky, in *Vtoraya Vses. Konferentsiya po Metodam Kontsentrirovaniya v Analiticheskoi Khimii, Tezisy Dokladov (2nd All-Union Conference on Concentration Methods used in Analytical Chemistry, Abstracts of Lectures)*, Nauka, Moscow, 1977, p. 78.
60 T. Yano, S. Ide, Y. Tobeta, H. Kobayashi and K. Ueno, *Talanta*, 23 (1976) 457.
61 A. Miyazaki, A. Kimura and Y. Umezaki, *Anal. Chim. Acta*, 138 (1982) 121.
62 B.M. Vanderborght and R.E. van Grieken, *Anal. Chem.*, 49 (1977) 311.
63 J. Smith and R. van Grieken, in *Int. Symp. Microchem. Techn., 1977, Abstracts*, Davos, 1977, p. 135.
64 L.I. Koval'chuk, V.P. Koryukova, L.V. Smirnova and E.V. Shabanov, *Zh. Anal. Khim.*, 34 (1979) 1136.
65 H.W. Nürnberg, *Fresenius' Z. Anal. Chem.*, 316 (1983) 557.
66 P.T. Kissinger and V.R. Heineman (Editors), *Laboratory Technique in Electroanalytical Chemistry*, Marcel Dekker, New York, Basel, 1984.
67 A.H.I. Ben-Bassat, J.-M. Blindermann, A. Salomon and E. Wakshal, *Anal. Chem.*, 47 (1975) 534.
68 B.H. Vassos, L.F. Hirsch and H. Letterman, *Anal. Chem.*, 45 (1973) 792.
69 A.A. Tikhomirova, S.A. Patkin and N.P. Morozov, *Zh. Anal. Khim.*, 31 (1976) 282.
70 S. Nakashima, *Anal. Chem.*, 51 (1979) 654.
71 M. Aoyama, T. Hobo and S. Suzuki, *Bunseki Kagaku*, 81 (1982) E99.
72 W. Faigle and D. Klockow, *Fresenius' Z. Anal. Chem.*, 306 (1981) 190.
73 W. Faigle and D. Klockow, *Fresenius' Z. Anal. Chem.*, 310 (1982) 33.
74 R. Winkler, *Fresenius' Z. Anal. Chem.*, 267 (1973) 261.
75 K. Oikawa, Y. Hashimoto and S. Yanagiawa, *J. Spectrosc. Soc. Jpn.*, 23 (1974) 111.
76 A. Alian and B. Sansoni, *J. Radioanal. Nucl. Chem.*, 89 (1985) 191.
77 B.H. O'Connor, G.C. Kerigan and W.M. Thomas, *X-Ray Spectrom.*, 4 (1975) 190.
78 M. Demuynck and R. Dams, *Bull. Soc. Chim. Belg.*, 90 (1981) 265.
79 T.R. Fogg and R.C. Seeley, *Am. Lab. (Fairfield, Conn.)*, 16 (1984) 36.
80 N.R. McQuaker and D.F. Brow, *Dev. Atom. Plasma Spectrochem. Anal. Proc. Int. Winter Conf., San Juan, January 7–11, 1980*, London, 1981, p. 404.
81 A. Sugimae and T. Mizoguchi, *Anal. Chim. Acta*, 144 (1982) 205.
82 K.Y. Chiou and O.K. Manuel, *Anal. Chem.*, 56 (1984) 2721.
83 G.R. Fuchs, E. Lisson, B. Schwarz and K. Bachmann, *Fresenius' Z. Anal. Chem.*, 320 (1985) 498.
84 G. Torsi, E. Desimoni, F. Palmisano and L. Sabbatini, *Anal. Chem.*, 53 (1981) 1035.
85 K. Kemp and M. Tscherning, *Nucl. Instrum. Methods*, 181 (1981) 481.
86 W. Thain, *Monitoring Toxic Gases in the Atmosphere for Hygiene and Pollution Control*, Pergamon Press, Oxford, 1980.
87 G. Choudhary (Editor), *Chemical Hazards in the Workplace (ASS Symp. Ser.)*, American Chemical Society, Washington, 1981.
88 J.E. Scott and J.M. Ottaway, *Analyst (London)*, 106 (1981) 1076.
89 T. Dumas and E.J. Bond, *J. Chromatogr.*, 206 (1981) 384.
90 E.A. Peregud, *Khimicheskii Analiz Vozdukha (Chemical Analysis of Air)*, Khimiya,

Leningrad, 1976 (in Russian).
91 W. de Jonghe, D. Chakraborti and F. Adams, *Anal. Chem.*, 52 (1980) 1974.
92 D.J. David, *Analyst (London)*, 85 (1960) 495.
93 R.A. Isaac, *J. Assoc. Off. Anal. Chem.*, 63 (1980) 788.
94 K. Tanabe, K. Matsumoto, H. Haraguchi and K. Fuwa, *Anal. Chem.*, 52 (1980) 2361.
95 R.L. Flannery and D.K. Markus, *J. Assoc. Off. Anal. Chem.*, 63 (1980) 779.
96 D.W. Hoult, M.M. Beaty and G.F. Wallace, *At. Spectrosc.*, 1 (1980) 157.
97 T.F. Borovik-Romanova and I.F. Gribovskaya, *Agrokhimiya*, No. 5 (1967) 126.
98 R.L. Mitchell, *The Spectrochemical Analysis of Soils, Plants, and Related Materials*, Commonwealth Agricultural Bureaux, London,1964, No. 44a.
99 I.T. Klimov, *Gidrokhim. Mater.*, 34 (1961) 131.
100 R. Bock, *A Handbook of Decomposition Methods in Analytical Chemistry*, International Textbook Company, Glasgow, 1984.
101 P. Tschöpel and G. Tölg, *J. Trace Microprobe Techn.*, 1 (1982) 1.
102 V.I. Rigin and A.O. Eryemina, *Zh. Anal. Khim.*, 39 (1984) 510.
103 V.I. Rigin, *Zh. Anal. Khim.*, 39 (1984) 648.
104 *Metody Analiza Chistykh Khimicheskikh Reaktivov (Methods of Analysing Pure Chemical Reagents)*, Khimiya, Moscow, 1984 (in Russian).
105 M. Thompson, *Analyst (London)*, 110 (1985) 229.
106 M. Sakata and O. Shimoda, *Bunseki Kagaku*, 31 (1982) T81.
107 W.E. Hillebrand, G.E. Lundell, H.A. Bright and J.I. Hoffman, *Applied Inorganic Analysis, With Special Reference to the Analysis of Metals, Minerals and Rocks*, Wiley, New York, Chapman and Hall, London, 2nd ed., 1953.
108 P.G. Jeffery, *Chemical Methods of Rock Analysis*, Pergamon Press, Oxford, New York, Toronto, Sydney, Braunschweig, 1970.
109 S.I. Ginzburg, N.A. Ezerskaya, I.V. Prokofyeva, N.V. Fedorenko, V.I. Shlenskya and N.K. Belsky, *Analiticheskaya Khimiya Platinobykh Metallov (Analytical Chemistry of Platinum Metals)*, Nauka, Moscow, 1978 (in Russian).
110 R.D. Reeves and R.R. Brooks, *Trace Elements Analysis of Geological Materials*, Wiley, New York, Toronto, Sydney, Braunschweig, 1978.
111 G.V. Ostroumov (Editor), *Metodicheskie Osnovy Issledovaniya Khimicheskogo Sostava Gornykh Porod, Rud i Mineralov (Methodical Basis of Studying the Chemical Composition of Rocks, Ores and Minerals)*, Nedra, Moscow, 1979 (in Russian).
112 G.V. Ostroumov (Editor), *Opredelenie Rudnykh i Rasseyannykh Metallov v Mineralnom Syrye (Determining Ore and Trace Elements in Mineral Products)*, Nedra, Moscow, 1982 (in Russian).
113 G.V. Ostroumov (Editor), *Opredelenie Redkikh i Radioaktivnykh Elementov v Mineralnom Syrye (Determining Rare and Radioactive Elements in Mineral Products)*, Nedra, Moscow, 1983 (in Russian).
114 I.A. Blyum and Yu. A. Zolotov, *Zh. Anal. Khim.*, 31 (1976) 159.
115 E.M. Dorokhova, O.P. Shvoeva, A.S. Cherevko and G.V. Myasoedova, *Zh. Anal. Khim.*, 34 (1979) 1140.
116 K. Kawabuchi, M. Kante, T. Muraoka and Y. Miwako, *Jpn. Anal.*, 25 (1976) 213.
117 Yu. A. Zolotov and N.M. Kuz'min, *Ekstraktsionnoe Kontsentripovanie (Extraction Concentration)*, Khimiya, Moscow, 1971 (in Russian).
118 N.M. Kuz'min, V.S. Vlasov, V.Z. Krasil'schchik and V.G. Lambrev, *Zavod. Lab.*, 43 (1977) 1.
119 I.G. Yudelevich and E.A. Startseva, *Atomno-Absorbtsionnoe Opredelenie*

Blagorodnykh Metallov (Atomic Absorption Determination of Noble Metals), Nauka, Novosibirsk, 1981 (in Russian).

120 Yu. A. Zolotov, V.A. Bodnya and A.N. Zagruzina, *CRC Crit. Rev. Anal. Chem.*, 14 (1982) 93.

121 S.J. Al-Bazi and A. Chow, *Talanta*, 31 (1984) 815.

122 E.E. Rakovskii, *J. Radioanal. Nucl. Chem.*, 88 (1985) 161.

123 N.M. Kuz'min, in *Teoriya i Praktika Ekstraktsionnykh Metodov (Theory and Practice of Extraction Methods)*, Nauka, Moscow, 1985, p. 186 (in Russian).

124 O.M. Petrukhin and G.I. Malofeeva, in *Teoriya i Praktika Ekstraktsionnykh Metodov (Theory and Practice of Extraction Methods)*, Nauka, Moscow, 1985, p. 246 (in Russian).

125 G.A. Vall, M.V. Usol'tseva, I.G. Yudelevich, I.V. Seryakova and Yu. A. Zolotov, *Zh. Anal. Khim.*, 31 (1976) 27.

126 P.P. Kish, B. Ya. Spivakov, V.V. Roman and Yu. A. Zolotov, *Zh. Anal. Khim.*, 32 (1977) 1942.

127 N.A. Borshch, O.M. Petrukhin and Yu. A. Zolotov, *Zh. Anal. Khim.*, 33 (1978) 1120.

128 G.A. Vall, L.P. Poddubnaya, I.G. Yudelevich and Yu. A. Zolotov, *Zh. Anal. Khim.*, 34 (1979) 885.

129 O.M. Petrukhin, Yu. A. Zolotov, V.N. Shevchenko, O.I. Kryuchkova, S.I. Gronskaya, V.N. Gushchin, G.I. Ramendik, V.V. Dunina and E.G. Rukhadze, *Zh. Anal. Khim.*, 34 (1979) 334.

130 Shan Xiao-Quan, Ni Zhe-Ming and Yuan Zhi-Neng, *Anal. Chim. Acta*, 171 (1985) 269.

131 N.K. Roy, D.K. De and A.K. Das, *At. Spectrosc.*, 5 (1984) 126.

132 D. Kolčava and J. Janáček, *Collect. Czech. Chem. Commun.*, 49 (1984) 370.

133 J.G. Viets, R.M. O'Leary and R.J. Clark, *Analyst (London)*, 109 (1984) 1589.

134 Yu. M. Yukhin, T.A. Udalova and V.G. Tsimbalist, *Zh. Anal. Khim.*, 40 (1985) 850.

135 A. Wyttenbach, S. Bajo and L. Tobler, *J. Radioanal. Chem.*, 78 (1983) 283.

136 G.V. Myasoedova and G.I. Malofeeva, *Zh. Anal. Khim.*, 34 (1979) 1626.

137 G.V. Myasoedova, I.I. Antokolskaja, O.P. Shvoeva, L.I. Bolshakova and S.B. Savvin, *Talanta*, 23 (1976) 866.

138 O.P. Shvoeva, G.P. Kuchava, G.V. Myasoedova, S.B. Savvin, L.N. Bannykh, N.G. Zhukova, O.N. Grishina and M.S. Mezhirov, *Zh. Anal. Khim.*, 40 (1985) 1606.

139 A.E. Watson and G.L. Moore, *S. Afr. J. Chem.*, 37 (1984) 81.

140 Yu. N. Kukushkin, S.A. Simakova, V.P. Alashkevich and D.F. Makarov, *Z. Prikl. Khim.*, 44 (1971) 1640.

141 A.P. Kuznetsov, Yu. N. Kukushkin and D.F. Makarov, *Izv. Vuzov, Khim. Khim. Tekhnol.*, 20 (1977) 65.

142 I. Yu. Davydova, A.P. Kuznetsov, I.I. Antokolskaya, N.N. Nikolskaya and Z.A. Ezhova, *Zh. Anal. Khim.*, 34 (1979) 1145.

143 V.G. Goryushina, S.B. Savvin and E.V. Romanova, *Zh. Anal. Khim.*, 18 (1963) 1340.

144 J.A.C. Broekaert and P.K. Hörmann, *Anal. Chim. Acta*, 124 (1981) 421.

145 *Trudy Komissii po Analiticheskoi Khimii AN SSSR, Vol. XV, Metody Kontsentrirovaniya Veshchestv v Analiticheskoi Khimii (Analytical Methods of Concentration of Substances)*, Nauka, Moscow, 1965 (in Russian).

146 *Metody Analiza Veshchestv Vysokoi Chistoty (Methods of Analysing High-purity Substances)*, Nauka, Moscow, 1965 (in Russian).

147 A. Mizuike, *Bunseki Kagaku*, 65 (1980) 314.

148 J.R. Moody, *Phil. Trans. R. Soc. London, Ser. A*, 305 (1982) 669.

149 L. Kosta, *Talanta*, 29 (1982) 985.
150 L. Licht, J. Birkhahn and B. Wagner, *Mitteilungsbl. Chem. Ges. DDR*, 31 (1984) 231.
151 A. Mizuike, *Fresenius' Z. Anal. Chem.*, 319 (1984) 415.
152 Yu. A. Zolotov, M. Grasserbauer, G.H. Morrison and Yu. A. Karpov, *Pure Appl. Chem.*, 57 (1985) 1133.
153 Yu. A. Karpov and I.P. Alimarin, *Zh. Anal. Khim.*, 34 (1979) 1402.
154 G. Tölg, *Talanta*, 21 (1974) 327.
155 G. Tölg, *J. Less-Common Metals*, 43 (1975) 157.
156 G. Tölg, *Pure Appl. Chem.*, 44 (1975) 645.
157 G. Tölg, *Naturwissenschaften*, 63 (1976) 99.
158 A.B. Blank, *Zh. Anal. Khim.*, 15 (1960) 359.
159 E.S. Zolotovitskaya and E.I. Yakovenko, *Zavod. Lab.*, 33 (1967) 1516.
160 N.M. Kuz'min, G.I. Zhuravlyev, I.A. Kuzovlev, A.N. Galaktionova and T.I. Zakharova, *Zh. Anal. Khim.*, 24 (1969) 429.
161 G. Baudin, *Prog. Anal. At. Spectrosc.*, 3 (1980) 1 and 57.
162 M.S. Chupakhin and S.U. Kreingold, *Z. Vses. Khim. Obs.*, 29 (1984) 700.
163 Yu. A. Zolotov, *Zavod. Lab.*, 50 (1984) 3.
164 I.G. Yudelevich, L.M. Buyanova and I.R. Shelpakova, *Khimikospektral'ny Analiz Veshchestv Vysokoi Chistoty (Atomic Emission Analysis of High-Purity Substances)*, Nauka, Novosibirsk, 1980 (in Russian).
165 I.R. Shelpakova, I.G. Yudelevich and B.M. Ayupov, *Posloiny Analiz Materalov Elektronnoi Tekhniki (Layerwise Analysis of Electronic Engineering Materials)*, Nauka, Novosibirsk, 1984 (in Russian).
166 Yu. A. Karpov, I.G. Yudelevich, E.N. Gil'bert, E.A. Startseva, D.V. Kormilitsyn and O.V. Shuvaeva, *Zh. Anal. Khim.*, 40 (1985) 373.
167 M. Pinta, *Pure Appl. Chem.*, 49 (1977) 893.
168 Kh. I. Zil'bershtein (Editor), *Spectrochemical Analysis of Pure Substances*, Adam Hilger Ltd., Bristol, 1977.
169 N.N. Semenov, *Spektralny Analiz Kvartsevogo Cyrya (Spectral Analysis of Quartz Materials)*, Promstroiizdat, Moscow, 1957 (in Russian).
170 B.D. Brodskaya, M.A. Notkina and N.P. Men'shova, *Zh. Anal. Khim.*, 27 (1972) 151.
171 L.E. Vakulenko, B. Ya. Kaplan, Yu. I. Merisov, A.I. Mikhailichenko and G.S. Skripkin, *Zavod. Lab.*, 39 (1973) 1342.
172 V.T. Mishchenko, L.I. Kovalchuk, L.P. Ponomarenko and L.V. Smirnova, *Ukr. Khim. Zurnal*, 47 (1981) 767.
173 Mo Shengjun and Zharg Zhuoyong, *Fenxi huaxue*, 12 (1984) 408.
174 M.A. Notkina, E.I. Petrova, T.V. Cherkashina and Yu. A. Chernikhov, *Tr. Kom. Anal. Khim. Akad. Nauk SSSR*, 15 (1965) 80.
175 I.G. Yudelevich, V.P. Shaburova, V.G. Torgov and O.I. Shcherbakova, *Zh. Anal. Khim.*, 28 (1973) 1049.
176 B.R. Bear, M.C. Edelson, B. Gopalan and V.A. Fassel, *Anal. Spectrosc. Proc. 26th Conf. Anal. Chem. Energy Technol., Knoxville, TN, October 11–13, 1983*, Amsterdam, 1984, p. 187.
177 A.A. Halouma, R.B. Farrar, E.A. Hester and R.W. Morrow, *Anal. Spectrosc. Proc. 26th Conf. Anal. Chem. Energy Technol., Knoxville, TN, October 11–13, 1983*, Amsterdam, 1984, p. 201.
178 A.K. Seshagiri, Y. Babu, K.M.L. Jayanth, G.I. Dalvi, M.D. Sastry and B.D. Joshi, *Talanta*, 31 (1984) 773.

179 Fang Qiying, Huang Zongzhi and Zhou Hongdi, *Nucl. Techn.*, No. 3 (1985) 57.
180 C.E. Michel and G.E. Brown, *Anal. Spectrosc. Proc. 26th Conf. Anal. Chem. Energy Technol., Knoxville, TN, October 11–13, 1983*, Amsterdam, 1984, p. 235.
181 R. Ko, *Appl. Spectrosc.*, 38 (1984) 909.
182 V.P. Belyaev, N.M. Kuz'min, V.R. Kalinachenko and L.M. Yakimenko, *Zavod. Lab.*, 28 (1962) 685.
183 N.M. Kuz'min, L.M. Yakimenko and V.R. Kalinachenko, *Metody Anal. Khim. Reakt. Prep.*, No. 15 (1968) 84.
184 N.M. Kuz'min, *Zh. Anal. Khim.*, 22 (1967) 451.
185 N.M. Kuz'min, T.P. Dubrovina and O.M. Shemshuk, *Zh. Anal. Khim.*, 28 (1973) 364.
186 N.M. Kuz'min, T.P. Dubrovina and O.M. Shemshuk, *Zavod. Lab.*, 35 (1969) 784.
187 N.M. Kuz'min, *Zavod. Lab.*, 32 (1966) 1349.
188 N.M. Kuz'min, V.L. Sabatovskaya and L.S. Khorkina, *Metody Anal. Galogenidov Shchelochnykh Shchelochnozemel'nykh Met. Vys. Chist.*, 1971, Part II (1971) 57.
189 A.K. Babko, N.M. Kuz'min, G.S. Lisetskaya, M.I. Ovrutsky and S.V. Freger, *Ukrainsk. Khim. Zh.*, 33 (1967) 828.
190 L.I. Pavlenko, O.M. Petrukhin, Yu. A. Zolotov, A.V. Karyakin, G.N. Gavrilina and I.E. Tumanova, *Zh. Anal. Khim.*, 29 (1974) 933.
191 V.V. Shaburova, I.G. Yudelevich, I.V. Seryakova and Yu. A. Zolotov, *Zh. Anal. Khim.*, 31 (1976) 255.
192 N.A. Borshch, O.M. Petrukhin and Yu. A. Zolotov, *Zh. Anal. Khim.*, 33 (1978) 1120.
193 Yu. A. Zolotov, N.G. Vanifatova, T.A. Chanysheva and I.G. Yudelevich, *Zh. Anal. Khim.*, 32 (1977) 317.
194 Yu. A. Zolotov, N.G. Vanifatova, T.A. Chanysheva and I.G. Yudelevich, *Zh. Anal. Khim.*, 34 (1979) 1720.
195 N.M. Kuz'min, I.A. Kuzovlev, S.V. Tsykunova, G.V. Krasnikova and A.N. Galaktionova, *Zavod. Lab.*, 34 (1968) 1058.
196 P.P. Kish, I.S. Balog, B. Ya. Spivakov and Yu. A. Zolotov, *Zh. Anal. Khim.*, 31 (1976) 1114.
197 N.M. Kuz'min, V.S. Solomatin, A.N. Galaktionova and I.A. Kuzovlev, *Zh. Anal. Khim.*, 24 (1969) 725.
198 N.M. Kuz'min, V.S. Solomatin and V.A. Bystrova, *Metody Anal. Khim. Reakt. Prep.*, No. 17 (1971) 44.
199 V.S. Solomatin, N.M. Kuz'min, D.N. Nishanov, L.T. Shatalina and N.I. Lakeeva, *Promst. Khim. Reakt. Osobo Chist. Veshchestv*, No. 27 (1973) 5.
200 N.M. Kuz'min, I.A. Kuzovlev, S.V. Meshchankina and V.A. Bystrova, *Metody Anal. Khim. Reakt. Prep.*, No. 17 (1971) 65.
201 N.M. Kuz'min, I.A. Kuzovlev and V.V. Chugunova, *Metody Anal. Khim. Reakt. Prep.*, No. 17 (1971) 34.
202 V.A. Fedorov, V.A. Krutikov, B.I. Kozyrkin, B.G. Gribov and N.M. Kuz'min, *Elektron. Tekh., Ser. 14, Materialy*, No. 7 (1971) 123.
203 L.A. Demina, O.M. Petrukhin and Yu. A. Zolotov, *Zh. Anal. Khim.*, 27 (1972) 593.
204 V.A. Orlova, B. Ya. Spivakov, V.M. Shkinev, T.I. Kirillina, V.A. Ivanova, T.I. Malyutina and Yu. A. Zolotov, *Zh. Anal. Khim.*, 33 (1978) 91.
205 N.M. Kuz'min, L.S. Khorkina and I.A. Kuzovlev, *Zh. Anal. Khim.*, 24 (1969) 589.
206 V.M. Shkinev, B. Ya. Spivakov, V.A. Orlova, T.M. Malyutina, T.I. Kirillina and Yu. A. Zolotov, *Zh. Anal. Khim.*, 33 (1978) 922.
207 N.M. Kuz'min, G.I. Zhuravlev, I.A. Kuzovlev, A.N. Galaktionova and T.I. Zakharova, *Metody Anal. Khim. Reakt. Prep.*, No. 17 (1971) 54.

346

208 T.A. Kul'beda, T.P. Tsvetkova and N.M. Kuz'min, *Elektron. Tekh. Ser. 14, Materialy*, No. 11 (1974) 103.
209 G.I. Zhuravlev, G.D. Popova, T.I. Zakharova and N.M. Kuz'min, *Metody Anal. Khim. Reakt. Prep.*, No. 17 (1971) 28.
210 N.M. Kuz'min, V.S. Vlasov and V.G. Lambrev, *Zh. Anal. Khim.*, 27 (1972) 1614.
211 N.M. Kuz'min, A.I. Lebedev, G.D. Popova and T.A. Bokova, *Elektron. Tekh. Ser. 14, Materialy*, No. 3 (1974) 94.
212 N.M. Kuz'min, S.V. Meshankina and I.A. Kuzovlev, *Metody Anal. Khim. Reakt. Prep.*, No. 17 (1971) 69.
213 N.M. Kuz'min, G.D. Popova and V.L. Sabatovskaya, *Zavod. Lab.*, 37 (1971) 658.
214 I.A. Kuzovlev, O.A. Sverdlina and N.M. Kuz'min, *Zh. Anal. Khim.*, 27 (1972) 199.
215 N.M. Kuz'min, A.V. Emelyanov, S.V. Meshankina, V.L. Sabatovskaya, I.A. Kuzovlev and T.I. Zakharova, *Zh. Anal. Khim.*, 26 (1971) 282.
216 T.A. Bokova and N.M. Kuz'min, *Zh. Anal. Khim.*, 29 (1974) 386.
217 N.M. Kuz'min, T.I. Zakharova and V.V. Chugunova, *Metody Anal. Khim. Reakt. Prep.*, No. 17 (1971) 49.
218 N.P. Krivenkova, L.I. Pavlenko, B. Ya. Spivakov, I.A. Popova, T.S. Plotnikova, V.M. Shkinev and I.P. Kharlamov, *Zh. Anal. Khim.*, 31 (1976) 514.
219 B. Ya. Spivakov, V.I. Lebedev, V.M. Shkinev, N.P. Krivenko, T.S. Plotnikova, I.P. Kharlamov and Yu. A. Zolotov, *Zh. Anal. Khim.*, 31 (1976) 757.
220 Yu. A. Zolotov, N.V. Shakhova, O.I. Kryuchkova, S.I. Gronskaya, B. Ya. Spivakov, G.I. Ramendik and V.N. Gushin, *Zh. Anal. Khim.*, 33 (1978) 1253.
221 Yu. A. Zolotov and I.P. Alimarin, *Dokl. Akad. Nauk SSSR*, 136 (1961) 603.
222 Yu. A. Zolotov and I.P. Alimarin, *Radiokhimiya*, 4 (1962) 272.
223 Yu. A. Zolotov, *Acta Chim. Acad. Sci. Hung.*, 32 (1962) 327.
224 B. Ya. Spivakov, V.M. Shkinev and Yu. A. Zolotov, *Zh. Anal. Khim.*, 30 (1975) 2182.
225 B. Ya. Spivakov, V.M. Shkinev and G.A. Vorobyeva, in *Theory and Practice of Extraction Methods*, Nauka, Moscow, 1985, p. 137 (in Russian).
226 E. Jackwerth, J. Lohmar and G. Wittler, *Fresenius' Z. Anal. Chem.*, 270 (1974) 6.
227 E. Jackwerth and H. Berndt, *Anal. Chim. Acta*, 74 (1975) 299.
228 E. Jackwerth, *Fresenius' Z. Anal. Chem.*, 271 (1974) 120.
229 M. Kimura, *Talanta*, 24 (1977) 194.
230 H. Berndt, E. Jackwerth and M. Kimura, *Anal. Chim. Acta*, 93 (1977) 45.
231 A.T. Kashuba and C.R. Hines, *Anal. Chem.*, 43 (1971) 1758.
232 V.A. Ryabukhin, N.G. Gatinskaya and A.N. Ermakov, *Zh. Anal. Khim.*, 32 (1977) 909.
233 R.L. Slyusareva and L.I. Kondratyeva, *Zavod. Lab.*, 41 (1975) 689.
234 W. Reichel and B.G. Bleakley, *Anal. Chem.*, 32 (1960) 909.
235 M. Kimura and S. Egawa, *Talanta*, 29 (1982) 329.
236 G. Henrion, J. Gelbrecht, T. Hoffmann and D. Marquardt, *Z. Chem.*, 21 (1981) 192.
237 Yang Chao-Yeuh and Yeh Yu-Chai, *Analyst (London)*, 107 (1982) 1492.
238 P. Heininger and G. Henrion, *Z. Chem.*, 25 (1985) 73.
239 Y. Harada, N. Kurata and Y. Goto, *Bunseki Kagaku*, 33 (1984) 71.
240 Y. Akama, M. Hayakawa, T. Nakai and F. Kawamura, *Anal. Chim. Acta*, 157 (1984) 207.
241 Z. Marczenko and J. Maruszak, *Chem. Anal. (PRL)*, 27 (1982) 391.
242 R. Bye and L. Engvik, *Anal. Chim. Acta*, 166 (1984) 289.
243 R. Koehler, H. Scharf, K. Licht, M. Klewc, R. Matschat and K. Richter, *Chem. Tech. (Leipzig)*, 37 (1985) 101.

347

244 V. Valković, *Trace Element Analysis*, Taylor and Francis, London, 1975.
245 T.T. Gorsuch, *The Destruction of Organic Matter*, Pergamon Press, Oxford, New York, Toronto, Sydney, Braunschweig, 1970.
246 S.K. Kyuregyan, *Atomny Spektralny Analiz Nefteproduktov (Atomic Emission Analysis of Petroleum Products)*, Khimiya, Moscow, 1985.
247 S.S.M. Hassan, *Organic Analysis using Atomic Absorption Spectrometry*, Ellis Horwood, New York, Chichester, Brisbane, Toronto, 1984, 384p.
248 M.S. Viglar and V.E. Gaylor, *Appl. Spectrosc.*, 28 (1974) 342.
249 A. Serbănescu, G. Bănăteanu, T. Fedin and S. Badea, *Rev. Chim. (RSR)*, 26 (1975) 863.
250 S. Oemar, in *Abstracts of the Pittsburgh Conference on Anal. Chem. Appl. Spectrosc., Cleveland, OH, 1977*, Pittsburgh, PA, 1977, p. 48.
251 A.I. Chernova, *Zavod. Lab.*, 34 (1968) 1072.
252 G. Kaiser, P. Tschöpel and G. Tölg, *Fresenius' Z. Anal. Chem.*, 253 (1971) 177.
253 M.A. Volodina, A.G. Buyanovskaya, T.A. Torshkova and A.S. Arutyunova, in *IV Vsesoyuznaya Konf. po Anal. Khim. Org. Soedinenii (IVth All-Union Conference on Analytical Chemistry of Organic Compounds), Moscow, January 7–9, 1980, Abstracts of Reports*, Nauka, Moscow, 1979, p. 237.
254 A.G. Buyanovskaya, M.A. Volodina, T.V. Bykova and N.K. Gavrilova, in *IV Vsesoyuznaya Konf. po Anal. Khim. Org. Soedinenii (IVth All-Union Conference on Analytical Chemistry of Organic Compounds), Moscow, January 7–9, 1980, Abstracts of Reports*, Nauka, Moscow, 1979, p. 240.
255 T.A. Gorshkova, M.A. Volodina, A.S. Arutyunova and T.M. Repetyuk, in *IV Vsesoyuznaya Konf. po Anal. Khim. Org. Soedinenii (IVth All-Union Conference on Analytical Chemistry of Organic Compounds), Moscow, January 7–9, 1980, Abstracts of Reports*, Nauka, Moscow, 1979, p. 246.
256 M. Gocłowski, M. Jaskóła, J. Ludziejewski and R. Matusiewicz, in *Instrum. Methods Phys. Res.*, B231 (1–3) (1984); *Particle Induced X-Ray Emission and Analytical Applications 3, Proc. 3rd Int. Conf., Heidelberg, July 18–22, 1983*, p. 163.
257 L. Vos and R. van Grieken, *Anal. Chim. Acta*, 164 (1984) 83.
258 L. Kotz, G. Henze, G. Kaiser, S. Pahlke and G. Tölg, *Talanta*, 26 (1979) 681.
259 U. Voellkopf and Z. Grobenski, *At. Spectrosc.*, 5 (1984) 115.
260 R.A. Jacob and L.M. Klevay, *Anal. Chem.*, 47 (1975) 741.
261 Z. Aneva and M. Ivancheva, *Anal. Chim. Acta*, 167 (1985) 371.
262 Z. Turlakiewicz, M. Jakubowski and J. Chmielnicka, *Br. J. Ind. Med.*, 42 (1985) 63.
263 H.A. Chandler and M. Scott, *At. Spectrosc.*, 5 (1984) 230.
264 S. Palágyi, *Isotopenpraxis*, 21 (1985) 130.
265 C.L. Lochmüller, J.W. Galbraith and K.H. Walter, *Anal. Chem.*, 46 (1974) 440.
266 I.K. Kressin, *Anal. Chem.*, 56 (1984) 2269.
267 S.A. Lewis, T.C. O'Haver and J.M. Harnly, *Anal. Chem.*, 57 (1985) 2.
268 D.R. Webb and D.E. Carter, *J. Anal. Toxicol.*, 8 (1984) 118.
269 E. de Olivera, J.W. McLaren and S.S. Berman, *Anal. Chem.*, 55 (1983) 2047.
270 B. Welz and M. Melcher, *Anal. Chem.*, 57 (1985) 427.
271 Z. Mianzhi and R.M. Barnes, *Appl. Spectrosc.*, 38 (1984) 635.
272 B. Welz, M. Melcher and J. Neve, *Anal. Chim. Acta*, 165 (1984) 131.
273 J. Piwonka, G. Kaiser and G. Tölg, *Fresenius' Z. Anal. Chem.*, 321 (1985) 225.
274 W. Maher, *Anal. Chim. Acta*, 138 (1982) 365.
275 A. Stracheim, D.J. Eve and R.M. Fourie, *J. South Afric. Chem. Inst.*, 12 (1959) 75.
276 J. Aznarez, F. Palacios and J.C. Vidal, *An. Estac. Exp. Aula Dei*, 16 (1982) 45.

348

277 W. Miketukova and K. Kack, *Arch. Toxicol.*, 20 (1964) 242.
278 P.P. Kish, G.M. Vitenko and V.I. Buletsa, *Lab. Delo*, No. 1 (1974) 29.
279 Yu. F. Korshunov, L.I. Zhuk, I.I. Orestova, E.S. Gureev and A.A. Kist, *Zh. Anal. Khim.*, 31 (1976) 1962.
280 A.V. Vanaeva, N.I. Kuznetsova and T.M. Moroshkina, *Vestn. Leningr. Univ.*, No. 16 (1971) 154.
281 V.I. Rigin, *Zh. Anal. Khim.*, 34 (1979) 261.

Subject index

thiourea (266), of diethyldithiocarbamate (274, 275), with Aliquat 336 (307), preconcentration by extraction of the matrix (212, 226, 319, 320, 323, 326).

Sorption with hydrated Sb_2O_5 (75), on cation exchanger KU-2 (61), dithiocarbamate derivatives of cellulose (66), polymeric thioether (243), of pyrrolidinedithiocarbamate on silica gel grafted with Bondapak Porasil B (293).

Precipitation as dibenzyldithiocarbamate (241), as sulphide in the presence of cellulose (241), coprecipitation with $Fe(OH)_3$ (88, 89, 330), $Al(OH)_3$ (88, 89), $Mn(OH)_4$ (88, 89), $Zr(OH)_4$ (88, 89), $La(OH)_3$ (329), preconcentration by precipitating the matrix (80).

Electrodeposition (100, 103–105, 263), cementation (106, 107), preconcentration with the use of mercury as chemical reagent (163).

Volatilization as elemental Sb (128), chlorides (128, 316), fluorides (128), iodides (131) and hydrides (128, 132, 134–135, 218, 219, 230, 234, 292) after chemical transformation, preconcentration by volatilization (118–120, 296) and sublimation (122) of the matrix, by volatilization of the matrix after chemical transformation (315), by dry (332, 333) and wet (335, 336) ashing of the organic matrix.

Zone melting (145).

Flotation (158, 295).

Filtration (160).

Arsenic

Extraction of molybdoarsenic heteropoly acid (236), AsO_4^{3-} with dialkyl tin dinitrates (33, 226, 228, 236, 258, 321), of diethyldithiocarbamate (275), with Aliquat 336 (307), preconcentration by extraction of the matrix (174, 212, 319, 323).

Sorption on Al_2O_3 (75), manganese or zinc sulphided (363), cation exchanger

KU-2 (61), sorption As(V) with dithiocarbamate derivatives of cellulose (67), cellulose filter containing functional groups of 2,2′-diaminediethylamine (67), on porous glass balls treating N-β-aminoethylaminopropyltrimethoxysilane (69), polymeric thioether (243), Rexyn 201 (257), sorption of AsH_3 on Chromosorb 102 (301).

Precipitation as sulphide in the presence of cellulose (89, 241), as arsenate (177), pyrrolidine-dithiocarbamate (82), dibenzyldithiocarbamate (82, 241), coprecipitation with $Al(OH)_3$ (88, 89), $Mn(OH)_4$ (89), $La(OH)_3$ (329), molybdenum sulfide (89); preconcentration by precipitating the matrix (80).

Electrodeposition (100, 105, 106), cementation (107).

Volatilization as As (128), AsF_3 (128), $AsCl_3$ (128, 129), oxides (128, 248, 476), AsH_3 (128, 132–135, 218, 230, 233, 234), by volatilization of the matrix after chemical transformation (121, 315), fractional volatilization (215), preconcentration by dry (122, 123, 125, 126, 230, 231, 333) and wet (218, 219, 230, 231, 335) ashing of the organic matrix.

Zone melting (145).

Flotation (157, 158, 295).

Filtration (160, 161).

Barium

Preconcentration by extraction of the matrix (212, 319, 320, 323, 326, 327).

Sorption on cation-exchangers KU-2-8 (60), Dowex 50W x 8 (76), Chelex-100 (78).

Precipitation of sulphate (80), co-precipitation with lead chromate (89).

Fractional volatilization (215), preconcentration by volatilization of the matrix (119, 120, 174, 292, 314), after chemical transformation (315), by dry (124–126, 332, 333) and wet (219) ashing of the organic matrix.

Beryllium

Extraction as trifluoroacetylacetonate (293), preconcentration by extraction of the matrix (212, 319, 320, 323).

Sorption on Chelex-100 (78), complex-forming sorbents based on aminopolystyrenepolyarsenazo and polypyridylazoresorcinol (304).

Co-precipitation with metal hydroxides (86), with $Mg(OH)_2$ and $Fe(OH)_3$ (86, 88, 89).

Fractional volatilization (215), preconcentration by volatilization of the matrix (119, 130, 174, 314), by dry (125, 126, 332) and wet (125, 126) ashing of organic matrix.

Bismuth

Extraction of 8-hydroxyquinolinate (324, 326), diethyldithiocarbamate (302, 319), of pyrrolidinedithiocarbamate with molten stearic alcohol (38), with diethyldithiophosphoric acid (266), thiourea (266), Aliquat 336 (307); preconcentration by extraction of the matrix (174, 212, 319, 320, 323, 324, 326).

Sorption on activated charcoal (52, 53), as chelate with xylenol orange (55), of hexamethylenedithiocarbamate (54), xanthate (55), diethyldithiophosphate (54), 8-hydroxyquinolinate (54, 55) on activated charcoals, sorption on metatitanic acid (298), manganese or zinc sulphides (244), on cation exchanger KU-2 (60), anion exchanger EDE-10M (60), cotton precoated with iminodiacetylcellulose (66), polymeric thioether (243), sorbents based on aminopolystyrenepolyarsenazo and polypyridylazoresorcinol (304).

Precipitation with alizarin blue (82) as sulphides in the presence of cellulose (89, 241), co-precipitation with $Fe(OH)_3$ (88, 264), $Al(OH)_3$ (88), $Mn(OH)_4$ (88, 89), $Mg(OH)_2$ and $Fe(OH)_3$ (87, 330), $La(OH)_3$ (329), AgI in the presence of 1,10-phenanthroline (89), of

dibenzyldithiocarbamate with Ag and Ni dibenzyldithiocarbamates (92, 96), thiocyanate or iodide, crystalline violet with 2,4-nitroaniline (93), preconcentration by precipitating the matrix (80, 81).

Electrodeposition (105, 242, 263, 264), cementation (106, 107), preconcentration with the use of mercury as chemical reagent (108, 109), electrodiffusion (114, 115).

Volatilization as Bi (128), $BiCl_3$ (128), BiF_3 (128), BiH_3 (128, 132, 135, 136, 218, 219, 230, 234), fractional volatilization (214, 229), preconcentration by volatilization (119, 121, 315) and sublimation (121) of the matrix as SiF_4 (130), $GeCl_4$ (130), $SbCl_3$ (130, 318), SeO_2 (130), $Ni(CO)_4$ (130), by dry ashing of the organic matrix (124).

Direct crystallization of the matrix (140, 142, 174), zone melting (145).

Fire assay (153), pyrometallurgical slag formation (153).

Preconcentration by partial dissolution of the matrix (110, 111).

Boron

Preconcentration by extraction of the matrix (320).

Volatilization as boronmethyl ether (131), fractional volatilization (215, 216), preconcentration by dry (230, 231) and wet (230, 231) ashing of the organic matrix.

Zone melting (144, 145).

Bromine

Sorption on activated charcoals (53).

Precipitation (176).

Volatilization as Br_2 after chemical transformations (128, 176), preconcentration by volatilization of the matrix (120).

Directed crystallization of the matrix (267).

353

Cadmium

Extraction of dithizonate (24, 247), diethyldithiocarbamate (266, 319), pyrrolidinedithiocarbamate (38, 44, 228, 244, 319), tetramethyldithiocarbamate and hexamethylenedithiocarbamate (220), 2-isopropyl-8-n-mercaptoquinolinate (297), with 2,2'-dichlorodiethyl ether (35), tridecylamine (43), trioctylamine (226), $(C_4H_9)_4NI$ (266), Aliquat 336 (307); preconcentration by extraction of the matrix (174, 212, 226, 319, 320, 323).

Sorption on activated charcoal (53), of diethyldithiocarbamate (54, 328), hexamethylenedithiocarbamate (54), xanthate (55), diethyldithiophosphate (54), 8-hydroquinolinate (55, 242, 298) on activated charcoals, sorption on manganese or zinc sulphide (244), metatitanic acid (298), sorption on cation exchangers KU-2-8 (60, 61) and Dowex A-1 (60), anion exchanger Bio-Rad Ag MP-1 (60), sorbent Chelex-100 (63, 78, 294), cotton precoated with iminodiacetylcellulose (65, 66); sorption on a modified cellulose containing 1-(2-hydroxyphenylazo)-2-naphthol or 4-(2-pyridylazo)resorcinol (58), on dithiocarbamate derivatives of cellulose (67), sorption with sodium diethyldithiocarbamate coated on Chromosorb W-DMCS (73), on transparent membrane (73, 74), anion exchanger representing Sephadex with diethylaminoethyl groups (67), small glass balls with complex-forming groups of 8-hydroxyquinoline (70), sorbent with functional groups of salicyclic acid (77), complex-forming sorbent Spheronoxine (243), polymeric thioether (243, 294), Rexyn 201 (257), anion exchanger JRA-900 (294), complex-forming cellulose sorbent Hyphan (294), on sorbents based on aminopolystyrenepolyarsenazo and polypyridylazoresorcinol (304), ion exchangers of the iminodiacetic type (304), sorbent Polyorgs XI-H (305).

Precipitation as dibenzyldithiocarbamate (241), preconcentration by precipitating the matrix (80), coprecipitation with metal hydroxides (86), $Bi(OH)_3$ (86), $Mg(OH)_2$ and $Fe(OH)_3$ (87), Bi_2S_3, $Bi(OH)_3$, In_2S_3 (212), coprecipitation of sulphide in the presence of cellulose (89, 241), with AgI in the presence of 1,10-phenanthroline (89), coprecipitation of thioxinate with 8,8'-diquinolyldisulfide (92), diethyldithiocarbamate with diethyldithiocarbamates Co, Cu, Fe and Zn (92), dibenzyldithiocarbamate with dibenzyldithiocarbamate of silver and nickel (92, 93, 294, 295), butylxantate with napthalene (92), 1-nitroso-2-naphtholate with 1-nitroso-2-naphthol (93), thiocyanate or iodide, crystalline violet with 2,4-nitroaniline (93), with $Fe(OH)_3$ or iron tetramethylenedithiocarbamate (231), co-precipitation with molybdenum pyrrolidinedithiocarbamate (241), CuS (298), 1-(2-pyridylazo)-2-naphthyl alcohol (295).

Electrodeposition (97, 100, 105, 263, 264, 298), cementation (107).

Fractional volatilization (215, 216, 229), volatilization as Cd (128), $CdCl_2$ (128), CdO (128) after chemical transformation, preconcentration by volatilization of the matrix (118, 119, 121, 174, 292, 296, 314, 318), as borate ester (130), $Ni(CO)_4$ (130), by dry (124, 230, 231, 333) and wet (126, 219, 230, 231, 336) ashing of the organic matrix.

Directed crystallization of matrix (142, 174, 240), zone melting (145).

Flotation (158, 159, 232, 295).

Preconcentration by partial dissolution of the matrix (110, 111).

Dialysis (164).

Calcium

Preconcentration by extraction of the matrix (174, 212, 319, 320, 323, 324, 327).

Sorption on cation exchanger KU-2-8 (60), anion exchanger Bio-Rad Ag MP-1 (61), complex-forming sorbent Wofatit MC-50 (330).

Preconcentration by precipitating the matrix (80).

Electrodialysis (111, 112).

Preconcentration by volatilization (119, 120, 174, 292, 315) and sublimation (122) of the matrix, volatilization of the matrix as borate ester (130), TiF_4 (130), SiF_4 (130), $GeCl_4$ (130), by dry (123–126, 230, 231, 232) and wet (216, 217, 230, 231) ashing of the organic matrix.

Dialysis (164).

Filtration (300).

Carbon

Absorption of CO_2 (176).

Precipitation as carbonate (176).

Preconcentration by electrodissolution of the matrix (110).

Volatilization after chemical transformations (311).

Chlorine

Precipitation and co-precipitation as Cl^- (176) and ClO_4^- (177), co-precipitation Cl^- with $BaCrO_4$ (89).

Volatilization as Cl_2 after chemical transformations (128), preconcentration by dry (230, 231) and wet (230, 231) ashing of the organic matrix.

Preconcentration by directed crystallization of the matrix (142, 144).

Filtration (300).

Chromium

Extraction of hydroxyquinolinate (320, 326), with molten 8-hydroxyquinoline (38), molten diethylammonium diethyldithiocarbamate (38), of diethyldithiocarbamate (44, 235),

pyrrolidinedithiocarbamate (273, 296, 297), tetramethyldithiocarbamate and hexamethylenedithiocarbamate (220); preconcentration by extraction of the matrix (174, 212, 226, 319, 320, 323, 324, 326, 327).

Sorption on oxidized charcoals (53, 56), of 8-hydroxyquinolinate on the anion exchangers AV-17 and AV-18 (73), on activated charcoals (54, 55, 297, 298), sorption on metatitanic acid (298), hydrated Sb_2O_5 (75), on cation exchanger KU-2-8 (60), anion exchanger AV-17 (61), on modified cellulose with different functional groups (66, 67), sorption of $Cr_2O_7^{2-}$ on porous glass balls treating N-β-aminoethylaminopropyltrimethoxy-silane (69), ion exchangers of the iminodiacetic type (304), sorbent Polyorgs XI-H (305).

Precipitation of thiooxinate (82), pyrrolidinedithiocarbamate (82), diethyldithiocarbamate (241), preconcentration by precipitating of the matrix (80, 81), co-precipitation of diethyldithiocarbamate with diethyldithiocarbamates of Co, Cu, Fe and Zn (92), pyrrolidinedithiocarbamate of cadmium or cobalt (294), 1-(2-pyridylazo)-2-naphthol alcohol (295), $Bi(OH)_3$ (86), $Al(OH)_3$ (88, 89), $Ni(OH)_2$ (88, 89), Bi_2S_3, $Bi(OH)_3$ and In_2S_3 (212), $Ti(OH)_4$ (241).

Electrodeposition (242, 298).

Volatilization of chloride after chemical transformation (128), fractional volatilization (215), preconcentration by volatilization (118, 120, 121, 174, 315, 318) and sublimation (122) of the matrix, as TiF_4 (130), As_2O_3 (130), by dry (123–125, 230, 231, 233) and wet (126, 230, 231, 336) ashing of the organic matrix.

Directed crystallization of matrix (142, 143, 174), zone melting (145).

Flotation (159).

Filtration (160, 161, 300).

355

Cobalt

Extraction with molten 8-hydroxyquinoline (37), of pyrrolidinedithiocarbamate with molten stearic alcohol (38, 244), pyrrolidinethiocarbamate (44, 45, 296, 319), diethyldithiocarbamate (44, 45, 235, 245, 302, 319), tetramethyldithiocarbamate and hexamethylenedithiocarbamate (220), 8-hydroxyquinolinate (322–324, 326), 2-isopropyl-8-mercaptoquinolinate (297), preconcentration by extraction of the matrix (174, 212, 226, 319, 320, 323, 324, 326, 327).

Sorption on activated charcoal (53), of the chelates with diethyldithiocarbamate (54, 328), hexamethylenedithiocarbamate (54), xanthate (55), 8-hydroxyquinoline (55, 298) on activated charcoals, sorption with hydrated Sb_2O_5 (75), manganese or zinc sulphide (244), metatitanic acid (298), on cation exchanger KU-2-8 (60), Dowex A-1 (60), anion exchanger AV-17 (61), sorbent Chelex-100 (63, 78, 229) cotton precoated with iminodiacetylcellulose (66), sorption on modified cellulose containing 1-(2-hydroxyphenylazo)-2-naphthol or 4-(2-pyridylazo)-resorcinol (66), dithiocarbamate derivatives of cellulose (67), cellulose containing azorhodanine groups (67), small glass balls with complex-forming groups of 8-hydroxyquinoline immobilized on their surface (70), polyester polyurethane foam containing 1-(2-pyridilazo)-2-naphthol (74), Dowex 1x12 (239), on polymeric thioether (243), sorbents based on aminopolystyrenepolyarsenazo and polypyridylazoresorcinol (304), ion exchangers of the iminodiacetic type (304), sorbent Polyorgs XI-H (305).

Preconcentration by co-precipitation with $Bi(OH)_3$ (86), $Al(OH)_3$ (88, 89), sulphide in the presence of cellulose (89), AgI in the presence of 1,10-phenantroline (89), Bi_2S_3, $Bi(OH)_3$ and In_2S_3 (212),

$Ti(OH)_4$ (241), KF (330), with alizarin blue (82), of diethyldithiocarbamate and thiooxinate (82), diethyldithiocarbamate (82, 241), dibenzyldithiocarbamate with dibenzyldithiocarbamates of silver and nickel (92, 295), 2-naphthol (92), $Fe(OH)_3$ and iron tetramethylenedithiocarbamate (231), dibenzyldithiocarbamate (241), molybdenum pyrrolidinethiocarbamate (241), cupferrate (82), 8-hydroxyquinolinate (82), of thiooxinate with 8,8'-diquinolildisulphide (92), tetrahydrofurfurylxantate with naphthalene (93), 1-nitroso-2-naphtholate with 1-nitroso-2-naphthol (93), with 1-(2-pyridylazo)-2-naphthyl alcohol (295).

Fractional volatilization (215), preconcentration by volatilization of the matrix (118–121, 174, 292, 296, 314, 318), dry (123–125, 230, 231, 332, 333) and wet (126, 230, 231) ashing of the organic matrix.

Electrodeposition (97, 103, 104, 242, 298), electrodialysis (111), electroosmosis (113).

Preconcentration by directed crystallization of matrix (142, 143, 174).

Flotation (158).

Filtration (160, 161, 295, 300).

Dialysis (164).

Preconcentration by partial dissolution of the matrix (110, 111).

Copper

Extraction of complex with 3-(4-phenyl-2-pyridyl)-5,6-diphenyl-1,2,4-triazine and tetraphenylborate with molten naphthalene (38), diethyldithiocarbamate (44, 235, 245, 250, 266, 302, 319), pyrrolidinedithiocarbamate (45, 228, 296), tetramethyldithiocarbamate and hexamethylenedithiocarbamate (220), pyrrolidinedithiocarbamate with molten stearic alcohol (245), with O-isopropyl-N-methylthiocarbamate (226), O,O-dialkyl-S-thiocarbomoildithiophosphate (266), of dithizonate (247), 8-hydroxyquinolinate

(322–324, 326), with neocupreine (268), as pyridinesalicyalate complex (269), with 2-isopropyl-8-mercaptoquinoline (297), trioctylamine (226), Aliquat 336 (307), diphenylthiourea (28, 226, 323), tetrazamacrocyclic extractants (228), preconcentration by extraction of the matrix (174, 212, 320, 323, 324, 326, 327).

Sorption on activated charcoal (53, 330), of a chelate with diethyldithiocarbamate (54, 328), diethyldithiophosphate (54), 8-hydroxyquinolinate (54, 298), hexamethylenedithiocarbamate (54), xanthate (55), xylenol orange (55) on activated charcoals, sorption on manganese or zinc sulphide (244), metatitanic acid (298), cation exchanger KU-2-8 (60, 61), Dowex A-1 and 1x12 (60, 77, 239), anion exchanger Bio-Rad Ag MP-1 (60), on a cotton precoated with iminodiacetylethylcellulose (65, 66), Chelex-100 (78, 294), on cellulose filters containing chromotropic acid group (66), sorption on modified cellulose containing 1-(2-hydroxyphenylazo)-2-naphthol or 4-(2-pyridylazo)resorcin (66), dithiocarbamate derivatives of cellulose (67), cellulose containing azorhodanine groups (67), complex-forming cellulose sorbent Hyphan (294), anion exchanger representing Sephadex with diethylaminoethyl groups (67), silica gels with complex-forming groups of β-diketones immobilized on their surface (70), of 8-hydroxyquinolinate on silica gel grafted with Bondapak Porasil B (293), modified 8-hydroxyquinoline silica (77), small glass balls with complex-forming groups of 8-hydroxyquinoline immobilized on their surface (70), glass packing immobilized with 8-hydroxyquinoline (76, 77), of phosphate, dimethylglyoximate, 8-hydroxyquinolinate on the anion exchangers AV-17 and AV-18 (73), complex-forming sorbent Spheronoxine (243), on transparent membranes (73, 74),

sorption on a sorbent obtained from titanium hydroxide, acrylamide and N,N'-methylenediacrylamide (75), sorbent with inoculated functional groups of salicylic acid (76, 77), 2-mercaptobenzothiazole coated with the aid of collodion (294), ion exchanger Amberlite CG-400 (239), Chromosorb W-HP (243), polymeric thioether (243, 294), ion exchangers of the iminodiacetic type (304), sorbent Polyorgs XI-H (305), Wofatit MC-50 (330).

Precipitation with alizarin blue (82), of pyrrolidinedithiocarbamate (82) diethyldithiocarbamate (82, 241), dibenzyldithiocarbamate (241), cupferrate (82), thiooxinate (82), 8-hydroxyquinolinate (82), preconcentration by precipitating the matrix (80), co-precipitation with $Bi(OH)_3$ (86), Bi_2S_3, $Bi(OH)_3$ and In_2S_3 (212), $Ti(OH)_4$ (241), KF (330), $Mg(OH)_2$ and $Fe(OH)_3$ (87, 241), $Ni(OH)_3$ (88), sulphide in the presence of cellulose (89, 241), AgI in the presence of 1,10-phenanthroline (89), of thiooxinate with 8,8'-diquinolildisulphide (92), diethyldithiocarbamate with diethyldithiocarbamates of Co, Fe and Zn (92), pyrrolidinedithiocarbamate with pyrrolidinedithiocarbamate Co(III) (92), dibenzyldithiocarbamate with dibenzyldithiocarbamates of silver and nickel (92, 295), $Fe(OH)_3$ or iron tetramethylenedithiocarbamate (231), with pyrrolidinedithiocarbamate of cadmium or cobalt (295), with molybdenum pyrrolidinedithiocarbamate (241).

Electrodeposition (98, 100, 104, 105, 242, 263, 264, 298), cementation (106, 107), preconcentration by electrodissolution of the matrix (110), electrodialysis (112), electroosmosis (113, 114), electrodiffusion (115).

Fractional volatilization (215, 216, 229), preconcentration by volatilization (119, 120, 174, 216, 292, 296, 314, 315, 318) and sublimation (122) of the matrix, by volatilization of the matrix as TiF_4 (130), SiF_4 (130), by dry (123–126, 333) and wet (125, 126, 219) ashing of the organic matrix.

Preconcentration by directed crystallization of the matrix (140, 142, 144, 174, 240) and zone melting (145).

Pyrometallurgical slag formation (153).

Flotation (158, 159, 295).

Preconcentration by partial dissolution of the matrix (110).

Filtration (173, 302), dialysis (164).

Fluorine

Volatilization as F_2 after chemical transformation (128, 176).

Preconcentration by directed crystallization (267).

Gallium

Extraction of 8-hydroxyquinolinate (323, 324), with molten 8-hydroxy-quinoline (38), $HGaCl_4$ with 2,2′-dichlordiethylether (33, 35, 43), methylisobutylketone (226), preconcentration by extraction of the matrix (212, 323, 324, 326, 327).

Sorption on TiO_3 (74), metatitanic acid (298), cation exchanger KU-2 (61), anion exchanger AV-17 (61), polyurethane foam (74), sorbents based on aminopolystyrenepolyarsenazo and polypyridylazoresorcinol (304).

Precipitation of pyrrolidinedithiocarbamate (82), of diethyldithiocarbamate (82), preconcentration by precipitating the matrix (80), co-precipitation with $Mn(OH)_4$ (88, 89), with induline hydrochlorine (88).

Electrodeposition (263).

Volatilization as $GaCl_3$ (128, 316) after chemical transformation, fractional volatilization (215), preconcentration by volatilization of the matrix (110, 121, 174,

314), by volatilization of the matrix after chemical transformation (130, 315).

Zone melting (145).

Germanium

Extraction of thiocyanate complex of diantipyrilmethane (38), molybdogermanium heteropoly acid (236).

Sorption on TiO_2 (74) or metatitanic acid (298).

Co-precipitation with $Al(OH)_3$ (88).

Electrodeposition (263).

Volatilization as $GeCl_4$ and GeF_4 (128), GeH_4 (134, 136, 218), preconcentration by volatilization of the matrix (121).

Flotation (158).

Gold

Extraction of the cationic Au(III) 8-mercaptoquinolinate in the presence of trichloracetate (26, 324), of diphenylthiourea (27, 28), with organic sulphides (28–30, 42–44, 256, 258, 259, 306), O-isopropyl-N-methylthiocarbamate (35), $AuCl(Ph_3P)$ (29, 65), $HAuX_4$ with 2,2′-dichlordiethyl ether (33), of 8-hydroxyquinolinate (38), thiocyanate complex of diantipyrilmethane (38), preconcentration by extraction of the matrix (212, 226, 327), by a mixture of petroleum sulphides and O-isopropyl-N-methylthiocarbamate (227), with petroleum sulphides (266), with dihexylsulphide (267), di-2-ethylhexyldithiophosphoric acid and *n*-octylamine (307).

Sorption on activated charcoals or sulphonated coal (53), cation exchanger KU-2-8 (60), polyurethane foam (74), ion exchanger Srafyon NMRR (76), CuS and complex-forming sorbent (212), on Monivex ion exchanger (221, 305), polymeric thioether (243), anion exchanger JRA-900 (294), sorbent Polyorgs XI-H (305).

Precipitation of dithizonate (82), with

diphenylthiourea (82), co-precipitation with $Bi(OH)_3$ (86), elemental S (89), thiocyanate or iodide, crystalline violet with 2,4-nitroaniline (93).

Electrodeposition (105, 263), cementation (107).

Volatilization as $AuCl_3$ (128), preconcentration by volatilization (119, 120) and sublimation (121, 122) of the matrix, by volatilization of the matrix as $GeCl_4$ (130, 315), chlorides of antimony (174), preconcentration by wet ashing of the organic matrix (219).

Fire assay (124, 147, 149–153).

Flotation (159).

Preconcentration by partial dissolution of the matrix (110, 111).

Hafnium

Sorption of 8-hydroxyquinolinate on activated charcoals (54, 55, 297).

Co-precipitation of 8-hydroxyquinolinate with indifferent organic co-precipitants (92).

Volatilization as $HfCl_4$ and HfF_4 after chemical transformation (128), preconcentration by volatilization of the matrix (314).

Indium

Extraction of $HInBr_4$ with 2,2'-dichlorodiethyl ether (33, 35), of $InBr_4^-$ with brilliant green (36), with tridecylamine (43), methylisobutylketone (226), of 8-hydroxyquinolinate (326), of diethyldithiocarbamate (319), preconcentration by extraction of the matrix (174, 212, 319, 324, 327, 328).

Sorption on activated charcoal (53), on TiO_2 (74), on metatitanic acid (298), of diethyldithiocarbamate (54, 328), hexamethylenedithiocarbamate (54), diethyldithiophosphate (54), 8-hydroxyquinolinate (55), of chelate with xylenol orange (55), on polymeric thioether (243, 294).

Precipitation with alizarine blue (82), preconcentration by precipitating the matrix (80), co-precipitation with metal hydroxides (86), $Al(OH)_3$ (88), $Mn(OH)_4$ (88).

Electrodeposition (105, 263), cementation (107).

Volatilization as $InCl_3$ after chemical transformation (128), fractional volatilization (215), preconcentration by volatilization of the matrix (121, 174, 216, 314), by volatilization of the matrix after chemical transformation (130, 315).

Preconcentration by directed crystallization of the matrix (140, 174, 240) and zone melting (144, 145).

Fire assay (152, 153).

Preconcentration by partial dissolution of the matrix (110, 111).

Iodine

Extraction of I_2 (177).

Sorption on activated charcoals (53).

Precipitation as I^- (177).

Volatilization as I_2 after chemical transformation (128).

Iron

Extraction of $HFeCl_4$ (33, 327), with molten 8-hydroxyquinoline (38), of pyrrolidinedithiocarbamate with molten stearic alcohol (38), of thiocyanate complex of diantipyrylmethane (38), diethyldithiocarbamate (44, 45, 235, 245, 319), pyrrolidinethiocarbamate (44, 228, 296), tetramethyldithiocarbamate and hexamethylenedithiocarbamate (220), preconcentration by extraction of the matrix (212, 226, 319, 320, 323, 324, 326, 327).

Sorption on activated charcoal (53, 330), of diethyldithiocarbamate (54), 8-hydroxyquinolinate (55, 298), xanthate (55), of a chelate Fe(III) with xylenol orange (55), on activated charcoals, on a sorbent obtained from $Ti(OH)_4$,

359

acrylamide and N,N'-methylenediacrylamide (75), metatitanic acid (298), on anion exchanger AV-17 (61), anion exchanger Bio-Rad Ag MP-1 (61), of phosphate and 8-hydroxyquinolinate on the anion exchangers AV-17 and AV-18 (73), Chelex-100 (63, 78, 294), cation exchanger KU-2-8 (60, 61), Dowex 50 W x 12 (239), complex-forming sorbent Spheronoxine (243), polymeric thioether (243), ion exchangers of the iminodiacetic type (304), sorbent Polyorgs XI-H (305), polyester polyurethane foam containing 1-(2-pyridylazo)-2-naphthol (74), cotton coated with iminodiacetylethylcellulose (65, 66), cellulose filter containing chromotropic acid group (66), modified cellulose containing 1-(2-hydroxy-phenylazo)-2-naphthol or 4-(2-pyridylazo)resorcin (66), dithiocarbamate derivatives of cellulose (66, 67) cellulose containing azorhodanine groups (67), silica gel with complex-forming groups of β-diketones immobilized on its surface (70), precipitation with alizarin blue (82), of pyrrolidinedithiocarbamate (82), dibenzyldithiocarbamate (241), of diethyldithiocarbamate (82, 241), thiooxinate (82), diethyldithiocarbamate, cupferrate and 8-hydroxyquinolinate (82), preconcentration by precipitating the matrix (80), co-precipitation with Ni, Co, Zn and Cu phosphates (84), Cd, Mn, Pb and Mg phosphates (84), AgI in the presence of 1,10-phenantroline (89), $Ti(OH)_4$ (241), $La(OH)_3$ (329), KF (330), diethyldithiocarbamate with diethyldithiocarbamates of Co, Cu and Zn (92), hexamethylenedithiocarbamate with hexamethyleneammonium hexamethylenedithiocarbamate (92), dibenzyldithiocarbamate with dibenzyldithiocarbamates of silver and nickel (92, 93, 96), molybdenum pyrrolidine dithiocarbamate (241) pyrrolidinedithiocarbamate of cadmium or cobalt (295).

Electrodeposition (97, 100, 105, 106), electrodialysis (112), electroosmosis (113), preconcentration by electrodissolution of the matrix (113).

Volatilization as $FeCl_3$ (128), fractional volatilization (325), preconcentration by volatilization (118–120, 174, 292, 296, 314) and sublimation of the matrix (122), by volatilization of the matrix after chemical transformation (130, 315, 318), dry (123–125, 230, 231, 333) and wet (126, 219, 230, 231) ashing of the organic matrix.

Preconcentration by directed crystallization of matrix (142, 143, 174, 240), and zone melting (144, 145).

Pyrometallurgical slag formation (153).

Filtration (173, 300).

Lanthanides

Extraction with di-(2-ethylhexyl)phosphoric acid (43), with thenoylthrifluoacetone and 1,10-phenanthroline (307), preconcentration by extraction of the matrix (320).

Sorption of lanthanides (55), Dy, Eu, Hf, Lu, Sm, Tb, and Yb (298) 8-hydroxyquinolate on activated charcoals, Ce, Tb and Yb on hydrated Sb_2O_5 (75), La, Ce and Pr on the anion exchanger Bio-Rad IX-10 (60), AV-17 (61), Dowex 1 x 8 (61), on the ion exchanger Dowex 50W x 12, Dowex 1 x 8 and Dowex 50W x 8 (239, 308), the cation exchangers AG50W x 8 (249), La and Yb on sorbents based on amino-polystyrenepolyarsenazo and polypyridylazoresorcinol (304).

Preconcentration of Ce and La by precipitating the matrix (81), co-precipitation of Ce with metal hydroxides (86), lanthanides with $Fe(OH)_3$ (88, 89), $Mg(OH)_2$ (88, 89), mixed oxalates (308).

Preconcentration of lanthanides (101) and La (101, 102) by electrodeposition of matrix, Eu, Lu, Sm, Tb and Tm by electrophoresis (114).

Volatilization of Ce as chloride (128), preconcentration of Ce, Eu and La (120), La (174), lanthanides (314) by volatilization of the matrix.
Dialysis of Eu, La and Lu (164).

Lead

Extraction with molten 8-hydroxyquinoline (38), thiocarbamate with molten stearic alcohol (38, 244), of pyrrolidinethiocarbamate (43–46, 228), diethyldithiocarbamate (245, 302, 319), tetramethyldithiocarbamate and hexamethylenedithiocarbamate (220), dithizonate (247), 8-hydroxyquinolinate (322–324, 326, 327), oximates (230, 231), with 18-crown-6 (237), O,O-dialkyl-S-thiocarbamoyldithiophosphate (266), Aliquat 336 (307), preconcentration by extraction of the matrix (174, 212, 319, 320, 323).

Sorption on activated charcoals (53, 330), of hexamethylenedithiocarbamate (54), diethyldithiocarbamate (54), diethyldithiophosphate (54), 8-hydroxyquinolinate (54, 55, 298), xanthate (55), of Pb chelate with Xylenol orange (55) on activated charcoals, metatitanic acid (298), sorption on cation exchanger KU-2-8 (60, 61), Dowex A-1 (60, 77), anion exchanger EDE-10M (61), cotton precoated with imino-diacetylcellulose (65, 66), modified cellulose containing 1-(2-hydroxy-phenylazo)-2-naphthol or 4-(2-pyridylazo)resorcin (66), dithiocarbamate derivatives of cellulose (67), anion exchanger representing Sephadex with diethylaminoethyl groups (67), small glass balls with complex-forming groups of 8-hydroxyquinoline immobilized on their surface (70, 77), sorption on phosphate and 8-hydroxyquinolinate on the anion exchangers AV-17 and AV-18 (73), sorbent with inoculated functional groups of salicyclic acid (76, 77), Chelex-100 (78, 294), complex-forming sorbent

Spheronoxine (243), polymeric thioether (243, 294), 2-mercaptobenzothiazole coated with the aid of collodion (294), complex-forming cellulose sorbent Hyphan (294), ion exchangers of the iminodiacetic type (304), Wofatit MC-50 (330).

Precipitation as sulphate (79, 80), nitrate (79, 80), diethyldithiocarbamate and thiooxinate (82), pyrrolidinedithiocarbamate (82), dibenzyldithiocarbamate (241), co-precipitation with $Bi(OH)_3$ (86), $Mg(OH)_2$ and $Fe(OH)_3$ (86, 87, 241, 264), $Al(OH)_3$ (88, 89), $Mn(OH)_4$ (88, 89), Bi_2S_3, $Bi(OH)_3$ and In_2S_3 (212), $Fe(OH)_3$ or iron tetramethylenedithiocarbamate (231, 232), CuS (298), $La(OH)_3$ (329), diethyldithiocarbamate with diethyldithiocarbamates Co, Cu, Fe and Zn (92), pyrrolidinedithiocarbamate Co(III) (92), dibenzyldithiocarbamate with dibenzyldithiocarbamates of silver and nickel (92, 295), molybdenum pyrrolidinedithiocarbamate (241), pyrrolidinedithiocarbamate of cadmium or cobalt (295), butylxantate with naphthalene (92).

Electrodeposition (97, 100, 104–106, 235, 263, 264, 298), cementation (106, 107), electrodiffusion (115), preconcentration by electrodissolution of the matrix (109).

Volatilization as chlorides (128), Pb (128), hydrides (128, 132, 136, 218, 230), fractional volatilization (215, 216, 229), preconcentration by volatilization (118, 119, 121, 174, 292, 296) and sublimation (122) of the matrix, by volatilization of the matrix after chemical transformation (130, 315, 318), by dry (123–125, 230, 231, 333) and wet (126, 219, 231, 335, 336) ashing of the organic matrix.

Preconcentration by direct crystallization of matrix (140, 142, 143, 174) and zone melting (144, 145).

Flotation (158).

Preconcentration by partial dissolution of the matrix (110).

Filtration (160, 300), dialysis (164).

Magnesium

Preconcentration by extraction of the matrix (174, 212, 319, 320, 323, 324, 326, 327).

Sorption on the cation exchanger KU-2-8 (60), modified cellulose containing 1-(2-hydroxyphenylazo)-2-naphthol or 4-(2-pyridylazo)resorcin (66), complex-forming sorbent Wofatit MC-50 (330).

Preconcentration by precipitating the matrix (280), co-precipitation with $Ni(OH)_3$ (88).

Preconcentration by volatilization (119, 174, 292, 315, 318, 329) and sublimation (121, 122) of the matrix, by volatilization of the matrix as TiF_4 (130), SiF_4 (130), $AsCl_3$ and $AsBr_3$ (130) by dry (123, 124, 230, 231) and wet (218, 231) ashing of the organic matrix.

Preconcentration by direct crystallization of the matrix (143, 174).

Filtration (300).

Manganese

Extraction with molten 8-hydroxyquinoline (38), of 8-hydroxyquinolinate (322, 326), diethyldithiocarbamate (44, 235, 245, 302, 319), of tetramethylenedithiocarbamate (220), pyrrolidinedithiocarbamate (296, 297), preconcentration by extraction of the matrix (174, 212, 226, 319, 320, 323, 324, 326, 327).

Sorption on activated charcoal (53, 329), of 8-hydroxyquinolinate on activated charcoal (55, 298), metatitanic acid (298), on cation exchanger KU-2-8 (60), anion exchanger AV-17 (61), cotton coated with iminodiacetylethylcellulose (65, 66), modified cellulose containing 1-(2-hydroxyphenylazo)-2-naphthol or 4-(2-pyridylazo)resorcin (66), dithiocarbamate derivatives of cellulose (66, 67), complex-forming cellulose sorbent Hyphan (295), MnO_4^{2-} on porous glass balls treating by N-β-aminoethylaminopropyltrimethoxy-silane (69), transparent membranes (74),

polyester polyurethane foam containing 1-(2-pyridylazo)-2-naphthol (74), on sorbent Dowex-A-1 (77), Chelex-100 (78), complex-forming sorbent Spheronoxine (243), of 8-hydroxyquinolinate on silica gel grafted with Bondapak Porasil B (293), ion exchangers of the iminodiacetic type (304), sorbent Wofatit MC-50 (330).

Preconcentration of diethyldithiocarbamate and thiooxinate (82), diethyldithiocarbamate (82), dibenzyldithiocarbamate (241), preconcentration by precipitating the matrix (80), co-precipitation with $Bi(OH)_3$ (86), $Al(OH)_3$ (88, 89), $Ni(OH)_3$ (88, 89), $Mg(OH)_2$ (88, 89), Bi_2S_3, $Bi(OH)_3$ and In_2S_3 (212), $Fe(OH)_3$ (241), KF (330), with 8,8'-diquinolildisulphide (92), pyrrolidinedithiocarbamate Co(III) (92), of 1-nitroso-2-naphtholate with 1-nitroso-2-naphthol (93), 1-(2-pyridylazo)-2-naphthyl alcohol (295).

Electrodeposition (104, 263, 264), preconcentration by electrodissolution of the matrix (109).

Volatilization as chlorides after chemical transformation (128), fractional volatilization (216), preconcentration by volatilization (118–120, 174, 216, 292, 296, 314, 315, 318), and sublimation (122) of the matrix, by volatilization of the matrix as TiF_4 (130), SiF_4 (130), $AsBr_3$ and $AsCl_3$ (130), by dry (123–125, 333) and wet (219, 326, 327) ashing of the organic matrix.

Preconcentration by direct crystallization of the matrix (142–143, 174, 240).

Flotation (159).

Filtration (160, 173, 300).

Mercury

Extraction of HgI_2 (236), with triphenylphosphine (29), diphenylthiourea (28), O-isopropyl-N-ethylthiocarbamate (35), with macrocyclic Schiff bases (238), of dithizonate (245, 297), dibutyldithiophosphate (266), diethyldithiocarbamate (294), 2-isopropyl-8-mercaptoquinoline (297), thenoyltrifluoroacetonate (297), with astraphloxine FF (306).

Sorption on activated charcoals (53), of 8-hydroxyquinolinate (54, 55, 298), xanthate (55) on activated charcoal, on PbS (75), manganese or zinc sulpide (244), sorption of mercury vapours from air on a gold wire/plate on a gold-plated carrier (301), on cation exchanger KU-2-8 (61), cellulose filter containing chromotropic acid group (66), dithiocarbamate derivatives of cellulose (66, 67), silica gel with complex-forming groups of 2-amino-1-cyclopentane-1-dithiocarboxylic acid immobilized on their surface (70), sodium diethyldithiocarbamate coated on Chromosorb W-DMCS (73), dithizone and thenoyltrifluoroacetone gels coated on a polystyrene (2% divinylbenzene) (73), transparent membranes (74), polymeric thioether (243).

Precipitation of Hg(I) sulphate (79), dibenzyldithiocarbamate (241), co-precipitation sulphide in the presence of cellulose (89, 241), with CuS (298), of thiooxinate with 8,8′-diquinolildisulphide (92), diethyldithiocarbamate with diethyldithiocarbamates Co, Cu, Fe and Zn (92).

Electrodeposition (100, 105, 231, 235, 242, 298).

Volatilization as chlorides and fluorides (128), oxides (128), Hg (128, 130, 131, 216, 234) after chemical transformation, preconcentration by dry (122, 124, 125, 230) and wet (231, 335, 336) ashing of the matrix.

Flotation (158, 295).

Filtration (161, 300, 301).

Molybdenum

Extraction with trioctylamine N-oxide (33), trioctylamine (226), Aliquat 336 (307), of thiocyanate complex with diantipyrylmethane (38), Mo(VI) with propylene carboxylic acid (39), of tetramethyldithiocarbamate with hexamethylenedithiocarbamate (220), diethyldithiocarbamate (302), 4-methyl-8-mercaptoquinolate (266), with α-benzoinoxime (307), of MoO_4^{2-} with dialkyltin dinitrates (328), preconcentration by extraction of the matrix (226, 320).

Sorption on oxidized charcoals (53), of 8-hydroxyquinolinate on activated charcoals (55), on TiO_2 (74, 75), metatitanic acid (298), Chelex-100 (63), modified cellulose containing 1-(2-hydroxyphenylazo)-2-naphthol or 4-(2-pyridylazo)resorcin (66).

Precipitation of diethyldithiocarbamate and thiooxinate (82), diethyldithiocarbamate (82), cupferrate (82), 8-hydroxyquinolinate (82), precipitation and co-precipitation as molybdate (177), co-precipitation of dibenzyldithiocarbamate with dibenzyldithiocarbamates of silver and nickel (92), with pyrrolidinedithiocarbamate of cadmium or cobalt (294, 295).

Volatilization as chlorides (128, 316) and fluorides (128), oxides (128), preconcentration by volatilization of the matrix (118, 120, 292, 315), by dry (230, 231) and wet (230, 231) ashing of the matrix.

Preconcentration by directed crystallization of the matrix (142).

Flotation (159, 295).

Filtration (160, 177).

Nickel

Extraction with molten 8-hydroxyquinoline (38), of 8-hydroxyquinolinate (322, 324, 326), of pyrrolidinedithiocarbamate with molten stearic alcohol (38, 244), oximates with

363

molten naphthalene (237), of pyrrolidinedithiocarbamate (45, 297), of tetramethyldithiocarbamate and hexamethylenedithiocarbamate (220), diethyldithiocarbamate (245, 302, 319), of dimethylglyoximate (266), preconcentration by extraction of the matrix (174, 212, 226, 319, 320, 323, 324, 326, 327).

Sorption on activated charcoals (53), of diethyldithiocarbamate (54, 328), hexamethylenedithiocarbamate (54), 8-hydroxyquinolinate (55, 242, 298), and xanthate (55) on activated charcoals, on manganese or zinc sulphide (244), metatitanic acid (298), on cation exchanger Dowex A-1 (60), cation exchanger KU-2-8 (60), sorption of phosphate, dimethylglyoximate and 8-hydroxyquinolinate on the anion exchangers AV-17 and AV-18 (73), on sorbent Chelex-100 (63, 78, 294), the modified cellulose containing 1-(2-hydroxyphenylazo)-2-naphthol or 4-(2-pyridylazo)resorcin (66), cellulose containing azorhodamine groups (67), complex-forming cellulose sorbent Hyphan (294), small glass balls with complex-forming groups of 8-hydroxyquinoline immobilized on their surface (70, 77), on a sorbent obtained from titanium hydroxide, acrylamide and N,N-methylenediacrylamide (75), sorbent with inoculated functional groups of salicylic acid (77), complex-forming sorbent Spheronoxine (243), polymeric thioether (243, 294), ion exchangers of the iminoacetic type (304), sorbent Polyorgs XI-H (305), Wofatit MC-50 (330).

Precipitation of dibenzyldithiocarbamate (241), pyrrolidinedithiocarbamate (82), diethyldithiocarbamate, cupferrate and 8-hydroxyquinolinate (82), preconcentration by precipitating of the matrix (80), co-precipitation with $Bi(OH)_3$ (86), $Mg(OH)_2$ and $Fe(OH)_3$ (87), $Mn(OH)_4$ (88, 89), AgI in the presence of 1,10-phenantroline (88, 89), Bi_2S_3, $Bi(OH)_3$ and In_2S_3 (212), $Fe(OH)_3$ or iron tetramethylenedithiocarbamate (231, 232), KF (330), diethyldithiocarbamate with diethyldithiocarbamates of Co, Cu, Fe and Zn (92), hexamethylenedithiocarbamate with hexamethyleneammonium hexamethylenedithiocarbamate (92), dibenzyldithiocarbamate with dibenzyldithiocarbamate of silver (92, 295), pyrrolidinedithiocarbamate of cadmium or cobalt (295), tetrahydrofurfurolxantate with naphthalene (93).

Electrodeposition (100, 104, 242, 298), preconcentration by electrodissolution of the matrix (110), electroosmosis (113).

Fractional volatilization (216), volatilization as $Ni(CO)_4$ (316), preconcentration by volatilization (118, 121, 174, 216, 292, 296), and sublimation (121, 122) of the matrix, by volatilization of the matrix after chemical transformation (130, 315, 318), by dry (123, 124, 231, 332, 333) and wet (126, 231) ashing of organic matrix.

Preconcentration by directed crystallization of matrix (140, 142, 143, 174, 239), and zone melting (145).

Flotation (158).

Preconcentration by partial dissolution of the matrix (110, 111).

Filtration (160, 165).

Dialysis (164).

Niobium

Preconcentration by extraction of the matrix (320).

Sorption of 8-hydroxyquinolinate on activated charcoals (55).

Precipitation of diethyldithiocarbamate and thiooxinate (82).

Volatilization as chloride and fluoride (128), oxides (128) after chemical transformation.

Nitrogen
   Precipitation as nitrate (176).
   Volatilization after chemical
transformation (128, 176).
   Preconcentration by directed
crystallization of the matrix (142).
   Flotation (240, 299).
   Filtration (300).

Oxygen
   Volatilization as O_2 after chemical
transformation (128).

Phosphorus
   Extraction of PO_4^{3-} with dialkyltin
dinitrates (33, 227, 236, 325, 328), as
heteropoly acid (220, 236, 240, 275, 297).
   Sorption on Al_2O_3 (75, 78).
   Volatilization as chlorides and
fluorides (128), P (128), PH_3 (128, 218, 219,
239, 240) after chemical transformation,
preconcentration of dry (230, 231) and
wet (219, 230, 231) ashing of the organic
matrix, by volatilization of the matrix
after chemical transformation (315).

Platinum group metals
   Extraction of Pd, Pt, Rh, Ru and Ir
with diphenylthiourea (27), Os, Pd, Pt,
Rh and Ru with hexa-
methylenephenylthiourea (28, 248), Pd,
Pt, Th and Ir di-*o*-tolylthiourea (28), Pd
(29, 258, 259), Ir, Pd and Pt (29) with
organic sulphides, Pd with di-*n*-
amylsulphide (29), Pd (29) and Pd, Pt and
Os (30) with triphenylphosphine, Ir, Rh
and Ru with *p*-octylaniline (30), Pd, Pt,
Ir, Rh and Ru with *p*-alkylaniline (30),
platinum group metals with 2-
octylaminopyridine (31), Ir with 4-
octylaminopyridine (31), Pt and Pd with
O-isopropyl-N-ethylthiocarbamate (35),
Ir, Pd, Pt, Rh and Ru with di-2-
(ethylhexyl)dithiophosphoric acid (226,
306, 307, 323), preconcentration by

extraction of the matrix [Ir, Pd and Rh
(226), Ru (320)], Ir, Pd, Pt and Ru with
N,N'-hexamethylene-N'-phenylthiourea
(306).
   Sorption of 8-hydroxyquinolinate on
activated charcoals (55), platinum group
metals on P–D sorbent containing
inoculated dithizone (63), chelate sorbent
PVB-MP (64, 65) and Polyorgs® (64, 65),
Ir, Pd, Pt, Rh and Ru on complex-forming
sorbents Polyorgs XI-H (305), sorption on
fiberous sorbents Milton-T (65), silica gel
with complex-forming groups of 5-
methylene-2-(2'-thiazoleazo)-anizol (70), or
2-amino-1-cyclopentane-1-dithiocarboxylic
acid (70) immobilized on its surface,
platinum group metals on polymer
thioester or polymeric tertiary amine (71,
72), preconcentration of platinum group
metals by paper chromatography (73, 74),
Ir, Pd and Pt on ion exchanger Srafyon
NMRR (76), Ir, Pd, Pt, Rh and Ru on
CuS and complex-forming sorbent (212),
Ir, Pd, Pt, Rh and Ru on polymeric
thioether (221), Ir, Pd, Pt, Rh and Ru on
Monivex ion exchanger (221), Pd and Pt
on ion exchanger Bio-Rad AG-50W x 8
(221), platinum group metals on
polymeric tertiary amine (244), Ir and Pd
on anion exchanger JRA-900 (294), Ir, Rh
and Ru on complex-forming sorbent with
3(5)-methyl-pyrazole group (305).
   Precipitation of Pd with alizarin blue,
Pd dithizonate (82), Ir, Pd, Pt, Rh and Ru
with diphenylthiourea (82), silver
cyanide (88), co-precipitation of platinum
group metals with sulphides of a number
of metals (89, 305), Pd and Pt on S (89),
Pt, Pd, Rh and Ir with thiobensamide
and the diphenylamine precipitate (91),
tetrahydrofurfurylxanthate of Ir with
naphthalene (93),
acenaphthequinonedioxime of Pd with
naphthalene (93), of Pd with
pyrrolidinedithiocarbamate of cadmium
or cobalt (295).
   Electrodeposition of Pt (231),
cementation of Ir, Pd and Pt (107),

preconcentration of Pd and Pt by electrodialysis (113).

Volatilization of Os and Ru as chlorides (128), Ir, Os and Ru (128) as fluorides, Ir and Ru as oxides (128), Os and OsO_4 (316) after chemical transformation, fractional volatilization Os (229), preconcentration of Pt by volatilization of the matrix (174), preconcentration by dry [(platinum group metals (124), Rh and Ru (230, 231), and wet [Rh and Ru (230, 231)]ashing of the organic matrix.

Fire assay: Ru and Os (148), Ir, Rh and Ru (149), platinum group metals (149–152), Ir, Pd, Pt, Rh and Ru (150, 151, 153), Ir, Pd and Ru (150, 151, 153), Pd, Pr and Rh (150, 151, 153), Ir, Pd, Pt and Rh (150–153), Os (152), Pt (152), Ir (152), Ir, Os, Pd, Pt, Rh and Ru (152), Pd and Pt (152).

Flotation Os (160), Ir, Pd, Pt, Rh (159).

Preconcentration of Pd by a partial dissolution of a matrix (110, 111).

Plutonium

Sorption on anion-exchanger JRA-900 (294).

Polonium

Volatilization as chlorides (128) and oxides (128) after chemical transformation.

Radium

Ultrafiltration (161, 162).

Rhenium

Extraction trioctylamine oxide (33). Sorption of 8-hydroxyquinolinate on activated charcoals (55, 298).

Volatilization as chlorides and fluorides (128), oxides (128) after chemical transformation, evaporation (175), fractional volatilization (229).

Scandium

Sorption with hydrated Sb_2O_5 (75). Preconcentration by volatilization of the matrix (120). Flotation (158).

Selenium

Extraction of SeO_4^{2-} and SeO_3^{2-} with dialkyltin dinitrates (33, 321, 328), of diethyldithiocarbamate (275, 319), with 3,3'-diaminobenzidine (327), as piazoselenole (271–273), preconcentration by extraction of the matrix (226).

Sorption of 8-hydroxyquinolinate on activated charcoals (55), manganese or zinc sulphide (244), on dithiocarbamate derivatives of cellulose (66, 67), cellulose filter containing functional groups of 2,2'-diaminediethylamine (67), SeO_4^{2-} on porous glass balls treated with N-β-aminoethylaminopropyltrimethoxy-silane (69), polymeric thioether (243), of pyrrolidine dithiocarbamate on silica gel grafted with Bondapak Porasil B (293).

Precipitation and co-precipitation as selenate and selenite (177), precipitation as dibenzyldithiocarbamate (241), co-precipitation with $Fe(OH)_3$ (88, 89, 274), As (89), $La(OH)_3$ (329), sulphide in the presence of cellulose (89, 241), diethyldithiocarbamate with diethyldithiocarbamate Co, Cu, Fe and Zn (92).

Electrodeposition (231, 330), cementation (107).

Volatilization as $SeCl_4$ and $SeBr_4$ (128), oxides (128, 234, 316), Se (128), H_2Se (128, 132, 134–136, 218, 219, 230, 233, 234, 245, 292), preconcentration by volatilization of the matrix (120), by dry (230, 231, 333) and wet (230, 231, 335, 336) ashing of the organic matrix.

Fire assay (150–153).

Flotation (158, 159).

Filtration (300).

Non-electrolytic concentration on a stationary dropping mercury electrode (108, 109).

Silicon

Extraction as molybdosilicic heteropoly acid (176, 236), preconcentration by extraction of the matrix (319, 320).

Precipitation of silicic acid (80, 176). Volatilization as $SiCl_4$ (128, 129), SiF_4 (128), SiH_4 (132), preconcentration by volatilization of the matrix (119, 315), dry ashing of the organic matrix (123). Filtration (300).

Silver

Extraction by diphenylthiourea (27, 28, 226, 306, 307, 323), organic sulphides (29, 44), petroleum sulphides (266), dihexylsulphide (267), by a mixture of petroleum sulphides and O-isopropyl-N-methylthiocarbamate (227), with triphenylphosphine (29), O-isopropyl-N-ethylthiocarbamate (35), O-isopropyl-N-methylthiocarbamate (35, 226), of diethyldithiocarbamate (251, 302, 319), 8-hydroxyquinolinate (38, 326), dithizonate (266), of thiocyanate complex with diantipyrylmethane (38), with nitrogen-containing macrocycles (238), with Aliquat 336 (307), di-2-ethylhexyldithiophosphoric acid and n-octylamine (307), preconcentration by extraction of the matrix (174, 212, 326, 327).

Sorption on activated charcoals or sulphonated coals (53, 330), of diethyldithiocarbamate (54, 328), hexamethylenedithiocarbamate (54), 8-hydroxyquinolinate (55, 242), xanthate (55), on activated charcoals, sorption with hydrated Sb_2O_5 (75), on manganese or zinc sulphide (244), cation exchanger KU-2-8 (60), dithiocarbamate derivatives of cellulose (67), silica gel with complex-forming groups of 2-amino-1-cyclopentane-1-dithiocarboxylic acid (70) or formazines (70), immobilized on its surface, on CuS and complex-forming sorbent (212), on Chromosorb W-HP

(242), polymeric thioether (243), sorbent Polyorgs XI-H (305).

Precipitation with alizarin blue (82), of pyrrolidinedithiocarbamate (82), with diphenylthiourea (82), preconcentration by precipitating the matrix (80), co-precipitation with $Bi(OH)_3$ (86), Bi_2S_3, $Bi(OH)_3$ and In_2S_3 (212), thiooxinate with 8,8'-diquinolildisulphide (92), diethyldithiocarbamate with diethyldithiocarbamates of Co, Cu, Fe and Zn (92), dibenzyldithiocarbamate with dibenzyldithiocarbamates of nickel (92, 96), with pyrrolidinedithiocarbamate of cadmium or cobalt (295), thiocyanate or iodide, crystalline violet with 2,4-nitroaniline (93).

Electrodeposition (106), cementation (106–108).

Fractional volatilization (229), preconcentration by volatilization (118, 119, 174, 296, 314, 315) and sublimation (122) of the matrix, by volatilization of the matrix as SiF_4 (130), $GeCl_4$ (130), $SbCl_3$ (130, 318), SeO_2 (130), pre-concentration by dry (230, 231) and wet (230, 231) ashing of the organic matrix.

Preconcentration by directed crystallization of the matrix (142, 174), by zone melting (145).

Fire assay (147, 150, 152).

Flotation (157).

Filtration (300, 301).

Preconcentration by partial dissolution of the matrix (110, 111).

Strontium

Preconcentration by extraction of the matrix (320, 324).

Sorption with hydrated Sb_2O_5 (75), cellulose filter containing chromotropic acid group (66), cation exchanger Dowex 50 W x 8 (76).

Precipitation of sulphate (79, 80).

Electrodisposition (100).

Preconcentration by volatilization of the matrix (292), by dry (230, 231, 332,

333) and wet (219, 231) ashing of the organic matrix.
Filtration (161, 165, 166).

Sulphur
Extraction of SO_4^{2-} with dialkyltin dinitrates (33, 328).
Precipitation as sulphur (80), sulphate (177), co-precipitation of SO_4^{2-} with $BaCrO_4$ (89).
Electrodeposition of SO_4^{2-} and SCN^- (105).
Volatilization as S (128, 311, 316), H_2S (128) after chemical transformation.
Filtration (300, 301).

Tantalum
Preconcentration by extraction of the matrix (320).
Precipitation of diethyldithiocarbamate and thiooxinate (82).
Volatilization as $TaCl_4$ and TaF_4 (128) after chemical transformation, preconcentration by volatilization of the matrix after chemical transformation (315).

Technetium
Sorption on anion exchanger JRA-900 (294).
Volatilization as chlorides and fluorides (128), oxides (128) after chemical transformation.

Tellurium
Extraction with tributylphosphate (43), butylrhodamine B (326), of diethyldithiocarbamate (275), preconcentration by extraction of the matrix (174, 212, 220, 319, 323, 326, 327).
Sorption on manganese or zinc sulphide (244), dithiocarbamate derivatives of cellulose, polymeric thioether (243).

368

Precipitation of tellurium (80), co-precipitation with $Fe(OH)_3$ (88, 264), $Al(OH)_3$ (88), $Mn(OH)_4$ (89), $La(OH)_3$ (329).
Electrodeposition (263), cementation (107).
Volatilization as chlorides and fluorides (128), oxides and Te (128), hydride (128, 132, 134, 135, 218, 230, 234) after chemical transformation, preconcentration by volatilization of the matrix after chemical transformations (315), by dry ashing (333) of the organic matrix.
Fire assay (151).
Non-electrolytic concentration on a stationary dropping mercury electrode (108, 109).
Flotation (295).
Filtration (300, 301).

Thallium
Extraction with diphenylthiourea (28, 226, 323), of $HTlCl_4$ with 2,2'-dichlordiethyl ether (33), diethyldithiocarbamate (319), preconcentration by extraction of the matrix (212, 319, 323, 327).
Sorption of diethyldithiocarbamate (54, 328), hexamethylenedithiocarbamate (54), 8-hydroxyquinolinate (54, 55, 242) and xanthate (55) on activated charcoals, polymeric thioether (294).
Preconcentration by precipitation of the matrix (80); co-precipitation with AgI in the presence of 1,10-phenanthroline (89).
Electrodeposition (103, 263), cementation (107).
Volatilization of Tl after chemical transformation (128, 315), fractional volatilization (229), preconcentration by volatilization (119, 121, 174) and sublimation (122) of the matrix, by volatilization of the matrix as SiF_4 (130), TlH_3 (136), by dry ashing of the organic matrix (123).

Pyrometallurgical slag formation (153).
Preconcentration by partial dissolution
of the matrix (110, 111).

Thorium
Extraction as a thiocyanate complex
with diantipyrilmethane (38),
preconcentration by extraction of the
matrix (320).
Precipitation with alizarin blue (82).
Flotation (157).

Tin
Extraction of thiocyanate complex
with diantipyrilmethane (38), with
molten 8-hydroxyquinoline (37), of 8-
hydroxyquinolinate (326),
diethyldithiocarbamate (302),
preconcentration by extraction of the
matrix (212, 320, 323, 327).
    Sorption of 8-hydroxyquinolinate on
activated charcoals (55), manganese or
zinc sulphide (244), metatitanic acid
(298), on polymeric thioether (243).
    Precipitation of stannous acid (80), of
diethyldithiocarbamate (82),
preconcentration by precipitating the
matrix (81), co-precipitation with $Bi(OH)_3$
(86), $Al(OH)_3$ (88), $Mn(OH)_4$ (89), $Bi(OH)_3$,
Bi_2S_3 and IN_2S_3 (212), $Fe(OH)_3$ (264, 330),
$La(OH)_3$ (329), sulphide in the presence of
cellulose (89, 241),
hexamethylenedithiocarbamate with
hexamethyleneammonium
hexamethylenedithiocarbamate (92),
dibenzyldithiocarbamate with
dibenzyldithiocarbamates of silver and
nickel (92, 93, 294, 295),
pyrrolidinedithiocarbamate of cadmium
or cobalt (295).
    Electrodeposition (104, 263),
cementation (107), preconcentration by
electrodissolution of the matrix (110).
    Volatilization as chlorides (128, 129,
316), fluorides (128), SnI_4 (128, 131), Sn
(128), SnH_4 (128, 132–136, 218, 230),

fractional volatilization (215, 216),
preconcentration by volatilization of the
matrix after chemical transformation
(130, 315), by volatilization (121, 292) and
sublimation (122) of the matrix, by dry
(123, 124, 230, 231) and wet (219, 230, 231)
ashing of the organic matrix.
    Preconcentration by directed
crystallization of the matrix (140, 142,
174), by zone melting (145).
    Preconcentration by partial dissolution
of the matrix (110, 111).
    Fire assay (153).
    Pyrometallurgical slag formation (153).
    Filtration (161).
    Flotation (295).

Titanium
Extraction of thiocyanate complex
with diantipyrilmethane (38), 8-
hydroxyquinolinate (326),
preconcentration by extraction of the
matrix (320, 324).
    Sorption of 8-hydroxyquinolinate on
activated charcoals (55), on cotton
precoated with iminodiacetylcellulose
(66).
    Precipitation with alizarin blue (82), of
diethyldithiocarbamate and thiooxinate
(82), cupferrate, diethyldithiocarbamate
and 8-hydroxyquinoline (82),
preconcentration by precipitating the
matrix (81), co-precipitation with $Al(OH)_3$
(88), $Mn(OH)_4$ (88).
    Volatilization as $TiCl_4$ (123, 128, 316),
TiF_4 (128, 129, 130), fractional
volatilization (215), preconcentration by
volatilization of the matrix after
chemical transformation (130, 315), by
sublimation of the matrix (122), dry
ashing of the organic matrix (124, 332,
333).
    Electrodeposition (105).
    Preconcentration by directed
crystallization of the matrix (142).
    Filtration (161, 300, 301).
    Flotation (295).

Tungsten

Extraction with trioctylamine N-oxide (33), of thiocyanate complex with diantipyrilmethane (38), of WO_4^{2-} with dialkyltin dinitrates (328), preconcentration by extraction of the matrix (320).

Sorption of 8-hydroxyquinolinate on activated charcoal (55), on TiO_2 (74, 75), Al_2O_3 (75), metatitanic acid (298), cellulose filter containing functional groups of 2,2'-diaminediethylamine (67), WO_4^{2-} and WO_3^{2-} on porous glass balls treating with N-β-aminoethyl-aminopropyltrimethoxysilane (69).

Precipitation as tungstate (177), tungstic acid (80), with alizarin blue (82), of diethyldithiocarbamate and thiooxinate (82).

Volatilization as chlorides (128, 316), fluorides (128), oxides (128) after chemical transformation, preconcentration by volatilization of the matrix after chemical transformation (315).

Flotation (295).

Vanadium

Extraction of tetramethyldithiocarbamate and hexamethylenedithiocarbamate (220), diethyldithiocarbamate (302), 8-hydroxyquinolinate (269), N-benzoyl-N-phenylhydroxylamine (307), with trioctylamine (226), preconcentration by extraction of the matrix (212, 323).

Sorption on oxidized charcoals (56), of 8-hydroxyquinolinate on activated charcoals (242), on TiO_2 (74), metatitanic acid (298), modified cellulose containing 1-(2-hydroxyphenylazo)-2-naphthol or 4-(2-pyridylazo)resorcin (66), cellulose filter containing functional groups of 2,2'-diaminediethylamine (67), on complex-forming sorbent Chelex-100 (294), sorbent Polyorgs XI-H (305).

Precipitation with alizarin blue (82), of

diethyldithiocarbamate and thiooxinate (82), diethyldithiocarbamate, cupferrate and 8-hydroxyquinolinate (82), pyrrolidinedithiocarbamate (82), co-precipitation with $Al(OH)_3$ (88), dibenzyldithiocarbamate with dibenzyldithiocarbamates of silver and nickel (92), with pyrrolidinedithiocarbamate of cadmium or cobalt (295), 1-phenyl-3-methyl-4-caprynoylpyrazolone-5 (330).

Electrodeposition (100).

Volatilization as chlorides (128, 136) and fluorides (128) after chemical transformation, preconcentration by volatilization (118, 216, 292, 296, 315) and sublimation (122) of the matrix, by dry (123–126, 230, 231) and wet (125–127, 230, 231) ashing of the organic matrix.

Filtration (161, 300).

Flotation (295).

Uranium

Extraction with dioctylsulphoxide (28), petroleum sulphoxides (266).

Sorption on TiO_2 (75), cotton precoated with iminodiacetylcellulose (66), cellulose containing 1-(2-hydroxyphenylazo)-2-naphthol or 4-(2-pyridylazo)resorcin (66), on discs of cellulose-phosphate paper Watman P-81 (67), silica gel with complex-forming groups of β-diketones immobilized on its surface (70).

Co-precipitation of dibenzyldithiocarbamate with iron dibenzyldithiocarbamate (92, 295), 1-nitroso-2-naphtholate with 1-nitroso-2-naphthol (93), 1-(2-pyridylazo)-2-naphthyl alcohol (295).

Preconcentration by volatilization of the matrix (120).

Flotation (157–159, 295).

Fire assay (153).

Yttrium

Preconcentration on complex-forming

sorbents based on aminopolystyrenepolyarsenazo and polypyridylazoresorcin (304).

Zinc

Extraction with molten 8-hydroxyquinoline (38), of pyrrolidinedithiocarbamate with molten stearic alcohol (38, 244), diethyldithiocarbamate (44, 235, 245), tetramethyldithiocarbamate and hexamethylenedithiocarbamate (220), pyrrolidinedithiocarbamate (297), 2-isopropyl-8-mercaptoquinolinate (297), 8-hydroxyquinolate (323, 324), of thiocyanate complex with diantipyrylmethane (38), with trioctylamine (226), Aliquat 336 (307), preconcentration by extraction of the matrix (174, 212, 226, 319, 320, 323, 324, 326, 327).

Sorption of diethyldithiocarbamate (54, 328), hexamethylenedithiocarbamate (54), 8-hydroxyquinolinate (55, 298) on activated charcoals, sorption on manganese or zinc sulphide (244), on metatitanic acid (298), cation exchangers KU-2-8 (60), Dowex A-1 (60), anion exchanger AV-17 (61), cotton precoated with iminodiacetylethylcellulose (65, 66), cellulose filter containing chromotropic acid group (66), modified cellulose containing 1-(2-hydroxyphenylazo)-2-naphthol or 4-(2-pyridylazo)resorcin (66), dithiocarbamate derivatives of cellulose (67), complex-forming cellulose sorbent Hyphan (294), small glass balls with complex-forming groups of 8-hydroxyquinoline immobilized on their surface (70, 77), sorption of phosphate and 8-hydroxyquinolinate on the anion exchangers AV-17 and AV-18 (73), with transparent membranes (73, 74), Chelex-100 (78), complex-forming sorbent Spheronoxine (243), polymeric thioether (243), Rexyn 201 (257), anion exchanger

JRA-900 (294), ion exchanger of the iminodiacetic type (304), sorbent Polyorgs XI-H (305), sorbent Wofatit MC-50 (330).

Precipitation of pyrrolidinedithiocarbamate (82), dibenzyldithiocarbamate (241), co-precipitation with $Bi(OH)_3$ (86), $Mg(OH)_2$ and $Fe(OH)_3$ (87), $Ni(OH)_3$ (88, 89), $Bi(OH)_3$, Bi_2S_3 and In_2S_3 (212), KF (330), sulphide in the presence of cellulose (89), AgI in the presence of 1,10-phenanthroline (89), thiooxinate with 8,8'-diquinolyl disulphide (92), diethyldithiocarbamate with diethyldithiocarbamates Co, Cu and Fe (92), pyrrolidinedithiocarbamate Co(III) (92), molybdenum pyrrolidinedithiocarbamate (241), pyrrolidinedithiocarbamate of cadmium or cobalt (295), with dibenzyldithiocarbamate of silver or nickel (295), 1-(2-pyridylazo)-2-naphthyl alcohol (295), 1-nitroso-2-naphtholate with 1-nitroso-2-naphthol (93), thiocyanate with thiocyanate of antipyrine (93).

Electrodeposition (97, 100, 105, 106, 242, 263, 264, 274, 275, 298), cementation (92), electrodialysis (111).

Volatilization as chlorides (128) and Zn (128) after chemical transformation, fractional volatilization (229), preconcentration by volatilization of the matrix (119–121, 174, 292), by volatilization of the matrix after chemical transformation (130, 315, 316, 318), by dry (123–125, 230, 231, 333) and wet (126, 127, 219, 230, 231, 334) ashing of the organic matrix.

Preconcentration by directed crystallization (174).

Flotation (158, 159).

Filtration (161, 165, 166, 300, 301), dialysis (164).

Preconcentration by partial dissolution of the matrix (111).

371

Zirconium

Extraction of thiocyanate complex with diantipyrylmethane (38), preconcentration by extraction of the matrix (212, 320, 321, 323).

Sorption of 8-hydroxyquinolinate on activated charcoals (55), on complex-forming sorbents based on aminopolystyrenepolyarsenazo and polypyridylazoresorcin (304).

Co-precipitation with $Al(OH)_3$ and $Fe(OH)_3$ (88, 89), metal arsenates (89), of 8-hydroxyquinolinate by indifferent organic co-precipitants (92), with pyrrolidinedithiocarbamate of cadmium or cobalt (295).

Volatilization as $ZrCl_4$ (128, 129, 316), ZrF_4 (128) after chemical transformation, fractional volatilization (229, 230).

Flotation (157, 160).

Filtration (161).